DEPARTMENT OF THE ARMY
TECHNICAL MANUAL

DEPARTMENT OF THE AIR
FORCE TECHNICAL ORDER

**TM 9-1819AC**

**TO 19-75CAJ-4**

# ORDNANCE MAINTENANCE

# 2 ½ TON 6x6 TRUCK

## TECHNICAL MANUAL

*BY DEPARTMENTS OF THE ARMY AND THE AIR FORCE*
*DECEMBER 1952*

©2013 Periscope Film LLC
All Rights Reserved
ISBN#978-1-940453-18-7
www.PeriscopeFilm.com

DISCLAIMER:

This document is a reproduction of a text first published by the Department of the Army, Washington DC. All source material contained herein has been approved for public release and unlimited distribution by an agency of the US Government. Any US Government markings in this reproduction that indicate limited distribution or classified material have been superseded by downgrading instructions promulgated by an agency of the US government after the original publication of the document No US government agency is associated with the publication of this reproduction. This manual is sold for historic research purposes only, as an entertainment. It contains obsolete information and is not intended to be used as part of an actual training program. No book can substitute for proper training by an authorized instructor.

©2013 Periscope Film LLC
All Rights Reserved
ISBN#978-1-940453-18-7
www.PeriscopeFilm.com

TM 9-1819AC/TO 19-75CAJ-4

---

ORDNANCE MAINTENANCE:

2½-TON 6x6
CARGO TRUCKS
M135 AND M211

DUMP TRUCK M215

GASOLINE
TANK TRUCK M217

TRACTOR TRUCK
M221

AND WATER TANK
TRUCK M222:

POWER TRAIN

---

*United States Government Printing Office*
*Washington : 1953*

DEPARTMENTS OF THE ARMY
AND THE AIR FORCE
WASHINGTON 25, D. C., *5 December 1952*

TM 9–1819AC/TO 19–75CAJ–4 is published for the information and guidance of all concerned.

[AG 461 (4 Nov 52)]

BY ORDER OF THE SECRETARIES OF THE ARMY AND THE AIR FORCE:

OFFICIAL:
    WM. E. BERGIN
    *Major General, USA*
    *The Adjutant General*

J. LAWTON COLLINS
*Chief of Staff, United States Army*

OFFICIAL:
    K. E. THIEBAUD
    *Colonel, USAF*
    *Air Adjutant General*

HOYT S. VANDENBERG
*Chief of Staff, United States Air Force*

DISTRIBUTION:
  *Active Army:*
    Tech Svc (1); Tech Svc Bd (2); AFF (2); AA Comd (2); OS Maj Comd (10); Base Comd (2); MDW (3); Log Comd (5): A (10); CHQ (2); Div (2); Regt 9 (2); Bn 9 (2); CO 9 (2); FT (2); Sch (5) except 9 (50); PMS & T 9 (1); Gen Dep (2); Dep 9 (10); POE (5); OSD (2); PRGR 9 (10); Ars 9 (10); Proc Dist 9 (10); Mil Dist (3).
  *NG:* Same as Active Army except one copy to each unit.
  *ORC:* Same as Active Army except one copy to each unit.
  For explanation of distribution formula, see SR 310–90–1.

# CONTENTS

| | | Paragraph | Page |
|---|---|---|---|
| **CHAPTER 1. INTRODUCTION** | | | |
| Section I. | General | 1–3 | 1 |
| II. | Description and data | 4–6 | 16 |
| **CHAPTER 2. PARTS, SPECIAL TOOLS, AND EQUIPMENT FOR FIELD AND DEPOT MAINTENANCE** | | 7–10 | 20 |
| **CHAPTER 3. TROUBLE SHOOTING** | | 11–14 | 27 |
| **CHAPTER 4. REMOVAL AND INSTALLATION OF MAJOR COMPONENTS** | | | |
| Section I. | Disassembly of vehicle into major components | 15–43 | 28 |
| II. | Assembly of vehicle from major components | 44–80 | 58 |
| **CHAPTER 5. TRANSFER ASSEMBLY** | | | |
| Section I. | Description and data | 81, 82 | 84 |
| II. | Disassembly of transfer into subassemblies | 83–91 | 89 |
| III. | Rebuild of input shaft and components | 92–94 | 99 |
| IV. | Rebuild of output gear and components | 95–97 | 103 |
| V. | Rebuild of output gear bearing retainer and related components | 98–100 | 107 |
| VI. | Rebuild of front axle output shaft components | 101–103 | 110 |
| VII. | Rebuild of forward rear axle output shaft components | 104–106 | 115 |
| VIII. | Rebuild of idler shaft components | 107–109 | 118 |
| IX. | Rebuild of shifting mechanism | 110, 111 | 123 |
| X. | Cleaning and inspection of transfer case and covers | 112, 113 | 125 |
| XI. | Assembly of transfer from subassemblies and components | 114–120 | 126 |
| **CHAPTER 6. POWER TAKE-OFF** | | | |
| Section I. | Description and data | 121, 122 | 133 |
| II. | Disassembly of power take-off | 123–128 | 137 |
| III. | Rebuild of power take-off components | 129–135 | 143 |
| IV. | Assembly of power take-off | 136–141 | 145 |
| V. | Description and replacement of accessory drive unit | 142–144 | 153 |
| VI. | Rebuild of tank truck accessory drive unit | 145–148 | 155 |
| VII. | Rebuild of dump truck accessory drive unit | 149–152 | 163 |

| | Paragraph | Page |
|---|---|---|
| **CHAPTER 7. FRONT AXLE** | | |
| Section I. Description and data | 153–154 | 171 |
| II. Front axle alinement | 155, 156 | 174 |
| III. Disassembly of front axle into subassemblies | 157–163 | 180 |
| IV. Rebuild of axle shaft and universal joint | 164–169 | 183 |
| V. Rebuild of steering knuckle, support, trunnions, seals, and bearings | 170–173 | 189 |
| VI. Rebuild of front axle housing | 174–179 | 194 |
| VII. Rebuild of tie rod | 180–184 | 198 |
| VIII. Assembly of front axle from subassemblies | 185–192 | 201 |
| **CHAPTER 8. REAR AXLES** | | |
| Section I. Description and data | 193, 194 | 209 |
| II. Disassembly of rear axle into subassemblies | 195, 196 | 209 |
| III. Rebuild of differential and carrier assembly | 197–202 | 211 |
| IV. Cleaning and inspection of rear axle housing and axle shafts | 203, 204 | 227 |
| V. Assembly of rear axle from subassemblies | 205–209 | 230 |
| **CHAPTER 9. PROPELLER SHAFTS, UNIVERSAL JOINTS, AND PILLOW BLOCK** | | |
| Section I. Description | 210–213 | 232 |
| II. Rebuild of axle propeller shaft and universal joints | 214–217 | 233 |
| III. Rebuild of transmission-to-transfer propeller shaft assembly | 218–221 | 237 |
| IV. Rebuild of pillow block assembly | 222–225 | 240 |
| **CHAPTER 10. SERVICE BRAKE SYSTEM** | | |
| Section I. Description and data | 226, 227 | 248 |
| II. Rebuild of brake shoes and drums | 228–235 | 251 |
| III. Rebuild of master cylinder | 236–238 | 265 |
| IV. Rebuild of wheel cylinders | 239–241 | 270 |
| **CHAPTER 11. PARKING BRAKE SYSTEM** | | |
| Section I. Description and data | 242, 243 | 272 |
| II. Rebuild of mechanical parking brake | 244–246 | 273 |
| III. Rebuild of temporary (electric) parking brake valve | 247–252 | 247 |
| **CHAPTER 12. WHEELS AND HUBS** | | |
| Section I. Description and data | 253, 254 | 283 |
| II. Rebuild of wheels and hubs | 255, 256 | 286 |
| **CHAPTER 13. STEERING GEAR AND DRAG LINK** | | |
| Section I. Description and data | 257, 258 | 293 |
| II. Rebuild of steering gear assembly | 259–264 | 295 |
| III. Rebuild of drag link | 265, 266 | 314 |
| **CHAPTER 14. FRONT SPRING SUSPENSION** | | |
| Section I. Description and data | 267, 268 | 315 |
| II. Rebuild of front spring suspension components | 269–272 | 316 |
| III. Rebuild of shock absorbers | 273–277 | 321 |

| | Paragraph | Page |
|---|---|---|
| **CHAPTER 15. REAR SPRING SUSPENSION** | | |
| Section I. Description and data | 278, 279 | 329 |
| II. Rebuild of rear spring suspension components | 280–283 | 331 |
| **CHAPTER 16. FRAME AND ASSOCIATED PARTS** | | |
| Section I. Description and data | 284, 285 | 334 |
| II. Rebuild of frame and associated parts | 286–289 | 336 |
| **CHAPTER 17. ELECTRICAL SYSTEM** | | |
| Section I. Description and data | 290–292 | 341 |
| II. Repair of cable connections | 293, 294 | 342 |
| III. Rebuild of horn | 295–297 | 344 |
| **CHAPTER 18. WINCH AND DRIVE LINE** | | |
| Section I. Description and data | 298, 299 | 346 |
| II. Disassembly of winch into subassemblies | 300–302 | 348 |
| III. Rebuild of end frame assembly | 303–305 | 352 |
| IV. Rebuild of gear case and automatic brake | 306–310 | 355 |
| V. Rebuild of drum assembly | 311, 312 | 362 |
| VI. Assembly of winch from subassemblies | 313–315 | 362 |
| VII. Rebuild of winch drive system | 316–319 | 365 |
| **CHAPTER 19. CAB AND ASSOCIATED PARTS** | | |
| Section I. Description and data | 320, 321 | 370 |
| II. General rebuild of cab | 322–329 | 371 |
| III. Rebuild of cab doors | 330–332 | 384 |
| IV. Rebuild of windshield assembly | 333, 334 | 391 |
| **CHAPTER 20. CARGO BODY** | | |
| Section I. Description and data | 335, 336 | 395 |
| II. Rebuild of cargo body | 337–341 | 400 |
| **CHAPTER 21. REPAIR AND REBUILD STANDARDS** | 342–357 | 403 |
| **APPENDIX REFERENCES** | | 425 |
| **INDEX** | | 431 |

# CHAPTER 1

# INTRODUCTION

## Section I. GENERAL

### 1. Scope

*a.* (1) These instructions are published for the information and guidance of personnel responsible for field and depot maintenance of this matériel. These instructions contain information on maintenance which is beyond the scope of the tools, equipment, or supplies normally available to using organizations. This manual does not contain information which is intended primarily for the using organization, since such information is available to ordnance maintenance personnel in the pertinent operators technical manual or field manual.

(2) This first edition is being published in advance of complete technical review of all concerned. Any errors or omissions will be brought to the attention of Chief of Ordnance, Washington 25, D. C., ATTENTION: ORDFM-Pub.

*b.* This manual contains a description of and procedures for removal, disassembly, inspection, repair, rebuild, and assembly of the units of the power train for 2½-ton 6x6 cargo truck M135 (figs. 1 and 2) and M211 (figs. 3 and 4), dump truck M215 (figs. 5 and 6), gasoline tank truck M217 (figs. 7 and 8), tractor truck M221 (figs. 9 and 10), and water tank truck M222 (figs. 11 and 12). The appendix contains a list of current references, including supply catalogs, technical manuals, and other available publications applicable to the matériel.

*c.* TM 9-819A contains operating and lubricating instructions for the matériel and contains all maintenance operations allocated to using organizations in performing maintenance work within their scope.

*d.* TM 9-1819AA contains service information on the engine, radiator, and air compressor.

*e.* TM 9-1819AB contains service information on the Hydra-Matic transmission (GMC Model 302M).

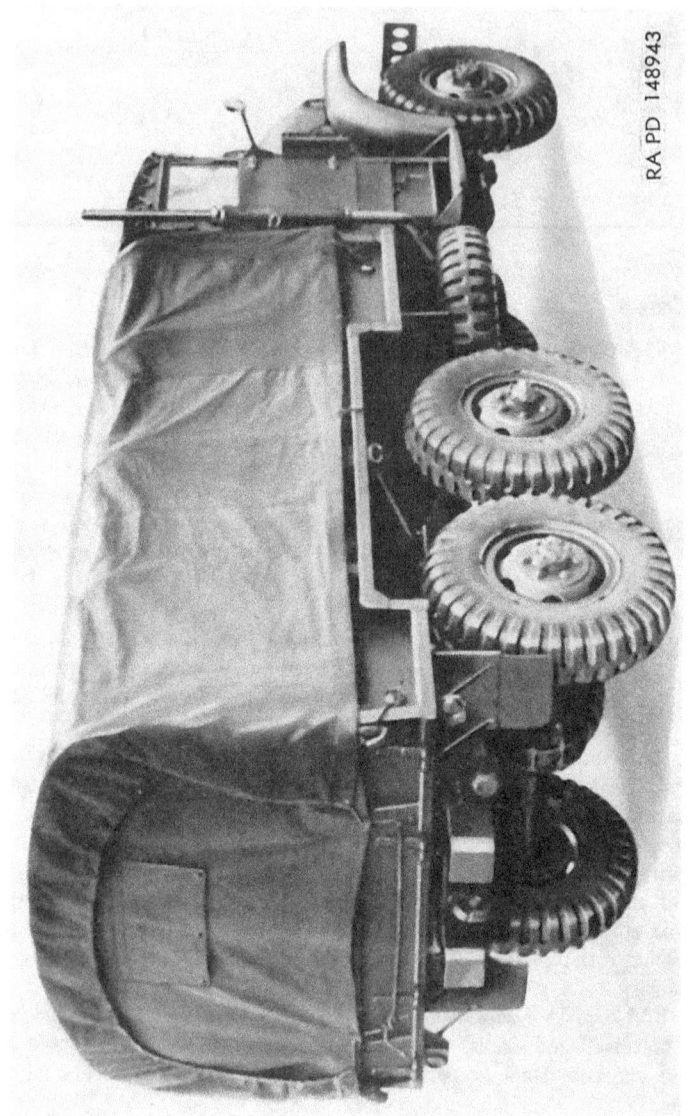

*Figure 1. Three-quarter right rear view of 2½-ton 6x6 cargo truck M135.*

*Figure 2. Three-quarter left front view of 2½-ton 6x6 cargo truck M135.*

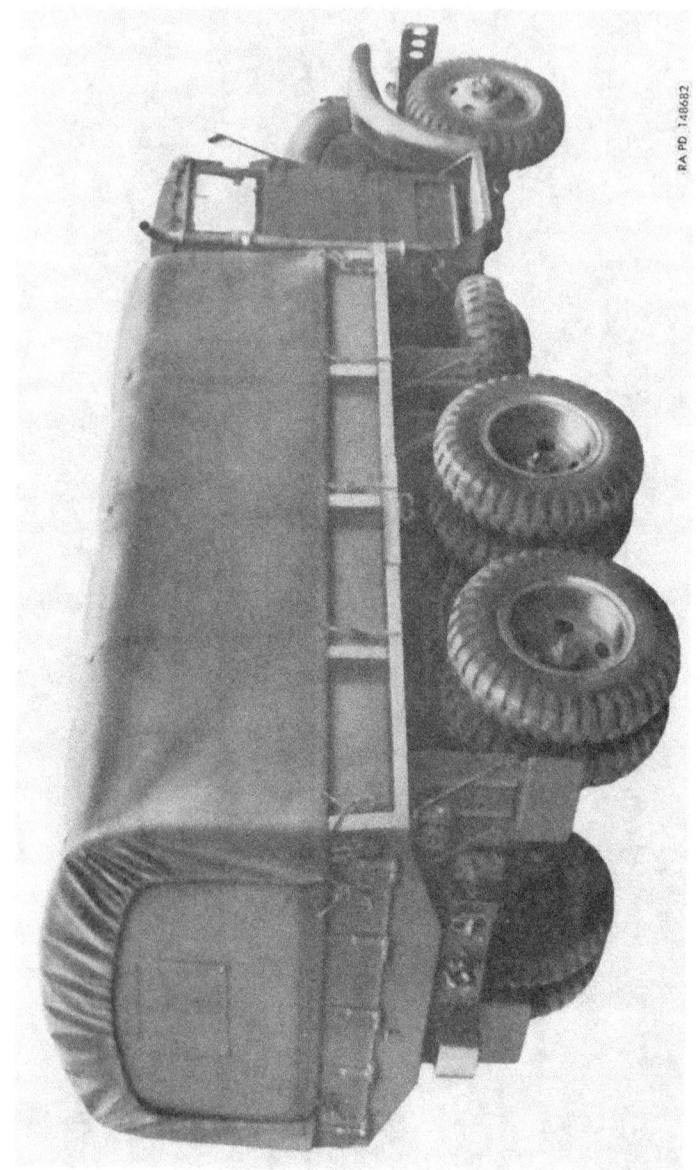

*Figure 3. Three-quarter right rear view of 2½-ton 6x6 cargo truck M211.*

*Figure 4. Three-quarter left front view of 2½-ton 6x6 cargo truck M211.*

*Figure 5. Three-quarter right rear view of 2½-ton 6x6 dump truck M215.*

*Figure 6. Three-quarter left front view of 2½-ton 6x6 dump truck M215.*

*Figure 7. Three-quarter right rear view of 2½-ton 6x6 gasoline tank truck M217.*

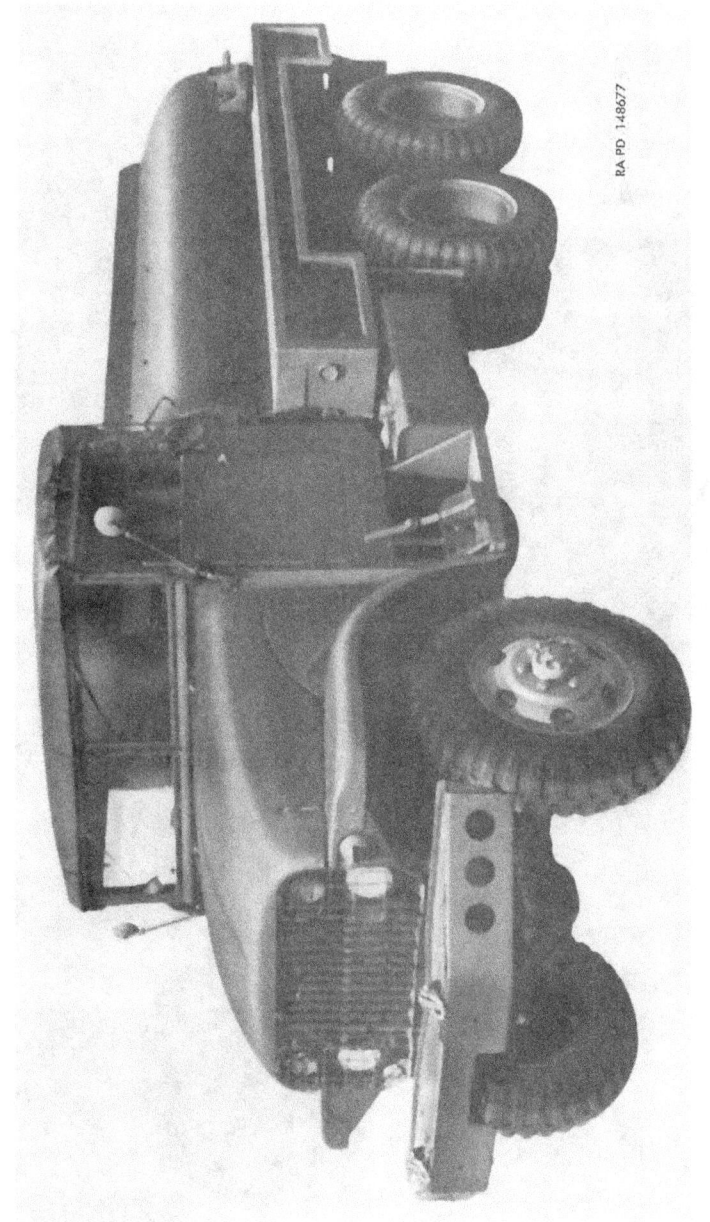

*Figure 8. Three-quarter left front view of 2½-ton 6x6 gasoline tank truck M217.*

*Figure 9. Three-quarter right rear view of 2½-ton 6x6 tractor truck M221.*

*Figure 10. Three-quarter left front view of 2½-ton 6x6 tractor truck M221.*

*Figure 11. Three-quarter right rear view of 2½-ton 6x6 water tank truck M222.*

*Figure 12. Three-quarter left front view of 2½-ton 6x6 water tank truck M222.*

*f.* TM 9-1819B contains service information on the Trico windshield wiper motor.

*g.* TM 9-1819C contains service information on the dump body used on dump truck M215.

*h.* TM 9-1819D contains service information on the gasoline tank body used on gasoline tank truck M217, water tank body used on water tank truck M222, and trailer attachments used on tractor truck M221.

*i.* TM 9-1825A contains service information on the Delco-Remy starter, generator, generator-regulator, and ignition unit.

*j.* TM 9-1825E contains service information on the Scintilla multiple cable connectors and light switch.

*k.* TM 9-1827A contains service information on the Bendix-Westinghouse low air pressure switch.

*l.* TM 9-1827B contains service information on the Bendix Products air-hydraulic power cylinder.

*m.* TM 9-1829A contains service information on the AC speedometer.

*n.* TM 9-1826D contains service information on the Holley carburetor.

*o.* TM 9-1828A contains service information on the Carter electric fuel pump.

## 2. Field and Depot Maintenance Allocation

The publication of instructions for complete disassembly and rebuild is not to be construed as authority for the performance by field maintenance units of those functions which are restricted to depot shops and arsenals. In general, the prescribed maintenance responsibilities will be reflected in the allocation of maintenance parts listed in the appropriate columns of the current ORD 8 supply catalog pertaining to 2½-ton 6x6 cargo truck M135 and M211, dump truck M215, gasoline tank truck M217, tractor truck M221, and water tank truck M222. Instructions for depot maintenance are to be used by maintenance companies in the field only when the tactical situation makes the repair functions imperative. Supply of parts listed in the depot guide column of ORD 8 supply catalogs will be made to field maintenance only when the emergency nature of the maintenance to be performed has been certified by a responsible officer of the requisitioning organization and upon express authorization by the chief of the service concerned. Those operations which can be performed as "emergency field maintenance" are specifically covered as such in this manual.

## 3. Forms, Records, and Reports

*a. General.*—Responsibility for the proper execution of forms, records, and reports rests upon the officers of all units maintaining this equipment. However, the value of accurate records must be fully appreciated by all persons responsible for their compilation, maintenance, and use. Records, reports, and authorized forms are normally utilized to indicate the quantity, and condition of matériel to be inspected, to be repaired, or to be used in repair. Properly executed forms convey authorization and serve as records for repair or replacement of matériel in the hands of troops and for delivery of matériel requiring further repair to ordnance shops in arsenals, depots, etc. The forms, records, and reports establish the work required, the progress of the work within the shops, and the status of the matériel upon completion of its repair.

*b. Authorized Forms.*—The forms generally applicable to units maintaining this equipment are listed in the appendix. No forms other than those approved for the Department of the Army will be used. For current and complete listing of all forms, refer to current SR 310-20-6. Additional forms applicable to the using personnel are listed in the operators manual. For instructions on use of these forms, refer to FM 9-10.

*c. Field Reports of Accidents.*—The reports necessary to comply with the requirements of the Army safety program are prescribed in detail in the SR 385-10-40 series of special regulations. These reports are required whenever accidents involving injury to personnel or damage to matériel occur.

*d. Report of Unsatisfactory Equipment or Materials.*—Any suggestions for improvement in design and maintenance of equipment, safety and efficiency of operation, or pertaining to the application of prescribed petroleum fuels, lubricants, and/or preserving materials, will be reported through technical channels as prescribed in SR 700-45-5 to the Chief of Ordnance, Washington 25, D. C., ATTN: ORDFM, using DA Form 468, Unsatisfactory Equipment Report. Such suggestions are encouraged in order that other organizations may benefit.

*Note.* Do not report all failures that occur. Report only **REPEATED** or **RECURRENT** failures or malfunctions which indicate unsatisfactory design or material. However, reports will always be made in the event that exceptionally costly equipment is involved. See also SR 700-45-5 and the printed instructions on DA Form 468.

## Section II. DESCRIPTION AND DATA

## 4. Description

*a. General.*—The power train units and systems covered in this manual include the transfer and propeller shafts, front and rear axles, spring suspension and frame, steering system, brake system, cab, and bodies on models itemized in paragraph 1*b*. Differences in the models are briefly described in paragraph 5.

*b. Transfer Assembly.*—The transfer assembly (fig. 47) is a two-speed unit which transfers power to front and rear axles. Complete description and tabular data on this component are described in chapter 5. Adjustment and replacement procedures of control linkage are contained in TM 9-819A.

*c. Power Take-Off Assembly* (figs. 85, 86, and 87).—The power take-off is single-speed and is mounted to the left side of transfer. Power take-off is used to drive winch, and to drive operating units of the dump, gasoline tank, and water tank body models. Control linkage adjustment and replacement procedures are contained in TM 9-819A. Complete description and tabular data on this component are included in chapter 6.

*d. Front Axle Assembly* (fig. 105).—The front axle is hypoid, single-reduction type using conventional differential and carrier assembly to transmit drive through constant-velocity universal joints. Power to front axle is automatically transmitted from transfer through conventional propeller shaft. Complete description and tabular data are included in chapter 7.

*e. Rear Axles* (fig. 131).—Rear axles are hypoid, single-reduction type using conventional differential and carrier assembly to transmit drive to rear wheels through full-floating axle shafts. Power is transmitted to both rear axles from transfer through conventional propeller shafts. Complete description and tabular data are included in chapter 8.

*f. Propeller Shafts and Universal Joints.*—Each axle propeller shaft (fig. 150) is tubular type, equipped with fixed yoke universal joint assembly at one end and slip yoke universal joint assembly at opposite end. Transmission-to-transfer propeller shaft consists of two universal point assemblies bolted together. All universal joint assemblies are needle roller bearing type. The pillow block assembly (fig. 155), mounted on forward rear axle, connects and supports the two propeller shafts required to transmit power from transfer to rear axle. Rebuild procedures are included in chapter 9.

*g. Service Brake System.*

    (1) The service brake system is air-hydraulic type consisting primarily of a pedal, interconnected to a hydraulic master cylinder; an air-hydraulic power cylinder; wheel cylinders

to transmit hydraulic pressure to the brake assemblies at each wheel; air compressor which maintains a supply of compressed air to operate the power cylinder; and interconnecting lines, fittings, and linkage.

(2) Replacement procedures on interconnecting lines, fittings, and linkage are contained in TM 9–819A. Service information on the air compressor is contained in TM 9–1819AA. Service information on the air-hydraulic power cylinder is contained in TM 9–1827B. Rebuild of brake assemblies, brake drum, hydraulic master cylinder, and wheel cylinders is contained in chapter 10.

*h. Parking Brake System.*

(1) Mechanical parking brake system includes an external-contracting one-piece band type brake, located at rear of transfer assembly. Parking brake hand lever, located at right of driver in cab, operates brake band through a relay lever and interconnecting rods. Description, tabular data, and service information are contained in chapter 11.

(2) In addition to the mechanical parking brake system, a temporary (electric) parking brake system is used only in the event of failure of mechanical parking brake system. This system includes a braking switch electrically connected to a solenoid valve connected into hydraulic master cylinder outlet hydraulic line. Service information on the parking brake valve is contained in chapter 11.

*i. Wheels and Hubs.* Service information on wheels, hubs, and bearings for both single and dual wheels is contained in chapter 12.

*j. Steering Gear and Drag Link.*—The steering gear assembly (fig. 186) is recirculating-ball type, flange-mounted on left frame side rail. The Pitman arm of the gear assembly is interconnected with left steering knuckle of front axle by a drag link. The drag link (fig. 204) is tubular type equipped with ball studs at both ends. Ball studs are mounted in bearings which require no lubrication. Service information and necessary tabular data are contained in chapter 13.

*k. Front Spring Suspension* (fig. 205).—Front springs are semi-elliptic, mounted with the arch down, shackled at both ends. Springs are mounted on front axle housing and held in place with "U" bolts and spring bumper blocks. Three torque rods, two lower and one upper, transmit driving and braking forces to frame. A double-acting shock absorber is mounted to frame side rail on each side. Absorber arms are connected to axle with links which attach to bumper blocks. Service information on front spring suspension is contained in chapter 14.

*l. Rear Spring Suspension* (fig. 214).—Rear spring suspension consists of an articulated main spring assembly and a fixed secondary spring assembly on each side. Both spring assemblies are semielliptic with slipper-type ends and mounted with the arch up. The main spring is mounted on a spring seat (fig. 39) which in turn is mounted on a shaft with tapered roller bearing. The secondary spring is mounted rigidly to frame side rail. Slipper ends of main springs are inserted in axle housing brackets, while secondary spring ends contact top of brackets under heavy loaded conditions. Three torque rods are used at each axle, two upper and one lower. Service information on rear spring suspension is contained in chapter 15.

*m. Frame.*—Service information on frame and associated parts is contained in chapter 16.

*n. Cab.*—Cab consists of an open-top structure enclosing driver's compartment. The all-steel cab structure includes several subassemblies bolted together into a unit assembly. The cab is equipped with an adjustable windshield, canvas top deck and rear curtains, and metal doors with regulated glass windows. Service information on the cab and associated parts is contained in chapter 19.

*o. Cargo Body.*—The cargo body, of all-steel construction, is mounted on frame side rails, attached by means of bolts to brackets on frame side rails. Removable wood side racks incorporate folding troop seats. Canvas top paulin, front end curtain, and rear end curtain are supported by five removable wood bows. The cargo bodies on the cargo truck M135 and M211 are the same except that body on cargo truck M211 (fig. 240) has a flat floor, while body on cargo truck M135 (fig. 239) has wheel housings built into each side of body. Service information on cargo bodies is contained in chapter 20.

*p. Winch* (fig. 222).—Winch is a worm-geared, jaw-clutch, drum type mounted at front of truck. The winch is operated through propeller shafts by power take-off mounted on left side of transfer. Service information on winch assembly and drive line is contained in chapter 18. Replacement procedures on power take-off control linkage are contained in TM 9-819A.

## 5. Differences Between Models

*a. Cargo Truck M135* (figs. 1 and 2).—The cargo truck M135 is a 2½-ton 6 x 6 truck equipped with a metal cargo type body mounted on frame independent of cab. Complete descriptions of chassis, body, and cab items are contained in TM 9-819A. This truck is furnished with or without winch.

*b. Cargo Truck M211* (figs. 3 and 4).—The cargo truck M211 is similar to the cargo truck M135 (*a* above) except it is equipped with flat-floor steel cargo body and dual rear wheels. This truck is furnished with or without winch.

*c. Dump Truck M215* (figs. 5 and 6).—The dump truck M215 is similar to the cargo truck M135 (*a* above) except it is equipped with a metal dump body and dual rear wheels. The power take-off is equipped with an accessory drive which actuates the body hoist equipment when controls are properly positioned by the driver. An auxiliary governor is also used in conjunction with the engine governor to properly limit engine speed when operating dump hoist mechanism. The dump body can also be fitted with side stake racks and folding troop seats. Paulin and bows for body protection may be installed in same manner as on the cargo truck M135. This truck is furnished with or without winch.

*d. Gasoline Tank Truck M217* (figs. 7 and 8).—The gasoline tank truck M217 is similar to the cargo truck M135 (*a* above) except it is equipped with a 1,200-gallon gasoline tank body and dual rear wheels. An accessory drive on power take-off actuates the pumping equipment when controls are properly positioned by driver. An auxiliary governor is used in conjunction with the engine governor to properly limit engine speed when operating the pump equipment. The body may be fitted with bows and top paulin for camouflage purposes. This truck is furnished without winch only.

*e. Water Tank Truck M222* (figs. 11 and 12).—The water tank truck M222 is identical to the gasoline tank truck M217 (*d* above) except it is equipped with a 1,000-gallon water tank body. The power take-off accessory drive and the auxiliary governor are similar to gasoline tank truck M217. This truck is furnished without winch only.

*f. Tractor Truck M221* (figs. 9 and 10).—The tractor truck M221 is a tractor type truck with chassis and power plant construction similar to cargo truck M135 (*a* above), except vehicle is a shorter wheelbase with dual rear wheels, and is equipped with a conventional type fifth wheel and ramp. The spare tire is mounted on frame at rear of cab. This truck is furnished with or without winch.

## 6. Data

*a.* Reference should be made to TM 9–819A for tabular data pertaining to the general characteristics and performance of the models described in paragraph 5.

*b.* Pertinent tabular data on the various components or systems included in the power train will be found in the applicable chapters of this manual.

# CHAPTER 2

# PARTS, SPECIAL TOOLS, AND EQUIPMENT FOR FIELD AND DEPOT MAINTENANCE

## 7. General

Tools and equipment and maintenance parts over and above those available to the using organization are supplied to ordnance field maintenance units and depot shops for maintaining, repairing, and/or rebuilding the matériel.

## 8. Parts

Maintenance parts are listed in Department of the Army Supply Catalog ORD 8 SNL G-749 which is the authority for requisitioning replacements. Parts not listed in the ORD 8 catalog, but required by depot shops in rebuild operations may be requisitioned from the listing in the corresponding ORD 9 catalog and will be supplied if available. Requisitions for ORD 9 parts will contain a complete justification of requirements.

## 9. Common Tools and Equipment

Standard and commonly used tools and equipment having general application to this matériel are listed in ORD 6 SNL J-8, Sections 7, 12, 13, and 18; ORD 6 SNL J-9, Sections 1, 2, 3, 8, and 10; ORD 6 SNL J-10, Sections 4, 7, 8, 11, 12, and 15; and are authorized for issue by T/A and T/O & E.

## 10. Special Tools and Equipment

The special tools (figs. 13, 14, and 15) and equipment tabulated in table I will be listed in the Department of the Army Supply Catalog ORD 6 SNL J-16, Section 42. This tabulation contains only those special tools and equipment, with the exception of those special tools listed in TM 9-819A, necessary to perform the operations described in this manual. It is included for information only, and is not to be used as a basis for requisitions.

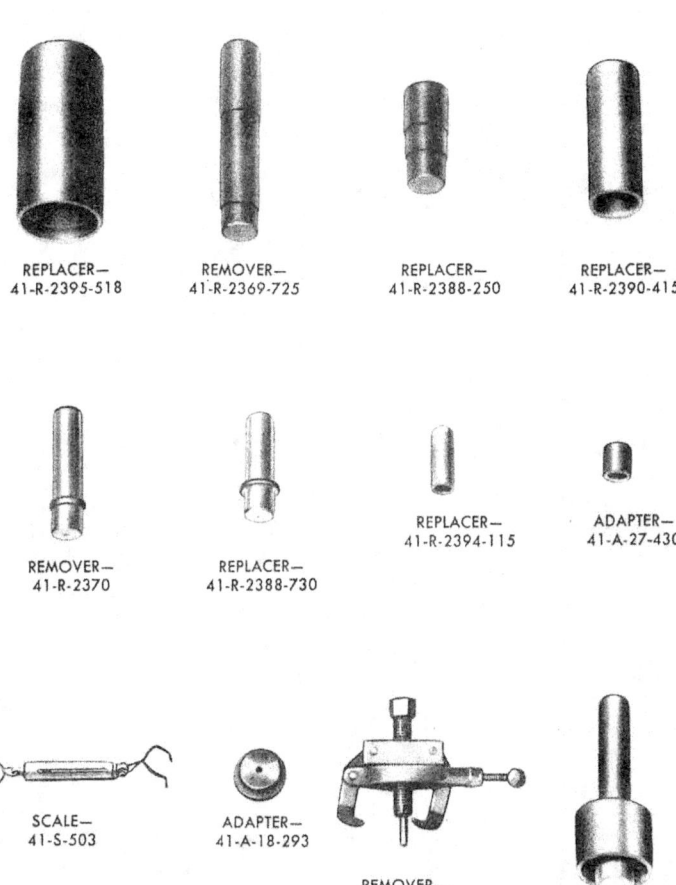

*Figure 13. Special tools.*

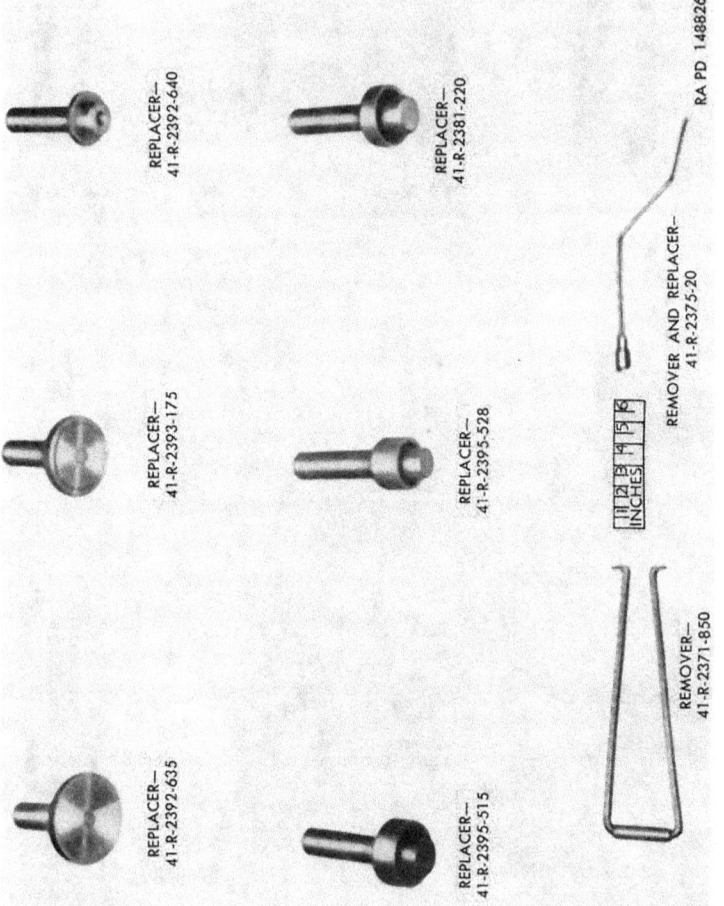

*Figure 14. Special tools.*

WRENCH—41-W-3634

TOOL—41-T-3215-910

WRENCH—41-W-3838-75

WRENCH—
41-W-545-5

WRENCH—
41-W-3825-66

WRENCH—7950690
RA PD 148827

*Figure 15. Special tools.*

Table I. *Special Tools and Equipment for Field and Depot Maintenance*

| Item | Identifying number | References | | Use |
|---|---|---|---|---|
| | | Fig. | Par. | |
| ADAPTER, puller, 1¼ in. diam, 1⅛ in. long. | 41-A-27-430 | 13, 187 | 21, 259c(3) | Use with PULLER 41-P-2954 to pull steering wheel. |
| ADAPTER, remover, drive pinion flange and differential side bearing. | 41-A-18-293 | 13, 134 | 198c, d | Use with REMOVER 41-R-2367-950 to remove differential side bearing. |
| REMOVER, bearing and drive pinion flange, differential side. | 41-R-2367-950 | 13, 134, 135 | 198c, d | For removing drive pinion flange and differential side bearing using ADAPTER 41-A-18-293. |
| REMOVER, bushing, 1³⁄₁₆ in. OD, 1¼ in. ID, 10 in. long. | 41-R-2369-725 | 13, 117 | 173a | For removing steering knuckle bushing type bearing. |
| REMOVER, bushing and oil seal retainer, 1½ in. diam, 6 in. long. | 41-R-2370 | 13, 192 | 262a(2) | For removing steering gear Pitman shaft bushing and oil seal retainer. |
| REMOVER, oil seal | 41-R-2371-850 | 14, 122 | 175d | For removing axle shaft oil seal in front axle housing. |
| REMOVER AND REPLACER, spring, ²⁵⁄₃₂ in. diam, 1¾ in. long, w/handle. | 41-R-2375-20 | 14, 162, 167 | 233h, 230b | For removing and replacing brake shoe springs. |
| REPLACER, bearing, 2⅜ in. diam, 7¼ in. long. | 41-R-2381-220 | 14, 143 | 202e(3) | For installing differential side bearings. |
| REPLACER, bushing, 2 in. diam, 5 in. long. | 41-R-2388-250 | 13, 117 | 173a(2) | For installing steering knuckle bushing type bearing. |
| REPLACER, bushing and oil seal retainer 1¾ in. diam, 5 in. long. | 41-R-2388-730 | 13, 193 | 262a(2) | For installing steering gear Pitman shaft bushing and oil seal retainer. |
| REPLACER, gear and bearing, 2⅜ in. OD, 2⁵⁄₁₆ ID, 7 in. long. | 41-R-2390-415 | 13, 66, 82 | 97a, 114c, 116c, 116d | For installing transfer ball type bearings and speedometer drive gear. |

| Item | Stock No. | Page | Para | Purpose |
|---|---|---|---|---|
| REPLACER, oil seal, 4 13/16 in. diam, 5/8 in. wide, w/hdl. | 41-R-2392-635 | 14, 183, 184 | 256d(4), 256e(2) | For installing oil seal in front and rear wheel hubs. |
| REPLACER, oil seal, 2 3/4 in. diam, 1 1/8 in. wide, w/hdl. | 41-R-2392-640 | 14, 123 | 179a | For installing axle shaft oil seal in front axle housing. |
| REPLACER, oil seal, 4 in. diam, 5/8 in. wide, w/hdl. | 41-R-2393-175 | 14, 139 | 201a | For installing oil seal in drive pinion outer bearing retainer. |
| REPLACER, oil seal, 1 in. diam, 2 7/8 in. long | 41-R-2394-115 | 13, 84 | 118e | For installing transfer shifter shaft oil seal. |
| REPLACER, oil seal sleeve, 3 3/4 in. OD, 3.280 in. ID, 8 3/4 in. long. | 41-R-2395-518 | 13, 120 | 173c(2) | For installing front and rear wheel hub inner oil seal sleeves. |
| REPLACER, oil seal sleeve, 3 1/8 in. diam, 7 1/2 in. long. | 41-R-2395-515 | 14, 140 | 201b | For installing oil seal sleeve on front and rear axle propeller shaft flanges. |
| REPLACER, oil seal sleeve, 2 1/2 in. diam, 7 7/8 in. long. | 41-R-2395-528 | 14, 156 | 224b | For installing oil seal sleeve on transfer output and pillow block propeller shaft flanges. |
| REPLACER, oil seal sleeve, 3 in. diam, 8 1/4 in. long. | 41-R-2395-535 | 13, 61, 153 | 220b, 93c(2) | For installing oil seal sleeve on transmission output and transfer input propeller shaft flanges. |
| SCALE, steering gear checking | 41-S-503 | 13, 201 | 264c(2) | For checking steering gear adjustment. |
| STAND, shock absorber rebuilding, complete. | 41-S-4977-5 | 209, 212 | 275, 277 | For holding shock absorber during rebuilding operations. |
| TOOL, holding, 3/8 in. thk, 6 1/2 in. wide, 38 in. long. | 41-T-3215-910 | 15, 51 | 84b, 90a, 114f, 116d, 198d, 202d, 223b | For holding propeller shaft flange while removing or installing retaining nut. |
| WRENCH, adjusting | 7950690 | 15 | 256f(1) | For adjusting rear wheel bearings. |
| WRENCH, bearing adj nut, socket (detachable) oct, 3 3/16 in. opng, 3/4 in. sq-drive, 1 7/8 in. lg. | 41-W-545-5 | 15, 40 | 49b(5), 256f(1) | For adjusting rear spring seat bearings. |

25

Table I. Special Tools and Equipment for Field and Depot Maintenance—Continued

| Item | Identifying number | References | | Use |
|---|---|---|---|---|
| | | Fig. | Par. | |
| WRENCH, torque indicating, ¾ in. sq-drive, 300 ft-lb, cap. | 41–W–3634 | 15, 40 | 49b(5), 256f(1) | For adjusting wheel and spring seat bearings. |
| WRENCH, wheel stud nut, dble-hd socket, 1 33/64 in. hex opng, 0.817 in. sq opng, 22½ in. long. | 41–W–3838–75 | 15 | 34a, 60b | For removing or installing wheel stud nuts. |
| WRENCH, wheel bearing nut, tubular, oct, 3 7/16 in. opng, ¾ in. female sq-drive, with 2.072 in. diam tubular pilot, 5 in. long. | 41–W–3825–66 | 15 | 256f(1) | For adjusting front wheel bearings. |

# CHAPTER 3

# TROUBLE SHOOTING

## 11. Purpose

*Note.* Information in this chapter is for use of ordnance maintenance personnel in conjunction with and as a supplement to the trouble shooting section in the pertinent operators manual (TM 9-819A). It provides the continuation of instructions where a remedy in the operators manual refers to ordnance maintenance personnel for corrective action.

Operation of a deadlined vehicle without a preliminary examination can cause further damage to a disabled component and possible injury to personnel. By careful inspection and trouble shooting, such damage and injury can be avoided and, in addition, the causes of faulty operation of a vehicle or component can often be determined without extensive disassembly.

## 12. Excessive Front Axle Caster

If excessive front axle caster is indicated by hard steering, refer to paragraph 155*b* for method of checking and causes of excessive axle caster.

## 13. Incorrect Front Wheel Alinement

If incorrect front wheel alinement is indicated by front wheel shimmy, refer to paragraphs 155 and 156 for front end alinement information.

## 14. Bent Brake Backing Plate

When a bent brake backing plate is indicated by noisy brakes, replacement of backing plate is necessary. Refer to paragraphs 228 through 235 for removal, disassembly, inspection, assembly, and installation of brake assembly.

# CHAPTER 4

# REMOVAL AND INSTALLATION OF MAJOR COMPONENTS

## Section I. DISASSEMBLY OF VEHICLE INTO MAJOR COMPONENTS

### 15. General

This section contains information for the guidance of personnel performing major rebuild work on the 2½-ton 6x6 cargo trucks M135 and M211, dump truck M215, gasoline tank truck M217, tractor truck M221, and water tank truck M222. It provides an assembly line procedure for the disassembly of the vehicle into its major components. It designates what constitutes a major component, illustrates the points of connection between major components, and states briefly what must be done. The illustrations shown in this section are keyed with reference letters which indicate disconnect points. These key letters are referred to as such in the various procedures.

*Note.* Procedures covering removal of cargo body (par. 16) apply only to cargo trucks M135 and M211; all other procedures generally apply to all of the vehicles after their respective bodies have been removed.

### 16. Removal of Cargo Body

*Note.* Key letters in following text indicate disconnect points and refer to figure 46.

a. *General*.
  (1) The cargo body, which includes racks, roof bows, top paulin, rear curtain, tail lights, marker lights, reflectors, and splash shields, is removed from chassis as an assembly.
  (2) A suitable chain fall with chain hooks must be provided to lift body high enough to clear chassis. The sequence of procedures for removing cargo body from chassis are listed in logical sequence, permitting the use of more than one mechanic. However, the sequence of procedures can be changed to meet existing conditions or facilities.

*b. Removal Procedures.*
(1) Remove six cap screws (C) and nuts (D), three on each side, which attach cargo body to brackets on chassis side members.
(2) Remove four bolts (A), nuts (B), plain washers (E), and four inner and four outer compression springs (F and G) which retain cargo body to brackets on chassis side members.
(3) Disconnect both tail light and marker light wiring cables from harness at bayonet type connectors. Body is now prepared for removal.
(4) Fasten chain fall hooks under each corner of body, then raise body to elevation required to allow removal of either chassis or body.
(5) Remove two wooden body support sills from chassis.
(6) Rebuild procedures for cargo body are contained in chapter 20.

## 17. Removal of Front Bumper Assembly

*Note.* Key letters in following text indicate disconnect points and refer to figure 16.

*a. General.*—The front bumper assembly, which includes gussets, is removed as an assembly.

*b. Removal Procedures.*
(1) Remove pin (B) retaining each tow hook (C) to bumper; then remove tow hooks.
(2) Remove two cap screws and nuts, one each side, under bumper attaching bumper to frame side members.
(3) Remove six cap screws (D) and nuts, three each side, which attach gusset and bumper to side members.
(4) Pull bumper and gussets forward and remove bumper.

## 18. Removal of Winch Assembly (When Used)

*Note.* Key letters in following text indicate disconnect points and refer to figure 17.

*a. General.*—The winch assembly, which includes support brackets and drive shaft universal joint, is removed as an assembly.

*b. Removal Procedures.*
(1) Support winch with hoist.
(2) Remove two nuts (D) from inside of frame attaching each front tie-down "U" bolt (E) to frame.
(3) Remove four cap screws (A, B, C, and F) and nuts which attach each winch support bracket to frame side member.

*Note.* It is not necessary to remove winch spacers from frame.

*Figure 16. Front bumper and fender disconnect points.*

(4) While supporting winch with hoist, pull winch, with support brackets attached, toward front and out of frame side members.

(5) Remove winch drive shaft universal joint from winch shaft by removing shear pin.

(6) Refer to chapter 18 for rebuild of winch assembly.

## 19. Removal of Hood Assembly

*Note.* Key letters in following text indicate disconnect points and refer to figure 18.

*a. General.*—The hood assembly, which includes horn and horn air supply line, is removed as an assembly.

*b. Preliminary Procedure.* Open drain cock at one air reservoir to exhaust air from system.

*Figure 17. Winch disconnect points.*

   c. *Removal Procedures.*
   (1) Disengage hood catches; then raise hood to vertical position.
   (2) Disconnect horn air supply line at air line fitting (E) on cowl.
   (3) Disconnect horn wiring cables at bayonet type connectors (J).
   (4) While supporting hood, remove two cap screws and nuts which attach each hood prop to hood support bracket (A).
   (5) Lower hood to horizontal position; then remove hinge cap screw and nut from each hinge. Remove hood.

## 20. Removal of Fender and Skirt Assemblies

   a. *General.*—Each fender and skirt assembly, which includes hood catch, fender support, fender brace, and blackout head light on left fender only, is removed as an assembly.
   b. *Removal Procedures.*
   (1) At left fender, disconnect blackout head light wiring connectors.

(2) Remove two cap screws which attach each fender support to brush guard and radiator side baffles ((A), fig. 16).
(3) Remove two cap screws which hold radiator side baffles ((E), fig. 16) to fender skirts.
(4) Remove cap screw and washer which attach each fender brace to bracket ((L), fig. 18) on cab cowl.
(5) Remove two cap screws, two nuts, and one rubber spacer which attach each fender to running board ((P), fig. 18). Remove fender assembly.

## 21. Removal of Cab Assembly

a. *General.*
(1) The cab assembly, which includes instruments, accelerator pedal, engine air cleaner, generator regulator, wiring cables, batteries, air and vent lines, parking brake lever, transfer lever, power take-off lever, muffler tail pipes, hood extension panels, hood props, running boards, tool box, fluid container bracket, gun mount bracket and gun mounting rear "U" bolts, is removed from chassis as an assembly.
(2) A suitable chain fall must be provided to lift cab high enough to clear steering column on chassis.
(3) The sequence of procedures for removing cab from chassis are listed in logical sequence, permitting the use of more than one mechanic; however, the sequence can be changed to meet existing conditions or facilities.

b. *Procedures at Front of Cab.*
(1) *Remove brake pedal cap screw.* Remove clamp cap screw and nut attaching upper brake pedal to lower brake pedal.
(2) *Disconnect choke and throtile controls.*
  (a) Disconnect choke control ((F), fig. 18) at carburetor.
  (b) Disconnect throttle rod ((H), fig. 21) linking accelerator lever ((GG), fig. 18) on cowl to accelerator lever on intake manifold.
(3) *Disconnect hoses and lines.*
  (a) Disconnect air cleaner hose ((G), fig. 18) at carburetor elbow ((B), fig. 19).
  (b) Disconnect vent line ((EE), fig. 18) connecting nipple ((D), fig. 19) on engine to cowl vent line.
  (c) Disconnect air hose ((Z), fig. 18) connecting air compressor governor to cowl air line fitting ((A), fig. 19) on governor.
(4) *Disconnect wiring harness and cables.*
  (a) From right side of engine, disconnect engine wiring harness at three bayonet type connectors ((H), fig. 19 and (DD), fig. 18).

*Figure 18. Disconnect points at front and bottom of cab.*

(b) Disconnect generator wiring cable ((Y), fig. 18) at generator ((C), fig. 19) using wrench 41-W-3249-900.

(c) Disengage battery-to-starter cable ((K), fig. 19) from clips ((FF), fig. 18) on cowl.

(5) *Remove batteries.*

(a) Disconnect all battery cables ((C), (D), (H), and (K), fig. 18) at batteries.

(b) Remove battery retainers from bolts ((B), fig. 18), then lift batteries off supports.

*Figure 19. Cab disconnect points on chassis.*

c. *Procedures Inside of Cab.*

(1) Fold companion seat bottom up and retain with latch to obtain maximum accessibility to interior of cab.

(2) Remove horn button from steering wheel by removing four screws attaching horn button retaining ring to steering wheel.
(3) Remove nut attaching steering wheel to steering shaft.
(4) Remove steering wheel from steering shaft, using wheel puller 41–P–2954 and puller adapter 41–A–27–430.
(5) Remove cap retaining steering column to dash bracket.
(6) Remove grommet from steering column.
(7) Disconnect horn cable from connector ((S), fig. 19) on steering column.
(8) Remove cab monting bolt hole covers from floor pan.
(9) Remove brake pedal upper half from brake pedal lower half.
(10) Remove upper and lower brake pedal plates from cowl.
(11) Lift rubber seal from shift control tower.
(12) Remove 14 cap screws attaching front floor pan to floor pan, then lift front floor pan from around shift control tower and remove from cab.
(13) Through cab floor opening at right side of transmission, disconnect three rubber lines ((CC), fig. 18 and (E), fig. 19—vent line), ((AA), fig. 18 and (G), fig. 19—air exhaust line), and ((BB), fig. 18 and (F), fig. 19—air line).
(14) Under driver's seat, disconnect power take-off control cable at hand control lever by removing yoke pin ((P), fig. 19).
(15) Remove power take-off control cable grommet at floor pan, using screw driver.

*d. Procedures Underneath Cab.*
(1) Disconnect speedometer shaft at transfer case ((T), fig. 18).
(2) Disconnect parking brake lever rod ((V), fig. 18) at relay lever ((N), fig. 19 or (G), fig. 36) on frame cross member.
(3) Disconnect transfer control lever rod ((U), fig. 18) at cross shaft ((F), fig. 36) on cross member.
(4) Remove clamp bolt which retains power take-off control cable to support bracket underneath cab floor.
(5) Disconnect rubber vent line ((R), fig. 18) at brake master cylinder ((R), fig. 19).
(6) Disconnect starter ground cable ((HH), fig. 18) from starter ((F), fig. 22).
(7) Remove cab front mounting bolts, nuts, washers, and one upper and two lower cushions ((M), fig. 18) which attach cab to support brackets ((J), fig. 19) on frame.

*e. Procedures at Rear of Cab.*
(1) Remove cab rear center mounting cap screw nut, steel washer, and lower cushion ((W), fig. 18) which mount cab to frame ((M), fig. 19).

(2) Remove two cap screws, nuts, and washers ((S), fig. 18) which attach cab rear mounting spring to frame cross member ((L), fig. 19).

*f. Procedures at Sides of Cab.*
  (1) At left side of cab, disconnect two wiring harnesses ((Q), fig. 19) at multiple plug and receptacle connectors ((N), fig. 18) located under cab floor directly above running board, using wrench 41-W-3249-900; then disengage harnesses from clips ((Q), fig. 18) on running board rear support.
  (2) Disconnect tail pipe ((X), fig. 18) from muffler by opening clamp.

*g. Removal of Cab from Chasis.*
  (1) Attach hoist to four lifting ring nuts on cab.

*Figure 20. Cab assembly removed.*

(2) Inspection should be made to see if all disconnect operations have been completed before cab is raised off mountings.
(3) Raise cab slowly, using several short lifts until free from mountings and past steering column; then remove cab.
(4) Remove cab mounting cushions and rear center cap screw spacer from chassis.
(5) Refer to chapter 19 for rebuild of cab.

## 22. Removal of Power Plant Assembly

*a. General.*—The power plant assembly, which includes engine and accessories, radiator, radiator baffles, brush guards, head lights, engine front exhaust pipe, transmission, universal joint, and transmission shift control tower is removed as an assembly.

*Note.* It is not necessary to drain cooling system, engine crankcase, or transmission when removing power plant assembly.

*b. Equipment.*—A chain fall and one special tool is required to accomplish power plant removal. Engine sling 41–S–3831–600 ((D), fig. 21) must be used to lift power plant out of vehicle.

*c. Removal Procedures.*
(1) *Operations at front of vehicle.*—Remove two bolts, washers, nuts, and springs attaching radiator to front cross member ((M), fig. 23).
(2) *Operations at left side of engine.*
 (*a*) Disconnect air compressor discharge line ((K), fig. 23) from elbow ((B), fig. 21) on air compressor; then detach line from clip ((J), fig. 21) at base of air compressor.
 (*b*) Remove one bolt and washer attaching power plant to rear mounting ((E), fig. 21 and (E), fig. 23).
 (*c*) Remove nut, washer, and rubber cushions attaching power plant at front mounting ((K), fig. 21 and (L), fig. 23).
 (*d*) Disconnect rod ((G), fig. 23) linking transmission lever and transfer reverse cross shaft lever at transmission lever ((F), fig. 21) by removing yoke pin.
 (*e*) Disconnect flexible fuel line ((H), fig. 23) at carburetor connector ((C), fig. 21).
 (*f*) Remove clamp ((J), fig. 23) retaining engine front exhaust pipe to engine rear exhaust pipe.
 (*g*) Remove radiator brace rod ((A), fig. 21) from clip on radiator, then swing rod into position at special cylinder head bolt. Install nut to hold lower end of rod at cylinder head.
(3) *Operations at right side of engine.*
 (*a*) Using wrench 41–W–3249–900, disconnect wiring harness ((N), fig. 23) at head light wiring connector ((B), fig. 22) at right side of radiator.

*Figure 21. Power plant disconnect points on left side.*

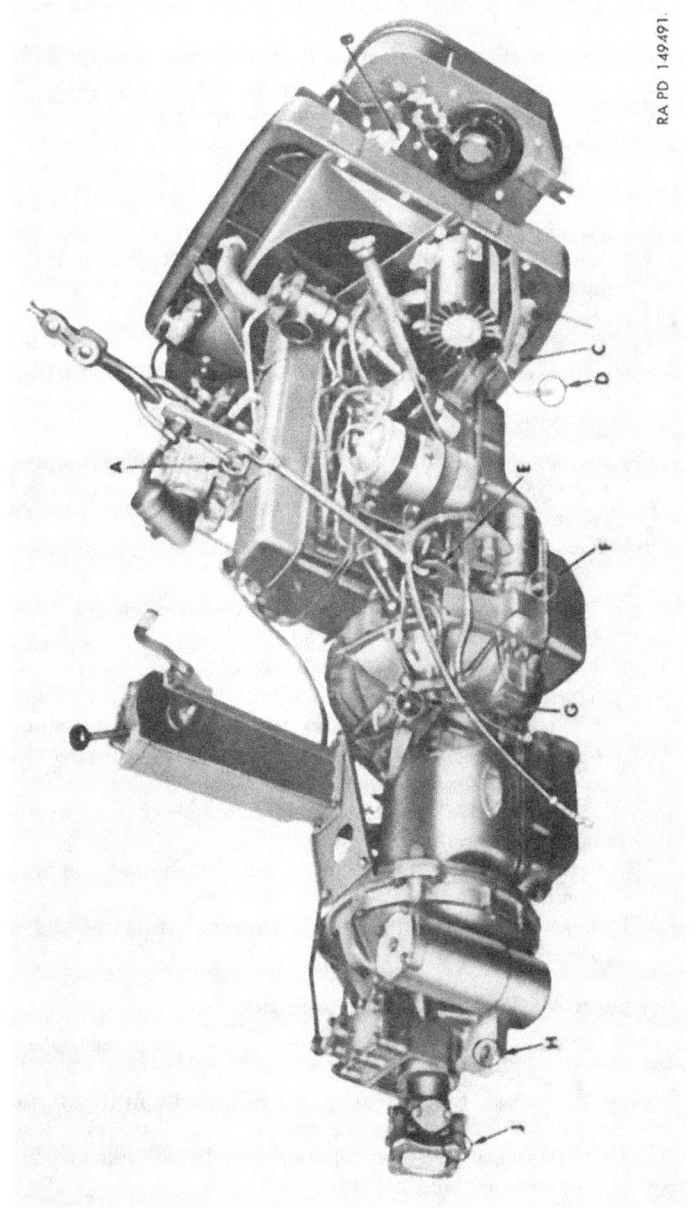

*Figure 22. Power plant removed showing disconnect points on right side.*

(*b*) Disconnect vent line ((D), fig. 23) at transmission filler neck ((G), fig. 22).
  (*c*) Remove four bolts connecting transmission-to-transfer universal joint flanges ((F), fig. 23 and (J), fig. 22).
  (*d*) Remove one bolt and washer attaching power plant to rear mounting ((H), fig. 22 and (E), fig. 23).
  (*e*) Remove bolt, nut, washers, and rubber cushions attaching power plant at front mounting ((C), fig. 22, and (P) and (L), fig. 23).
  (*f*) Remove engine ground strap ((D), fig. 22) bolt at frame side member ((Q), fig. 23).
 (4) *Removal of power plant from vehicle.*
  (*a*) Engage short hook of engine lifting sling 41–S–3831–600 with eye nut ((A), fig. 22) and long hook with lower lifting bracket ((E), fig. 22). Attach chain to engine sling cross bar.
  (*b*) Inspection should be made to see if all disconnect operations have been completed before power plant is raised off mountings.
  (*c*) Raise power plant slowly, using several short lifts until free from mountings.
   *Note.* As power plant is raised, be sure radiator lifting arms engage brackets on radiator support.
  (*d*) Pull power plant slowly forward, raising as necessary to clear front cross member.
  (*e*) When power plant is out of vehicle, support in manner which will permit access to drain plugs in transmission case and drain plug in engine oil pan cover.
   **Caution.** Do not permit weight of power plant to rest on oil pan or radiator support.
  (*f*) Remove radiator mounting cushion and shims from front cross member.
  (*g*) Refer to TM 9–1819AA for information on rebuild of power plant assembly.

## 23. Removal of Fuel Tank and Supports

*Note.* Key letters in following text indicate disconnect points and refer to figure 24.

 *a. General.* The fuel tank assembly, which includes filler cap, fuel gage, sending unit and fuel pump, is removed as an assembly. Fuel tank supports and straps are to be removed from frame side members as assemblies after removing fuel tank.
 *b. Removal Procedures.*
  (1) Remove drain plug and gasket from bottom of fuel tank and drain fuel.

*Figure 23. Power plant disconnect points on chassis.*

(2) After draining fuel from tank, install plug and gasket in tank.
(3) Disconnect two wiring harness connectors (H) at fuel pump.
(4) Disconnect wiring harness connector (G) at fuel gage sending unit.
(5) Disengage wiring harness from clip (D) at fuel gage sending unit.
(6) Disconnect fuel line (C) at shut-off cock (J) on tank.
(7) Disconnect vent line (E) at elbow (F) on tank.
(8) Remove nut from each strap holding fuel tank to support at (K). Raise straps and remove fuel tank assembly.

(9) Remove four cap screws (B) and nuts attaching each fuel tank support and strap assembly to side member; then remove support assemblies.

(10) Remove chassis wiring harness and clip (A) from fuel tank front support.

*Figure 24. Fuel tank disconnect points.*

## 24. Removal of Muffler and Rear Exhaust Pipe Assembly

*Note.* Key letters in following text indicate disconnect points and refer to figure 23.

*a. General.*—The muffler and rear exhaust pipe assembly, which includes muffler support bracket, muffler support, and rear exhaust pipe to transmission support hanger strap, is removed as an assembly.

*b. Removal Procedures.*

(1) Remove bolt (C), nut, washer, insulators, and spacer attaching rear exhaust pipe support strap to transmission rear support.

(2) Remove bolt (B) and nut attaching muffler supporting strap to transmission rear support.

(3) Remove four cap screws (A), washers, spacers, and eight insulators attaching muffler supporting bracket to side members; then remove muffler and rear exhaust pipe assembly from vehicle.

(4) Disconnect rear exhaust pipe and muffler bracket from muffler.

## 25. Removal of Spare Wheel and Carrier

*a. General.*—The spare wheel and tire is to be removed first from spare wheel carrier; then the spare wheel carrier, which includes swivel bracket and support bracket, is to be removed as an assembly.

*b. Removal Procedures.*
  (1) Remove two nuts attaching spare wheel swivel bracket to lock bracket on frame. Swing carrier and wheel out and away from frame and tip wheel and tire to upright position.
  (2) Remove four nuts attaching wheel to swivel bracket; then remove wheel and tire from bracket.
  (3) Remove four bolts and nuts attaching spare wheel carrier and support bracket to frame side member.

## 26. Removal of Steering Gear Assembly

*Note.* Key letters in following text indicate disconnect points and refer to figure 25.

*Figure 25. Steering gear disconnect points.*

43

*a. General.*—The steering gear assembly, which includes Pitman arm and steering drag link, is removed as an assembly.

*b. Removal Procedures.*

   (1) Disconnect drag link at front axle steering arm (D) by removing ball stud nut (C) and driving ball stud out of steering arm.

   (2) Remove four cap screws (A and B) and nuts which mount steering gear to frame side member; then remove steering gear and drag link from vehicle.

   (3) Disconnect drag link from steering gear Pitman arm by removing nut and driving out drag link ball stud.

   (4) Remove nut and washer which retain Pitman arm to Pitman shaft. With puller 41–P–2952, remove arm from shaft.

   (5) Refer to chapter 13 for rebuild of steering gear and drag link.

## 27. Removal of Brake Pedal Lower Half and Pedal Brace

*Note.* Key letters in following text indicate disconnect points and refer to figure 26.

*a.* Remove cap screw (B) and washer which attach brake pedal shaft brace (D) to side of brake master cylinder.

*b.* Disengage brake pedal pull back spring (G) from clip (F).

*c.* Remove cotter pin and clevis pin attaching master cylinder push rod yoke and clip (F) to brake pedal lower half.

*Figure 26. Master cylinder and brake pedal disconnect points.*

*d.* Remove cotter pin (E) retaining brake pedal lower half and pedal brace on pedal shaft, then remove pedal brace and pedal from chassis.

## 28. Removal of Brake Master Cylinder

*Note.* Key letters in following text indicate disconnect points and refer to figure 26.

*a.* Disconnect brake hydraulic line at rear of master cylinder by removing bolt (A) and washers from line connector.

*b.* Remove four cap screws (C) and nuts attaching brake master cylinder to frame bracket, then remove master cylinder.

*c.* Refer to paragraphs 236 through 238 for rebuild of brake master cylinder.

## 29. Removal of Winch Drive Line Assembly

*a. General.*—The winch drive line assembly, which includes both front and rear drive shafts, rear universal joint at power take-off end of rear shaft, pilot bearing, and pilot bearing bracket and shield, is removed as an assembly.

*b. Removal Procedures.*
   (1) At rear universal joint, loosen set screw securing joint to power take-off shaft.
   (2) At front of front shaft, loosen set screw on shaft stop and slide stop from end of shaft.
   (3) Remove two cap screws attaching pilot bearing bracket to front spring and torque rod bracket.
   (4) Pull complete drive line assembly toward front until rear universal joint yoke clears power take-off shaft; then pull drive line assembly to rear until end of front shaft clears hole in front cross member. Lower drive line assembly and remove from under vehicle.
   (5) Refer to paragraphs 316 through 319 for rebuild of winch drive line.

## 30. Removal of Air Reservoir Assemblies

*a. General.*—The air reservoir assemblies, which includes safety valve (on left reservoir only), air line fittings, mounting supports, and "U" bolts, are removed as assemblies.

*b. Removal Procedures.*
   (1) Disconnect all air lines at each air reservoir.
   (2) Remove two nuts from each "U" bolt retaining air reservoirs to frame side members. Remove air reservoirs.
   (3) Remove air line fittings, drain cocks, and safety valve from reservoirs.

## 31. Removal of Air-Hydraulic Cylinder Assembly

*Note.* Key letters in following text indicate disconnect points and refer to figure 27.

*a. General.*—The air-hydraulic cylinder assembly, which includes rear mounting bracket, is removed as an assembly.

*b. Removal Procedures.*
   (1) Disconnect three top air lines (A, B, and C) and two lower hydraulic fluid lines (D and E) from cylinder.
   (2) Remove two bolts (F) and nuts attaching cylinder rear bracket (G) to frame side member.
   (3) Remove nuts from two bolts attaching cylinder to front mounting bracket.
   (4) Remove air-hydraulic cylinder assembly from vehicle.

*Figure 27. Air-hydraulic cylinder disconnect points.*

## 32. Removal of Propeller Shaft Assemblies

*a. General.*—The propeller shaft assemblies, which includes universal joints and flange yokes, are removed as assemblies.

*b. Removal Procedure.*
   (1) Remove four bolts from each propeller shaft universal joint flange. Lower propeller shaft assembly and remove from under vehicle.
   (2) Refer to chapter 9 for rebuild of propeller shafts.

## 33. Removal of Transfer Assembly

*Note.* Key letters in following text indicate disconnect points and refer to figure 28.

*a. General.*—The transfer assembly, which includes parking brake, power take-off, and power take-off control cable, is removed as an assembly. A suitable jack or a chain fall is required to remove transfer assembly.

  b. *Removal Procedures.*

  (1) Disconnect rod (G) linking transfer reverse cross shaft (F) and transfer lower shifter shaft (H) by removing yoke pin (E) at transfer end of rod.
  (2) Disconnect transfer upper shifter shaft (J) from transfer control cross shaft (C) by removing yoke pin (D) and flat washers.
  (3) Disconnect transfer vent line at vent line tee on frame No. 2 cross member.
  (4) Disconnect parking brake pull rod at parking brake cam levers by removing pin.
  (5) Bend back transfer mounting bolt locks (B). With jack or hoist, raise transfer just enough to remove weight of transfer from mounting bolts. Remove mounting bolts (A) from each side of transfer; then lower transfer assembly and remove from under vehicle.
  (6) Disconnect vent line from transfer.
  (7) Refer to TM 9-819A for removal of parking brake components, power take-off, and power take-off control cable from transfer.

*Figure 28. Transfer disconnect points.*

(8) Refer to chapter 5 for rebuild of transfer, chapter 6 for rebuild of power take-off, and paragraphs 244 through 246 for rebuild of parking brake components.

## 34. Removal of Wheels

*a.* Loosen wheel stud nuts, using wheel stud nut wrench 41-W-3838-75 with handle 41-H-1541-10.

*b.* Raise complete chassis until tires are clear of floor or ground and place suitable supports under frame.

*c.* Remove wheel stud nuts; then remove wheels.

## 35. Removal of Shock Absorber Assemblies

*Note.* Key letters in following text indicate disconnect points and refer to figure 29.

*a. General.*—Each shock absorber assembly, which includes link, is removed as an assembly.

*b. Removal Procedures.*
   (1) Remove nut (F) and disconnect shock absorber link from spring bumper block (E).
   (2) Remove two cap screws (A) and nuts attaching each shock absorber to frame side member. Remove shock absorber assembly.

*Figure 29. Shock absorber and front axle disconnect points.*

(3) Remove nut attaching link pin to shock absorber arm; then drive out link pin and remove link.

(4) Refer to paragraphs 273 through 277 for rebuild of shock absorber assembly.

## 36. Removal of Front Axle Assembly

*a. General.*—The front axle assembly, which includes hubs, torque rods, brake lines, and steering tie rod, is removed as an assembly. A suitable jack or chain fall is required to remove front axle.

*b. Removal Procedures.*

(1) At axle, disconnect flexible hydraulic brake and axle vent lines by removing bolts ((A) and (B), fig. 30) and washers attaching each line connector ((C), fig. 30) to axle.

(2) Remove two clips ((D), fig. 30) retaining brake and vent lines in shield on upper torque rod.

(3) At frame end of each torque rod, remove nut ((B), fig. 29 and (F), fig. 30), and drive torque rod ((C), fig. 29 and (G), fig. 30) from frame brackets with soft metal hammer.

(4) Using jack or chain fall, raise front axle just enough to permit removal of "U" bolt nuts ((D), fig. 29), spring bumper block ((E), fig. 29), and spring "U" bolts ((G), fig. 29).

(5) Lower jack or chain fall until axle clears under side of chassis; then withdraw axle from under vehicle.

(6) Remove nut ((H), fig. 29) and washer from torque rod tapered pins. With soft metal hammer, drive pins out of axle bracket to remove torque rods from axle.

(7) Refer to chapter 7 for rebuild of front axle assembly.

*Figure 30. Front axle upper torque rod and lines disconnect points.*

## 37. Removal of Rear Axle Assemblies

*a. General.*—Each rear axle assembly, which includes brakes, hubs, metal brake lines, and propeller shaft pillow block on forward rear

49

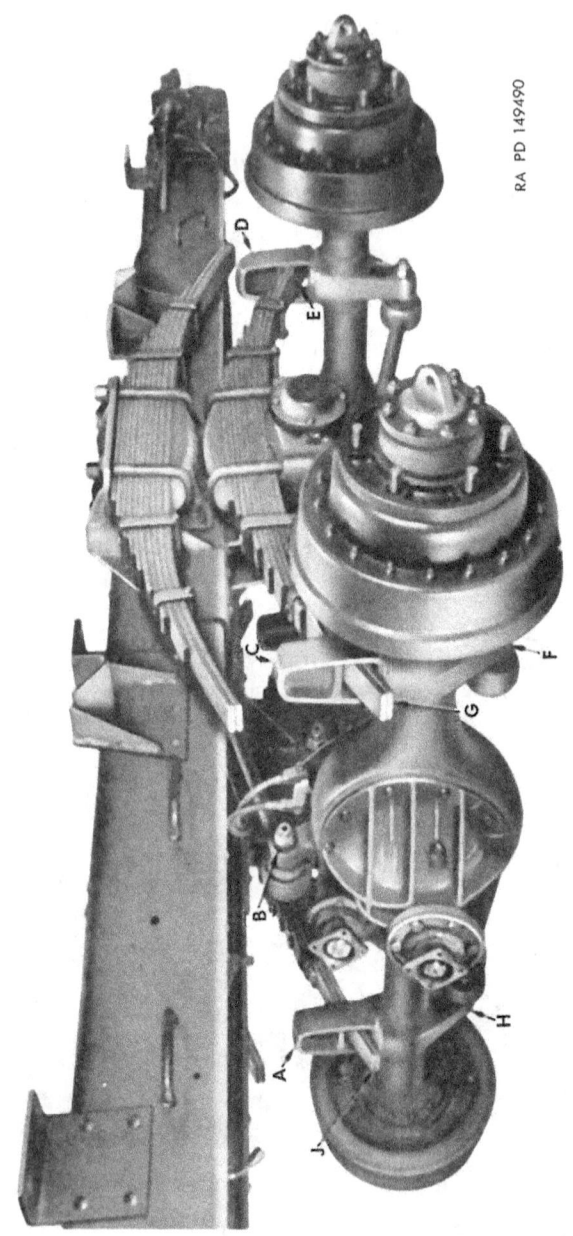

*Figure 31. Rear axles installed showing disconnect points.*

*Figure 32. Forward rear axle upper torque rod and lines disconnect points.*

51

axle only, is removed as an assembly. A suitable dolly type jack or a chain fall is required to remove rear axle assemblies from vehicle.

*b. Removal Procedures.*

 (1) At axles, disconnect hydraulic brake and axle vent lines by removing bolt ((D), fig. 32 or (G), fig. 33) and washers attaching each line connector ((E), fig. 32 or (H), fig. 33) to axle.

 (2) Remove tow clips ((A), fig. 32 or (F), fig. 33) retaining brake and vent lines in shield on each upper torque rod.

 (3) Place jack under axle or attach chain fall to axle and raise assembly just enough to remove tension from torque rods.

 (4) At axle end of each torque rod, remove nuts ((B), fig. 31, (C), fig. 32, or (D), fig. 33) and washer ((B), fig. 32); then drive torque rod pins from axle brackets ((F), fig. 32 or (J), fig. 33) with soft metal hammer.

 (5) Lower axle and at the same time move axle out and away from ends of springs; then withdraw axle from under vehicle.

 (6) Remove torque rods from frame brackets in same manner as described for removal of torque rod pins from axle brackets in (4) above.

 (7) Remove pillow block assembly from forward rear axle by removing four mounting stud nuts and dowels attaching pillow

*Figure 33. Rear rear axle upper torque rod and lines disconnect points.*

block to axle bracket. Refer to paragraphs 222 through 225 for rebuild of pillow block assembly.

(8) Refer to chapter 8 for rebuild of rear axle assemblies.

### 38. Removal of Front Spring Assemblies

*Note.* Key letters in following text indicate disconnect points and refer to figure 34.

*a. General.*—Each front spring assembly, which includes spring shackles, is removed as an assembly.

*b. Removal Procedures.*
  (1) Remove lubrication fittings (B) from shackle pins and bolts.
  (2) Support spring assembly adequately before removing shackle bolts.
  (3) Remove nut from bolt (A) attaching each spring shackle (D) to spring bracket (C and E) on frame.
  (4) Drive spring shackle bolts from spring brackets and shackles with soft metal hammer; then lower and withdraw spring from under chassis.
  (5) Remove spring shackles from spring by removing clamp bolt which secures shackle pin at inner side of shackle. From

*Figure 34. Front spring disconnect points.*

53

inner side of spring, drive pin out of shackle and spring eye at each end of spring.
(6) Refer to paragraphs 269 through 272 for rebuild of front spring suspension components.

### 39. Removal of Rear Main Spring Assemblies

*a. General.*—Each rear main spring assembly, which includes spring seat, spring seat bearing cups, spring "U" bolts, and "U" bolt spacers, is removed as an assembly.

*b. Removal Procedures.*
(1) Remove four cap screws and washers ((G), fig. 35) which attach each spring seat dust cap ((F), fig. 35) to seat. Remove cap.
(2) Bend tangs of nut lock away from lock nut retaining spring seat to spring seat shaft.
(3) Remove lock nut, nut lock, adjusting nut, washer, and outer bearing cone from spring seat shaft.
(4) Slide spring assembly off spring seat shaft and remove from vehicle.
(5) Remove oil seal flange and inner bearing from spring seat.
(6) Remove spring seat bearing seal from spring seat shaft.
(7) Refer to chapter 15 for rebuild of rear spring suspension components.

### 40. Removal of Rear Secondary Spring Assemblies

*Note.* Key letters in following text indicate disconnect points and refer to figure 35.

*a. General.*—Each rear secondary spring assembly is removed as an assembly.

*b. Removal Procedures.*
(1) Remove nuts (A) and spacers (B) from "U" bolts (C) which attach spring to spring seat (D) and frame bracket.
(2) While supporting spring, remove four bolts (E) and nuts which attach each spring seat to frame brackets; then remove spring assembly.

### 41. Removal of Control Shafts and Brackets

*a. Removal of Transfer Control Cross Shaft and Brackets Assembly.*—Remove two cap screws ((A) and (D), fig. 36) and nuts which attach each bracket to No. 2 cross member; then remove cross shaft and bracket assembly.

*b. Removal of Transfer Reverse Cross Shaft, and Brackets and Linkage Assembly.*—Remove three cap screws and nuts which attach the right cross shaft bracket to frame No. 2 cross member, and remove

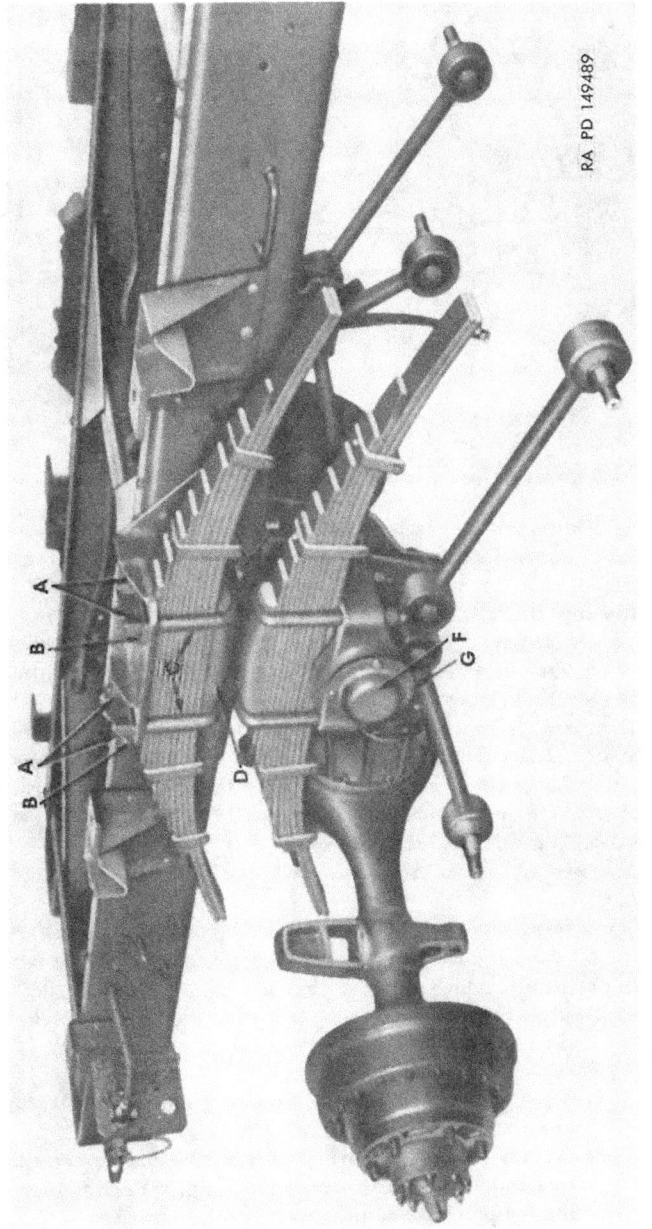

*Figure 35. Rear spring disconnect points.*

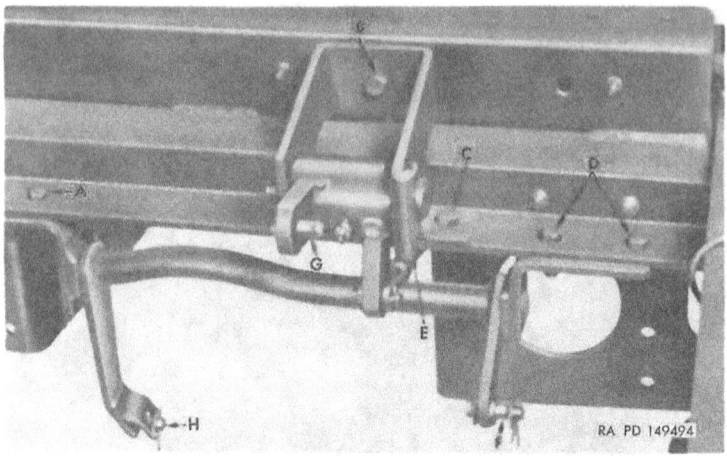

*Figure 36. Transfer cross shaft and parking brake relay lever disconnect points.*

two cap screws and nuts which attach the left cross shaft bracket to transmission rear support; then remove cross shaft and brackets assembly.

*c. Removal of Parking Brake Relay Lever Shaft, Bracket, and Rod Assembly.*—Remove three cap screws ((B) and (C), fig. 36) attaching relay lever shaft bracket ((E), fig. 36) to frame No. 2 cross member; then remove brake relay lever shaft, bracket, and rod assembly.

*d. Removal of Transmission Rear Support, Support Brackets, and Support Cushions Assembly.*—Remove four cap screws which attach support and bracket to each side member; then remove support and brackets. Support cushions can be removed from support by removing two cap screws from each cushion.

*e. Removal of Trailer Air Connection and Tail Light Bracket Assemblies.*

    (1) *General.*—The right bracket assembly, which includes trailer air connection cut-out cock and coupling, and the left bracket assembly, which includes chassis wiring harness trailer connection receptacle, receptacle cover, air cut-out cock, and trailer air coupling, are to be removed as assemblies.

    (2) *Removal Procedures.*

        (*a*) Disengage chassis wiring harness from clip ((C), fig. 37 and (E), fig. 38) on each bracket.

        (*b*) At left bracket, remove chassis wiring harness receptacle by removing four cap screws ((B), fig. 38) and nuts retaining receptacle cover and ground cable to bracket.

        (*c*) Disconnect trailer air supply line from cut-out cock ((D), fig. 37 and (D), fig. 38) on each bracket.

(*d*) Remove two cap screws ((B), fig. 37 and (A), fig. 38) and nuts which attach each trailer connection and tail light bracket to frame side member; then remove brackets.

*f. Removal of Rear Upper Torque Rod Brackets.* Remove six cap screws ((A), fig. 33) which attach each upper torque rod bracket (B), bracket reinforcement (C), and brake and vent line support (E) to cross member and remove brackets, reinforcements, and line supports.

*Figure 37. Right trailer air connection and tail light bracket disconnect points.*

*Figure 38. Left trailer air connection and tail light bracket disconnect points.*

## 42. Removal of Rear Bumpers, Tow Hooks, and Pintle

*a. Removal of Rear Bumpers.*—Remove six cap screws ((A), fig. 37) and nuts attaching each bumper to frame side member and rear cross member; then remove bumpers.

*b. Removal of Rear Tow Hooks.*—Remove cotter pin holding shackle pin to bracket; then remove shackle pin and shackle.

*c. Removal of Pintle and Pintle Brackets.*
   (1) Remove cotter pin securing pintle nut to pintle shaft. Insert bar through pintle jaw to prevent its turning; then remove nut and washer at end of pintle shaft. Remove pintle.
   (2) Remove two cap screws and nuts retaining inner and outer pintle brackets to rear cross member and remove brackets.

## 43. Removal of Lines and Wiring

*a. Removal of Air, Hydraulic, and Vent Lines From Frame.*—Disconnect all lines from each other. Disengage lines from clips and grommets that retain lines to frame; then remove lines from frame.

*b. Removal of Wiring Harnesses From Frame.*—Disengage wiring harnesses from clips and grommets that retain each wiring harness to frame; then remove harnesses from frame.

### Section II. ASSEMBLY OF VEHICLE FROM MAJOR COMPONENTS

## 44. General

This section provides an assembly line procedure for assembling the vehicle from its major components. The sequence of procedures for assembling vehicle from its major components are listed in logical sequence, permitting the use of more than one mechanic. However, the sequence of procedures can be changed to meet existing conditions or facilities. The illustrations shown in this section and in section I of this chapter are keyed with reference letters which indicate connect points. These key letters are referred to as such in the various procedures.

*Note.* Procedures covering installation of cargo body (par. 79) apply only to cargo trucks M135 and M211; all other procedures generally apply to all vehicles.

## 45. Installation of Lines and Harnesses on Frame

*a. General.*—The starting point of assembly procedures on this vehicle is with the frame assembly. The frame assembly consists of all brackets and reinforcements which are welded or riveted to frame side members and cross members. Support and position frame assembly in suitable manner. If an overhead hoist with a stationary type track is used, place frame assembly directly under track.

*b. Installation of Air Vent Lines on Frame.*—Install air vent lines on frame, using required clips and short pieces of loom which serve as insulation between lines and clips. Do not tighten attaching clips until vent lines are connected to their respective units.

*c. Installation of Air Lines on Frame.*—Install air lines on frame, using required clips, clip spacers, and short pieces of loom which serve as insulation between lines and clips. Do not tighten attaching clips until air lines are connected to their respective units.

*d. Installation of Hydraulic Lines on Frame.*—Install hydraulic lines on frame, using required clips and short pieces of loom which serve as insulation between lines and clips. Do not tighten attaching clips until hydraulic lines are connected to their respective units.

*e. Installation of Wiring Harnesses on Frame.*—Install wiring harnesses on frame, using required rubber coated type clips. Do not tighten attaching clips until wiring harnesses are connected to their respective units.

### 46. Installation of Rear Upper Torque Rod Brackets

*Note.* Key letters in following text indicate connect points and refer to figure 33.

Position each torque rod bracket (B) to cross member and install six $7/16$–20 x 2 cap screws (A). On opposite side of cross member install torque rod reinforcement (C) (metal plate with six screw holes) over bracket cap screws, then install six $7/16$–20 nuts. Tighten nuts to torque of 33 to 43 pound-feet.

### 47. Installation of Front Spring Assemblies

*Note.* Key letters in following text indicate connect points and refer to figure 34.

*a. General.*—Each front spring assembly, which includes shackles, can be installed as an assembly.

*b. Installation Procedures.*

    (1) Position spring assembly at spring brackets (C and E).

    (2) Install shackle bolt (A) through spring bracket and shackle. Shank of bolt at outer end is serrated. Drive shackle bolt in from outer side, seating bolt firmly in spring bracket. Install nut on bolt at each shackle and tighten until shackle binds; then loosen nuts just enough to relieve binding.

    (3) Install lubrication fittings (B) in shackle bolts and pins; then lubricate shackles.

### 48. Installation of Rear Secondary Spring Assemblies

*Note.* Key letters in following text indicate connect points and refer to figure 35.

*a.* Position spring under upper ledge of frame mounting bracket, with spring center bolt engaging hole in frame bracket. Install spring seat (D) under spring with center bolt properly located in seat; then attach seat to frame bracket with two $\frac{7}{16}$–20 x $1\frac{3}{4}$ and two $\frac{7}{16}$–20 x 2 cap screws (E) and four $\frac{7}{16}$–20 nuts. Tighten nuts to torque of 33 to 43 pound-feet.

*b.* Install "U" bolts (C), spacers (B), and nuts (A). Push spring against frame bracket and parallel to frame. Tighten nuts on "U" blots to torque of 375 to 400 pound-feet.

### 49. Installation of Rear Main Spring Assemblies

*Note.* Key letters in following text indicate connect points and refer to figure 39.

*a. General.* Each rear main spring assembly, which includes spring seat, bearing cups, "U" bolts, and "U" bolt spacers, can be installed as an assembly.

*b. Installation Procedures.*

(1) Coat inner diameter of oil seal assembly (J) with plastic type gasket cement. Press seal over sleeve (F) with lip of seal toward frame bracket. Press on sleeve until inner edge of seal is $\frac{1}{4}$-inch from inner edge of sleeve.

(2) Lubricate tapered roller bearing cones thoroughly. Install tapered roller bearing cone (E) in main spring seat assembly (D).

(3) Press oil seal flange (H) into spring seat.

(4) Slide spring and seat assembly over main spring seat shaft (G), with oil seal flange over oil seal lip; then insert tapered roller bearing cone (C).

(5) Install adjusting nut washer (M) over spring seat shaft threads. Install adjusting nut (N) and tighten to torque of 60 to 75 pound-feet, using wrench 41–W–545–5 with torque wrench 41–W–3634 (fig. 40), while oscillating spring and seat to make sure bearings are properly seated; then back off adjusting nut one-quarter turn.

(6) Install adjusting nut lock (P) and second adjusting nut (Q), which serves as a lock nut. Tighten lock nut to torque of 100 to 150 pound-feet, using wrench 41–W–545–5 with torque wrench 41–W–3634; then bend tangs of nut lock over flats on adjusting nuts.

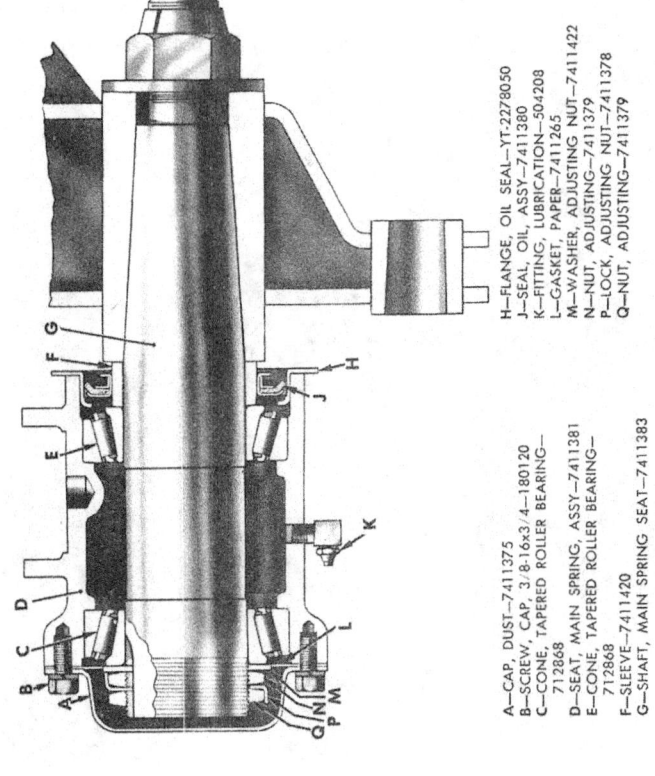

A—CAP, DUST—7411375
B—SCREW, CAP, 3/8-16x3/4—180120
C—CONE, TAPERED ROLLER BEARING—712868
D—SEAT, MAIN SPRING, ASSY—7411381
E—CONE, TAPERED ROLLER BEARING—712868
F—SLEEVE—7411420
G—SHAFT, MAIN SPRING SEAT—7411383
H—FLANGE, OIL SEAL—YT-2278050
J—SEAL, OIL, ASSY—7411380
K—FITTING, LUBRICATION—504208
L—GASKET, PAPER—7411265
M—WASHER, ADJUSTING NUT—7411422
N—NUT, ADJUSTING—7411379
P—LOCK, ADJUSTING NUT—7411378
Q—NUT, ADJUSTING—7411379

RA PD 149517

*Figure 39. Sectional view of rear main spring seat installed.*

61

(7) Install dust cap (A), using new paper gasket (L), and attach with four ⅜–16 x ¾ cap screws (B). Tighten cap screws firmly.

(8) Fill spring seat with lubricant through lubrication fitting (K) until lubricant appears at seal on inner side of seat. Refer to TM 9–819A or to lubrication order for type of lubricant.

*Figure 40. Using wrench 7950946 with torque wrench 41–W–3634 to tighten spring seat adjusting nuts.*

## 50. Installation of Front Axle Assembly

*a. General.*—The front axle assembly, which includes hubs, brakes, brake lines, and steering tie rod, can be installed as an assembly. A suitable dolly type jack or a chain fall is required to install front axle assembly.

*b. Preliminary Procedures.*—If metal hydraulic brake lines are not on axle, install brake lines as directed in paragraph 179*d*.

*c. Installation Procedures.*

(1) Place axle assembly into position under vehicle, using jack or chain fall.

(2) Lift axle until spring center bolts are located in holes of spring seats on top of axle. Install "U" bolts ((G), fig. 29) and spring bumper block ((E), fig. 29) with shock absorber link eye at front. Install "U" bolt nuts ((D), fig. 29) and tighten nuts to torque of 170 to 200 pound-feet.

(3) Install front torque rods (par. 51).

(4) Connect hydraulic brake and axle vent rubber lines at junction on top of axle housing with bolts ((A) and (B), fig. 30) and four plain copper washers.

*Note.* On each line, install one washer under bolt head and one washer between line connector ((C), fig. 30) and line junction on axle.

(5) On upper torque rod only, position brake and vent rubber lines in shield on top of torque rod. Secure lines in shield with two clips ((D), fig. 30).

## 51. Installation of Front Torque Rods

*Note.* Following procedures apply to the upper and both lower torque rods.

*a.* Place dust seal ((K), fig. 29) on both ends of torque rod bearing at frame bracket end. Position dust seal ((J), fig. 29) over end pin at axle end of rod.

*b.* Position torque rod tapered pin in axle bracket, being careful not to dislodge dust seal from bearing.

*c.* Position torque rod end with seals in frame bracket; then insert torque rod bolt ((C), fig. 29 or (G), fig. 30) in frame bracket, making sure torque rod bearing is alined with bolt hole. Drive bolt through bracket and bearing, using a soft metal hammer.

*Note.* Plain washer must be installed under bolt head at frame bracket on upper torque rod only. Upper torque rod bolt is installed from outer side and both lower bolts are installed from inner side. It may be necessary to turn bolts slightly when driving into place to aline serrations on bolt with serrations in bearing.

*d.* Install flat washer ((E), fig. 30) and nut ((B), fig. 29 or (F), fig. 30) on frame bracket bolt and tighten nut to torque of 190 to 250 pound-feet.

*e.* Install plain washer and nut ((H), fig. 29) on torque rod tapered pin at axle end; then tighten nut to torque of 350 to 400 pound-feet.

## 52. Installation of Shock Absorber Assemblies

*Note.* Key letters in following text indicate connect points and refer to figure 29.

*a. General.*—Each shock absorber assembly, which includes link, is installed as an assembly. Connect shock absorber link to shock absorber arm by inserting link pin in arm. Install 1/2–20 nut on link pin and tighten nut to torque of 48 to 64 pound-feet.

*b. Installation Procedures.*

(1) Position absorber assembly at frame; then insert two 9/16–18 x 4 cap screws (A) through shock absorber and frame side member.

(2) Place a reinforcing spacer at inner side of frame side member over cap screws. Install a 9/16–18 nut on each cap screw and tighten to torque of 63 to 84 pound-feet.

(3) Insert shock absorber link pin through eye on spring bumper block (E), install 1/2–20 nut (F) on link pin, and tighten to torque of 48 to 64 pound-feet.

### 53. Installation of Rear Axles

*a. General.*—The rear axle assemblies, which include hubs, brakes, brake lines, propeller shaft universal joint flanges, and pillow block on forward rear axle only, can be installed as assemblies. A suitable dolly type jack or a chain fall is required to install rear axles.

*b. Preliminary Procedures.*

(1) If metal brake lines are not on axle, install brake lines as directed in paragraph 208.

(2) Install propeller shaft pillow block on forward rear axle by positioning pillow block on inside (differential side) of mounting bracket with drain plug toward front. Install four tapered dowel wedges and four 1/2–20 nuts on mounting studs and tighten to torque of 48 to 64 pound-feet.

(3) Attach all torque rods to frame and spring seat shaft brackets by inserting tapered torque rod end pins through brackets, install plain washer and nut on each pin, and tighten to torque of 350 to 400 pound-feet.

*c. Installation Procedures.*

(1) Move each axle into position under chassis and engage main spring ends ((E), (G), and (J), fig. 31) with openings in brackets ((A), (C), and (D), fig. 31) at each end of axle housing.

(2) At each axle, attach three torque rods to axle brackets ((B), (F), and (H), fig. 31) by installing tapered pins in axle brackets. Install plain washer and nut on each pin and tighten nut to torque of 350 to 400 pound-feet.

(3) Connect hydraulic brake and axle vent rubber lines at junction on top of axle housing with two bolts ((G), fig. 33) and four plain copper washers.

*Note.* At each line connection, install one washer under bolt head and one washer between line connector and line junction on axle.

(4) On upper torque rod only, position brake and vent lines in shield on top of torque rod. Secure lines in shield with two clips ((A), fig. 32 and (F), fig. 33).

## 54. Installation of Transfer Assembly

*Note.* Key letters in following text indicate connect points and refer to figure 28.

*a. General.*—The transfer assembly, which includes parking brake, power take-off, and power take-off control cable, can be installed as an assembly. A suitable jack or a chain fall is required to install transfer assembly.

*b. Preliminary Procedures.*—Install power take-off, power take-off control cable, and parking brake on transfer. Refer to TM 9–819A for installation procedures and adjustments.

*c. Installation Procedures.*

(1) Using jack or chain fall, position transfer assembly under chassis.

(2) Install locks (B) on eight $\frac{1}{2}$–13 x $1\frac{3}{16}$ transfer mounting bolts (A) and install bolts attaching transfer to frame No. 2 cross member (four on each side of transfer). Tighten bolts to torque of 60 to 85 pound-feet. Bend bolt locks against bolt heads.

(3) Connect transfer vent line to tee on cross member and to transfer. Tighten line nuts firmly.

## 55. Installation of Transmission Rear Support, Support Brackets, and Cushions

*a.* Attach transmission support and brackets to underside of frame side member with four $\frac{1}{2}$–20 x $2\frac{1}{4}$ cap screws and $\frac{1}{2}$–20 nuts. Tighten nuts to torque of 48 to 64 pound-feet.

*b.* Attach each support bracket to side of frame side member with two $\frac{1}{2}$–20 x $1\frac{1}{4}$ cap screws and $\frac{1}{2}$–20 nuts. Tighten nuts to torque of 48 to 64 pound-feet.

## 56. Installation of Transfer Reverse Cross Shaft, Brackets, and Linkage Assembly

*a.* Position cross shaft assembly at frame No. 2 cross member and at transmission rear support.

*b.* Attach cross shaft right bracket to frame No. 2 cross member with two $\frac{3}{8}$–24 x 1 cap screws and $\frac{3}{8}$–24 nuts. Tighten nuts to torque of 20 to 27 pound-feet.

*c.* Attach cross shaft left bracket to transmission rear support with two $\frac{5}{16}$–24 x $\frac{7}{8}$ cap screws and $\frac{5}{16}$–24 nuts. Tighten nuts to torque of $9\frac{1}{2}$ to 13 pound-feet.

*d.* Connect transfer reverse cross shaft lever rod yoke to transfer lower shifter shaft by installing yoke pin. Secure yoke pin with cotter pin. Refer to TM 9–819A for adjustment procedures on transfer control linkage.

## 57. Installation of Transfer Control Cross Shaft and Brackets Assembly

*a.* Position cross shaft assembly to frame No. 2 cross member.

*b.* Attach each cross shaft bracket with two 3/8–24 x 1 cap screws ((A) and (D), fig. 36) and 3/8–24 nuts. Tighten nuts to torque of 20 to 27 pound-feet.

*c.* Install yoke pin ((D), fig. 28 or (H), fig. 36) and two plain washers connecting cross shaft lever ((C), fig. 28) to transfer upper shifter shaft ((J), fig. 28).

## 58. Installation of Parking Brake Relay Lever Shaft, Bracket, and Rod Assembly

*a.* Position shaft assembly on frame No. 2 cross member with parking brake rod placed at cam levers at brake.

*b.* Install three 3/8–24 x 1 cap screws ((B) and (C), fig. 36) and 3/8–24 nuts attaching bracket ((E), fig. 36) to frame No. 2 cross member. Tighten nuts to torque of 20 to 27 pound-feet.

*c.* Install parking brake rod between cam levers; then insert 3/8 x 15/16 clevis pin through cam levers and rod. Secure clevis pin with cotter pin.

## 59. Installation of Propeller Shaft Assemblies

Position propeller shaft between flanges of connecting units with lubrication fittings in same plane as other shafts; then install four bolts and nuts at each end.

*Note.* Propeller shafts must be installed with slip joints positioned as shown in figure 149.

## 60. Installation of Wheels

*a.* Position wheels on hub studs.

*b.* Install wheel stud nuts. Tighten nuts to torque of 300 to 350 pound-feet, using wheel stud nut wrench 41–W–3838–75 adapted to a torque wrench.

## 61. Installation of Winch Drive Line Assembly

*a. General.*—The winch drive line assembly, which includes front and rear drive shafts, rear shaft universal joint at power take-off, drive shaft pilot bearing, and pilot bearing bracket, can be installed as an assembly.

*b. Installation Procedures.*
  (1) Guide front end of front shaft through hole in frame front cross member; then position stop on shaft. Do not tighten stop set screw until after winch has been installed.
  (2) Push entire shaft assembly toward front until rear universal joint can be installed on power take-off shaft, alining key in shaft with keyway in universal joint yoke. Tighten set screw which secures joint yoke to power take-off shaft.
  (3) Position pilot bearing bracket against front spring and torque rod bracket; then install two 1/2–20 x 1 1/4 cap screws and 1/2–20 nuts. Tighten nuts to torque of 48 to 64 pound-feet.

## 62. Installation of Brake Master Cylinder

*Note.* Key letters in following text indicate connect points and refer to figure 26.

*a.* Position master cylinder to bracket on frame. Install four 3/8–24 x 1 1/4 cap screws (C) and 3/8–24 nuts which attach master cylinder to bracket. Tighten nuts to torque of 20 to 27 pound-feet.

*b.* Insert hydraulic brake line connector bolt (A) through connector, with washer on both sides of connector, and thread bolt into rear end of master cylinder. Tighten bolt to torque of 20 to 30 pound-feet.

## 63. Installation of Brake Pedal Lower Half and Pedal Brace

*Note.* Key letters in following text indicate connect points and refer to figure 26.

*a.* Place brake pedal lower half on pedal shaft.

*b.* Position brace (D) on pedal shaft and secure with cotter pin (E).

*c.* At master cylinder end of brace, install one 7/16–20 x 7/8 cap screw (B) and 7/16-inch lock washer which attach brace to master cylinder.

*d.* Position master cylinder push rod yoke and clip (F) to brake pedal lower half and secure with clevis pin and cotter pin.

*e.* Install brake pedal return spring (G) between clip (F) and winch drive shaft pilot bearing bracket.

## 64. Installation of Air-Hydraulic Cylinder Assembly

*Note.* Key letters in following text indicate connect points and refer to figure 27.

*a. General.*—The air-hydraulic cylinder assembly, which includes rear mounting bracket, can be installed as an assembly.

*b. Installation Procedures.*
  (1) Install rear mounting bracket (G) on stud at rear of air-hydraulic cylinder and secure with plain washer and 5/16–24 nut. Do not tighten nut.

(2) Position air-hydraulic cylinder at frame side member, with two end plate bolts inserted through front mounting bracket. Install 5/16–24 nut on each bolt. Tighten nuts to torque of 9½ to 13 pound-feet.

(3) Position rear bracket (G) at frame side member and secure with two 3/8–24 x 1 cap screws (F) and 3/8–24 nuts, attaching hydraulic brake line clip under front cap screw head. Tighten nuts attaching bracket to frame to torque of 20 to 27 pound-feet; then tighten nut attaching bracket to rear end of air-hydraulic cylinder to torque of 9½ to 13 pound-feet.

(4) Connect hydraulic line (D) to slave cylinder end fitting and hydraulic line (E) to bottom of control valve.

(5) Connect air line (A) to control valve inlet port, and connect air vent line (C) to exhaust port. Connect air line (B) to control valve trailer air outlet port. Tighten all line connections firmly.

## 65. Installation of Steering Gear Assembly

*Note.* Key letters in following text indicate connect points and refer to figure 25.

*a. General.*—The steering gear assembly, which includes Pitman arm and steering drag link, can be installed as an assembly.

*b. Preliminary Procedures.*

(1) Install Pitman arm on Pitman arm shaft, matching blank serration on shaft with blank serration in arm. Install lock washer and nut. Tighten nut to torque of 115 to 155 pound-feet.

(2) Connect drag link to steering gear Pitman arm by inserting link stud in Pitman arm; then install 5/8–18 nut. Tighten nut to torque of 75 to 100 pound-feet.

*c. Installation Procedures.*

(1) Position steering gear with drag link on frame side member and install three ½–20 x ½ cap screws (A), one ½–20 x 1⅞ cap screw (B), and four ½–20 nuts. Tighten nuts to torque of 48 to 64 pound-feet.

(2) Position steering drag link ball stud in hole in steering arm on front axle. Install 5/8–18 ball stud nut (C) and tighten to torque of 75 to 100 pound-feet.

## 66. Installation of Air Reservoir Assemblies

*a. General.*—The air reservoir assemblies, which includes air line fittings, mounting supports, "U" bolts, and safety valve on left reservoir only, can be installed as assemblies.

b. *Installation Procedures.*
  (1) Install air line connectors and drain cock on each tank, using plastic type gasket cement on all threads.
  (2) Install safety valve and elbow on left reservoir.
  (3) Position "U" bolts on reservoir, install reservoir supports on "U" bolts; then install assembly on frame side member. Install $5/16$–24 nuts on "U" bolts and tighten nuts to torque of $9\frac{1}{2}$ to 13 pound-feet.
  (4) Connect air lines to reservoirs and tighten connections firmly.

## 67. Installation of Muffler Assembly

*Note.* Key letters in following text indicate connect points and refer to figure 41.

  a. *General.*—The muffler assembly, which includes support bracket, strap, and brace can be installed as an assembly.
  b. *Installation Procedures.*
  (1) Position muffler supporting bracket (V) on muffler (N). Install muffler supporting strap (M) around muffler and attach strap to muffler supporting bracket (V) with four $3/8$–24 x $1\frac{1}{4}$ cap screws (W) and $3/8$–24 nuts. Tighten nuts just enough to support muffler until after exhaust pipe and tail pipe are connected.
  (2) Position four $5/16$–24 x $1\frac{5}{8}$ cap screws (S) in muffler, supporting bracket (V) which supports muffler to frame side member in the following manner: On each cap screw, place one washer (T), one spacer (U), and one rubber insulator (R) (over spacer) with smaller outside diameter of insulator facing threaded end of screw. Insert cap screw with washer, spacer, and insulator through muffler support bracket. Install another rubber insulator (Q) over spacer on cap screw with smaller outside diameter of insulator facing bracket.
  (3) Position muffler assembly, with cap screws installed in bracket, at frame side member. Install $5/16$–24 nuts (P) on cap screws and tighten to torque of $9\frac{1}{2}$ to 13 pound-feet.
  (4) Attach muffler support brace (L) to muffler with one $5/16$–24 x $5/8$ cap screw and $5/16$–24 nut, and attach support brace to transmission rear support with one $5/16$–24 x $3/4$ cap screw and $5/16$–24 nut. Tighten each nut to torque of $9\frac{1}{2}$ to 13 pound-feet.

## 68. Installation of Power Plant Assembly

  a. *General.*—The power plant assembly, which includes engine and accessories, radiator, radiator baffle, brush guards, head lights, engine

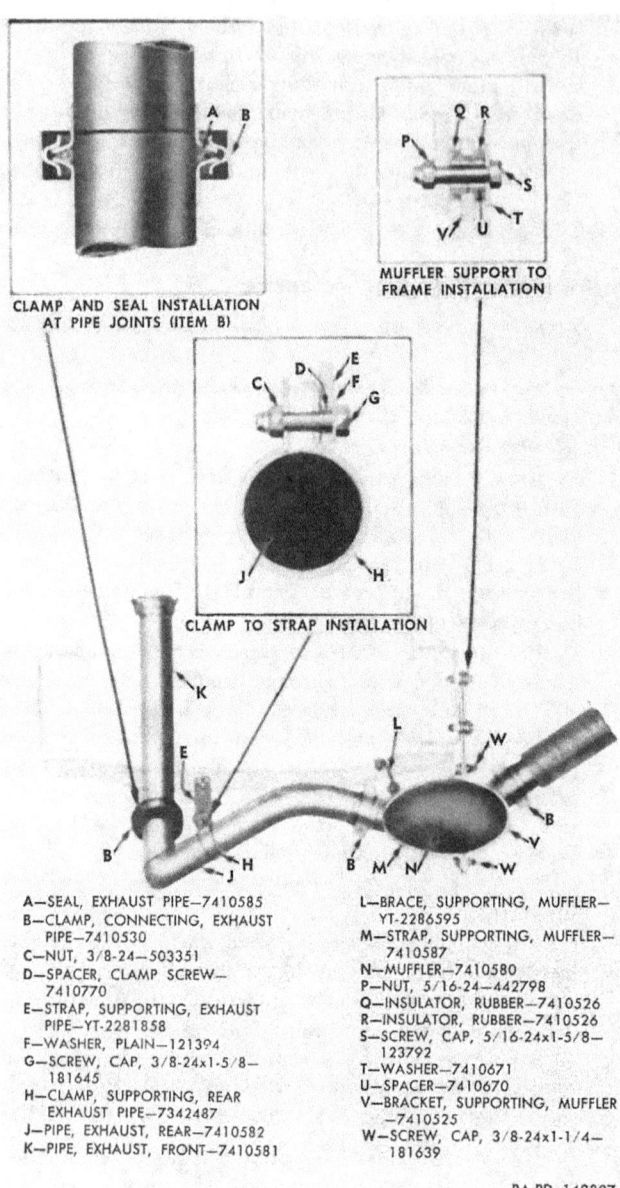

Figure 41. Rear exhaust pipe and muffler installation.

front exhaust pipe, transmission, transmission control tower, and universal joint, can be installed as an assembly.

*b. Equipment.*—A chain fall and one special tool are required to install power plant assembly. Engine lifting sling 41–S–3831–600 must be used to lift power plant into chassis.

*c. Installation Procedures.*
  (1) *Raise power plant into vehicle.*
    (a) Hook engine lifting sling 41–S–3831–600 into eye nut ((A), fig. 22) and lower lifting bracket ((E), fig. 22) on engine. Attach sling to chain fall.
    (b) Raise power plant assembly high enough to clear vehicle frame, then move into position above frame. Carefully lower power plant, taking care to avoid damage to lines, wiring, etc. Do not rest power plant solidly on mountings until two support cushion shims ((A), fig. 42) and support cushion ((D), fig. 42) at radiator support assembly ((E), fig. 42) are in place and bolt holes are alined. Drop engine mounting bolts ((J), fig. 42) with special washer ((M), fig. 42) through front engine mounting assemblies. Aline power plant rear mounting bolt holes; then lower power plant on mountings.
    (c) Remove engine lifting sling from power plant.
  (2) *Operations at front of vehicle.*—Assemble support spring washers ((G), fig. 42) and support springs ((F), fig. 42) on radiator support special bolts ((H), fig. 42), then insert bolts upward through bracket and install bolt spacer ((C), fig. 42) and ⅜–24 nuts ((B), fig. 42) on bolts. Tighten nuts firmly.
  (3) *Operations at right side of vehicle.*
    (a) Assemble engine mounting cushion ((K), fig. 42), special washer ((L), fig. 42), and ⅜–24 nut ((B), fig. 42) on front mounting bolt. Tighten nut firmly.
    (b) Install ½–20 x 1⅜ cap screw ((N), fig. 42) with ½-inch lock washer ((P), fig. 42) to anchor power plant rear mounting ((H), fig. 22) to support cushion ((Q), fig. 42).
    (c) Aline holes in universal joint flanges ((J), fig. 22 and (F), fig. 23) at rear of transmission and install four bolts.
    (d) Connect vent line ((D), fig. 23) at transmission filler neck ((G), fig. 22).
    (e) Attach engine ground strap ((D), fig. 22) to frame side member ((Q), fig. 23) with one ¼–28 x 1 cap screw, one 9/32-inch plain washer, one ¼-inch internal-external-teeth lock washer and one ¼–28 nut. Tighten nut to torque of 5 to 7 pound-feet.

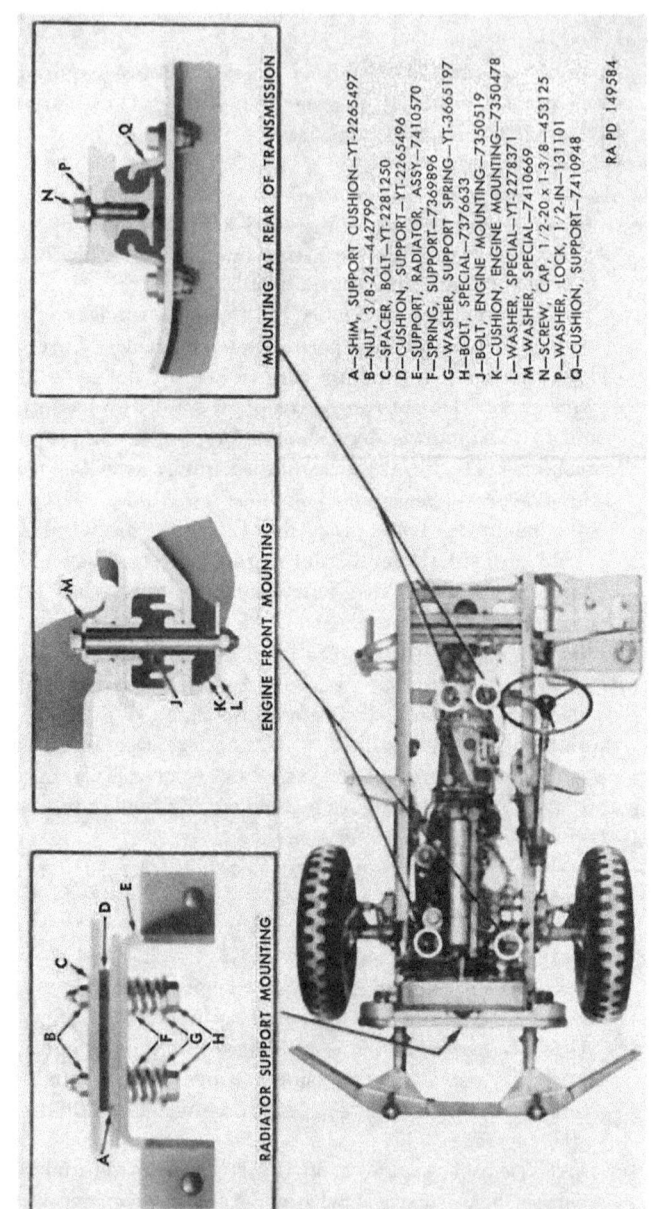

A—SHIM, SUPPORT CUSHION—YT-2265497
B—NUT, 3/8-24—442799
C—SPACER, BOLT—YT-2281250
D—CUSHION, SUPPORT—YT-2265496
E—SUPPORT, RADIATOR, ASSY—7410570
F—SPRING, SUPPORT—7369896
G—WASHER, SUPPORT SPRING—YT-3665197
H—BOLT, SPECIAL—7376633
J—BOLT, ENGINE MOUNTING—7350519
K—CUSHION, ENGINE MOUNTING—7350478
L—WASHER, SPECIAL—YT-2278371
M—WASHER, SPECIAL—7410669
N—SCREW, CAP, 1/2-20 x 1-3/8—453125
P—WASHER, LOCK, 1/2-IN—131101
Q—CUSHION, SUPPORT—7410948

RA PD 149584

*Figure 42. Power plant mountings.*

(f) Connect wiring harness ((N), fig. 23) to head light wiring connector ((B), fig. 22) using spanner wrench 41-W-3249-900.

(4) *Operations at left side of vehicle.*

(a) Assemble engine mounting cushion ((K), fig. 42), special washer ((L), fig. 42), and ⅜-24 nut on engine mounting bolt ((J), fig. 42) and tighten nut firmly.

(b) Install ½-20 x 1⅜ cap screw ((N), fig. 42) with ½-inch lock washer ((P), fig. 42) to anchor power plant rear mounting ((E), fig. 21) to rear support cushion ((Q), fig. 42).

(c) Connect flexible fuel line ((H), fig. 23) to carburetor ((C), fig. 21).

(d) Connect rod ((G), fig. 23) linking transmission lever and transfer reverse cross shaft by installing yoke pin at transmission end of rod ((F), fig. 21).

(e) Connect air compressor discharge line ((K), fig. 23) to elbow ((B), fig. 21) on air compressor and secure line to base of air compressor with clip ((J), fig. 21).

(f) Disconnect brace ((A), fig. 21) from engine cylinder head stud and position in clip on radiator support.

## 69. Installation of Rear Exhaust Pipe

*Note.* Key letters in following text indicate connect points and refer to figure 41.

a. Install exhaust pipe supporting strap (E) and rear exhaust pipe supporting clamp (H) on rear exhaust pipe (J) with ⅜-24 x 1⅝ cap screw (G), clamp screw spacer (D), plain washer (F), and ⅜-24 nut (C). Do not tighten nut.

b. Install new exhaust pipe seals (A) at exhaust pipe ends.

c. Position rear exhaust pipe (J) at muffler (N) and front exhaust pipe (K), then secure exhaust pipe connecting clamp (B) at each end of pipe.

d. Attach rear exhaust pipe hanger strap (E) to transmission rear support by installing ⁵⁄₁₆-24 x 1⅝ cap screw, flat washer, and spacer, with rubber insulator on each side of strap, through transmission rear support and secure with ⁵⁄₁₆-24 nut. Tighten nut firmly.

e. Tighten cap screw nut on rear exhaust pipe supporting clamp (H) to torque of 20 to 27 pound-feet.

## 70. Installation of Cab

*Note.* Key letters in following text indicate connect points and refer to figure 43.

*a. General.*
   (1) The cab assembly, which includes instruments, accelerator pedal, engine air cleaner, generator regulator, wiring, batteries, air lines, vent lines, parking brake lever, transfer lever, power take-off lever, muffler tail pipes, hood extension panels, hood props, running boards, tool box, fluid container bracket, gun mount bracket, and "U" bolts, can be installed as an assembly.
   (2) A suitable chain fall must be provided to lift cab onto chassis.
   (3) The sequence of procedures for installing cab on chassis are listed in logical sequence, permitting the use of more than one mechanic; however, the sequence can be changed to meet existing conditions or facilities.

*b. Raise Cab Onto Chassis.*
   (1) Attach hoist to four lifting ring nuts on cab.
   (2) Raise cab high enough to clear steering gear column and move cab to position above mountings on frame.
   (3) Lower cab, taking care to avoid damage to steering column, fuel tank, control linkage, etc. Do not rest cab solidly on supports until rubber cushions (D) at each front support are in place and bolt holes are alined.
   (4) Install steel washer (C) and cushion (D) on each front mounting bolt (B); then from inside of cab, insert bolts through cab, cushion (D), and support bracket.
   (5) Position rubber cushion (K) and steel spacer (G) into place under rear center of cab. Aline with center cap screw hole in cross member.
   (6) Install steel washer (H) and cushion (K) on $1/2$–20 x $4\frac{1}{2}$ cap screw (J); then insert cap screw through cab floor, steel spacer, mounting cushion, and frame cross member.
   (7) Install two $7/16$–20 x 1 cap screws (N) and $7/16$–20 nuts which attach each cab mounting spring to frame cross member.
   (8) Lower cab and remove hoist.
   (9) Install two lower cushions (D), steel washer (E), and $1/2$–20 nut (F) on each front mounting bolt (B). Tighten nut firmly against shoulder on bolt.
   (10) Install cushion (K), steel washer (L), and $1/2$–20 nut (M) on rear center cap screw. Tighten nut firmly.

*c. Procedures Underneath Cab.*
   (1) Connect speedometer shaft ((T), fig. 18) to transfer. Tighten connector nut firmly.
   (2) Connect parking brake lever rod ((V), fig. 18) to relay lever ((N), fig. 19 or (G), fig. 36) on frame cross member with yoke pin and cotter pin.

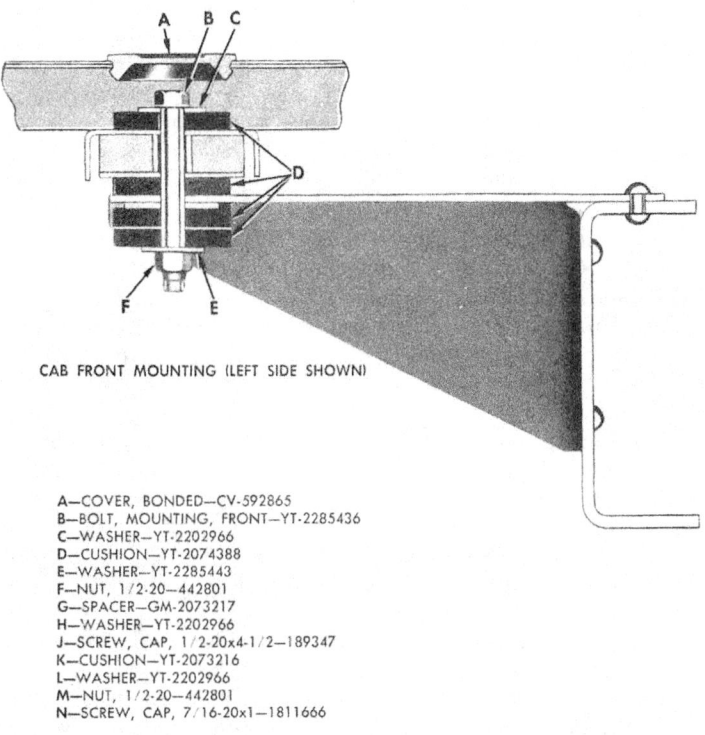

A—COVER, BONDED—CV-592865
B—BOLT, MOUNTING, FRONT—YT-2285436
C—WASHER—YT-2202966
D—CUSHION—YT-2074388
E—WASHER—YT-2285443
F—NUT, 1/2-20—442801
G—SPACER—GM-2073217
H—WASHER—YT-2202966
J—SCREW, CAP, 1/2-20x4-1/2—189347
K—CUSHION—YT-2073216
L—WASHER—YT-2202966
M—NUT, 1/2-20—442801
N—SCREW, CAP, 7/16-20x1—1811666

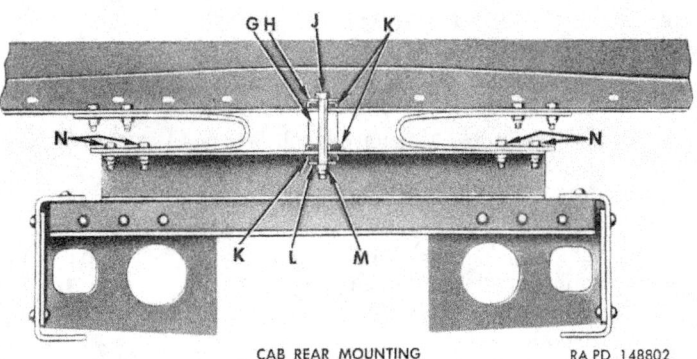

*Figure 43. Cab mountings.*

(3) Connect transfer control lever rod ((U), fig. 18) *to cross shaft* ((F), fig. 36) on frame cross member, with one yoke pin, two flat washers, and one cotter pin.
(4) Insert power take-off control cable ((P), fig. 19) up through hole in floor of cab; then secure cable to bracket under cab floor with clamp bolt and nut.
(5) Connect rubber vent line ((R), fig. 18) to brake master cylinder ((R), fig. 19).
(6) Connect starter ground cable ((H), fig. 18) to stud ((F), fig. 22) on starter. Tighten nut firmly.

*d. Procedures Inside of Cab.*
(1) Under drivers seat, connect power take-off control cable to power take-off hand control lever with yoke pin and cotter pin. Install rubber grommet around control cable at cab floor.
(2) Through cab floor opening at right side of transmission, connect three rubber lines ((BB), fig. 18 and (F), fig. 19—air line), ((CC), fig. 18 and (E), fig. 19—vent line), and ((AA), fig. 18 and (G), fig. 19—exhaust line). Tighten connections firmly.
(3) Install cab front floor pan over transmission control tower and secure to floor with fourteen $5/16$–24 x $5/8$ cap screws.
(4) Position rubber seal over shift control tower; then lower seal against floor pan.
(5) Raise rubber seal on steering column; then install upper brake pedal plate and lower brake pedal plate to cowl with seven $5/16$–24 x $5/8$ cap screws. Position steering column seal against top of pedal plates.
(6) Insert brake pedal upper ball through seal in upper brake pedal plate.

*Note.* Refer to e(1) below for attaching brake pedal upper ball to brake pedal lower half.

(7) Install cab front mounting bolt hole covers ((A), fig. 43) in floor pan.
(8) Position rubber grommet and bracket cap around steering column at dash. Install two $3/8$–24 x 2 cap screws and $3/8$–24 nuts which attach cap to dash bracket. Tighten nuts to torque of 20 to 27 pound-feet.
(9) Connect horn cable at connector ((S), fig. 19) on steering column.
(10) Install steering wheel on steering gear shaft; then install nut. Tighten nut to torque of 40 to 55 pound-feet.
(11) Position horn button contact and spring over steering shaft nut, and install horn button and retaining ring; then attach retaining ring to steering wheel with four No. 8 x $7/8$ screws.

*e. Procedures at Front of Cab.*

(1) *Connect brake upper pedal to lower pedal.* Insert brake pedal upper half into end of brake pedal lower half. Aline notch in pedal upper half with clamp screw hole in pedal lower half. Install $3/8$–24 x $1 5/8$ cap screw and $3/8$–24 nut. Tighten nut to torque of 20 to 27 pound-feet.

(2) *Connect choke and throttle controls.*

   (*a*) Connect choke control ((F), fig. 18) at carburetor.

   (*b*) Connect throttle rod ((H), fig. 21) linking accelerator lever ((GG), fig. 18) on cowl to accelerator lever on intake manifold with clevis pin and cotter pin.

(3) *Connect hoses and lines.*

   (*a*) Connect air cleaner hose ((G), fig. 18) to elbow ((B), fig. 19) on carburetor.

   (*b*) Connect engine vent line ((EE), fig. 18) to nipple ((D), fig. 19) on right side of engine.

   (*c*) Connect air compressor-to-cowl air line ((Z), fig. 18) to connector ((A), fig. 19) on air compressor governor.

(4) *Install batteries.*

   (*a*) Position battery on battery support. Install battery retainer over hold-down bolts. Make sure battery lifting handles are positioned in recess of battery to prevent battery retainer from resting on top of battery. Install $5/16$-inch lock washer and $5/16$–18 nut on each hold-down bolt; then tighten nuts firmly.

   (*b*) Install battery cables and terminals on battery posts and tighten clamp bolts. Refer to figure 44 for battery cable identification.

   (*c*) Engage starter cable ((K), fig. 19) in clips ((F), fig. 18) on cowl.

(5) *Connect wiring harness and cables.*

   (*a*) Connect engine wiring harness to instrument panel wiring harness at three bayonet type connectors ((DD), fig. 18 and (H), fig. 19).

   (*b*) Connect generator wiring cable ((Y), fig. 18) to generator, using spanner wrench 41–W–3249–900.

*f. Procedures at Sides of Cab.*

(1) At left side of cab, connect two wiring harnesses ((Q), fig. 19) at multiple plug and receptacle connectors ((N), fig. 18) located under cab floor directly above running board, using spanner wrench 41–W–3249–900, then engage harnesses in clips ((Q), fig. 18) on running board rear support.

(2) At right side of cab, install new seal on tail pipe and connect tail pipe to muffler with clamp.

*Figure 44. Battery cables installed.*

### 71. Installation of Fender and Skirt Assemblies

*a. General.*—Each fender and skirt assembly, which includes hood catch, fender support, fender brace, and blackout head light on left fender assembly only, can be installed as an assembly.

  *b. Installation Procedures.*
   (1) Position fender assembly to vehicle. Install one ½–20 x 1 cap screw and plain washer which attach each fender skirt to radiator side baffles ((E), fig. 16).
   (2) Install one ½–20 x 1 cap screw, one plain washer, and one lock washer which attach each fender brace to bracket ((L), fig. 18) on cowl.
   (3) Install two 5/16–24 x 1⅜ cap screws, one rubber spacer, two washers, and two 5/16–24 nuts which attach each fender to running board ((P), fig. 18).
   (4) Install two ½–20 x 1 cap screws, two plain washers, and two lock washers which attach each fender support to brush guard and radiator side baffle ((A), fig. 16).
   (5) At left fender, connect blackout head light wiring cables at two bayonet type connectors.

### 72. Installation of Hood Assembly

*a. General.*—The hood assembly, which includes horn and horn air supply line, can be installed as an assembly.

  *b. Installation Procedures.*
   (1) Position hood assembly on vehicle, with hinge cap screw holes alined. Install 5/16–24 x 2½ cap screw and 5/16–24 nut in each hinge; then, while supporting hood in upright position, connect each prop ((A), fig. 18) to hood with two ¼–28 x ⅝ cap screws.
   (2) Connect horn air supply line to air line fitting ((E), fig. 18) on cowl.
   (3) Connect horn wiring cables at two bayonet type connectors ((J), fig. 18).

## 73. Installation of Winch

*a. General.*—The winch assembly, which includes support brackets and drive shaft universal joint, can be installed as an assembly. A suitable chain fall is required to install winch assembly.

*b. Installation Procedures.*

(1) Install right and left support brackets ((D) and (A), fig. 45) on which with special bolts ((E), fig. 45) and $9/16$-inch lock washers.

(2) Install front universal joint ((B), fig. 45) on splines of winch shaft and insert shear pin ((C), fig. 45) through joint and shaft. Secure shear pin with two cotter pins.

(3) Supporting winch with chain fall, guide winch with support brackets attached (fig. 45) between frame side members. As winch is pushed toward rear, guide front universal joint splines onto front drive shaft splines.

(4) Position support brackets in place in frame. Install eight $5/8$–18 x $1 3/4$ bracket-to-frame cap screws ((A), (B), (C), and (F), fig. 17) and eight $5/8$–18 nuts. Tighten nuts to torque of 95 to 127 pound-feet.

(5) Position front drive shaft stop $3/8$-inch from winch universal joint yoke. Lock stop to shaft with $3/16$–16 x $1/2$ set screw.

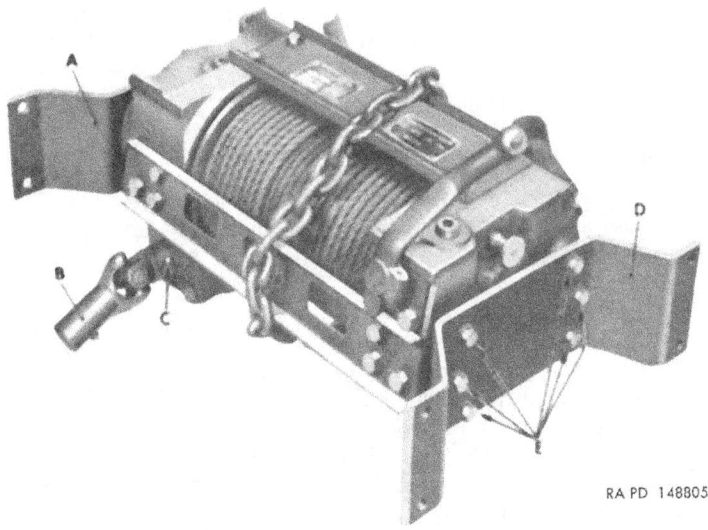

*Figure 45. Winch assembly.*

### 74. Installation of Front Bumper Assembly

*a. General.*—The front bumper assembly, which includes gussets, can be installed as an assembly.

*b. Installation Procedures.*
  (1) Position bumper assembly in place on front of frame side members with attaching cap screw holes in alinement.
  (2) Install eight cap screws ((D), fig. 16) and nuts. Tighten nuts firmly.
  (3) Position tow hook ((C), fig. 16) to bracket on bumper and secure with shackle pin ((B), fig. 16); then install cotter pin through hole in shackle pin.

### 75. Installation of Spare Wheel and Carrier

*a. General.*—The spare wheel carrier, less the spare wheel and tire, should be installed first on frame side member; then install spare wheel and tire on carrier.

*b. Installation Procedures.*
  (1) Position spare wheel carrier and bracket to frame side member. Install four ½–20 x 1¼ cap screws and ½–20 nuts which attach carrier to side member. Tighten nuts to torque of 48 to 64 pound-feet.
  (2) Position spare wheel and tire on carrier swivel bracket and attach with four ¾–16 nuts. Tighten nuts to torque of 250 to 300 pound-feet.
  (3) Tip wheel and tire to horizontal position and swing in toward frame, with swivel bracket in place against lock bracket. Install nuts on two attaching studs and tighten to torque of 250 to 300 pound-feet.

### 76. Installation of Fuel Tank and Supports

*Note.* Key letters in following text indicate connect points and refer to figure 24.

*a. General.*—The fuel tank, which includes filler cap, fuel gage, and fuel pump, can be installed as an assembly.

*b. Installation Procedures.*
  (1) Position each fuel tank support and strap to frame side member and install four ⅜–24 x 1 cap screws (B) and ⅜–24 nuts. Tighten nuts to torque of 20 to 27 pound-feet.
  (2) Position fuel tank on supports; then lower hold-down straps over tank. Secure each strap to support with ⅜–24 nut (K). Tighten each nut firmly.
  (3) Connect two bayonet type wiring harness connectors (H) at fuel pump.
  (4) Connect bayonet type wiring harness connector (G) at fuel gage sending unit.

(5) Engage wiring harness in clip (D) at fuel gage sending unit.
(6) Connect fuel line (C) to fuel tank shut-off cock (J).
(7) Connect vent line (E) to fuel tank elbow (F).
(8) Engage chassis wiring harness in clip (A) on fuel tank front support.

## 77. Installation of Rear Tow Hooks, Bumpers, Pintle, and Pintle Brackets

*a. Installation of Rear Tow Hooks.*—Position each tow hook at bracket on rear cross member and secure with shackle pin; then install cotter pin through hole in shackle pin.

*b. Installation of Rear Bumpers.*—Position each rear bumper against frame side member and rear cross member with cap screw holes in alinement. Install four $1/2$–20 x $15/8$ and two $1/2$–20 x $11/2$ cap screws ((A), fig. 37) and nuts in each bumper. Tighten nuts to torque of 48 to 64 pound-feet.

*c. Installation of Pintle and Pintle Brackets.*
  (1) Install two $3/4$–16 x 3 cap screws and two $3/4$–16 nuts which attach inner and outer pintle brackets to rear cross member. Tighten nuts to torque of 165 to 220 pound-feet.
  (2) Lubricate pintle shaft and insert shaft through brackets and cross member. Install one $11/2$–12 nut and one plain washer on shaft. Use bar through pintle jaw to prevent turning as nut is tightened. Tighten until pintle binds; then back off nut until cotter pin can be installed and pintle can be turned by hand.

## 78. Installation of Trailer Air Connection and Tail Light Bracket Assemblies

*a. General.* The right bracket assembly, which includes trailer air connection cut-out cock and coupling, and the left bracket assembly, which includes chassis wiring harness trailer connection receptacle, receptacle cover, air cut-out cock, and trailer air supply coupling, can be installed as assemblies.

*b. Installation Procedures.*
  (1) Install cut-out cock and trailer air supply coupling on each bracket.
  (2) Install each bracket assembly to frame side member with two $1/2$–20 x $11/2$ cap screws ((B), fig. 37 and (A), fig. 38) and two $1/2$–20 nuts. Tighten nuts to torque of 48 to 64 pound-feet.
  (3) On left bracket, install chassis wiring harness receptacle, receptacle cover ((C), fig. 38), and ground cable, using four $1/4$–28 x $7/8$ cap screws and four $1/4$–28 nuts ((B), fig. 38). Tighten nuts firmly.

(4) Connect air supply line to coupling ((D), fig. 37 and (D), fig. 38) on each bracket. Tighten line nut firmly.

(5) Engage chassis wiring harness in clip ((C), fig. 37 and (E), fig. 38) on each bracket.

### 79. Installation of Cargo Body

*Note.* Key letters in following text indicate connect points and refer to figure 46.

*a. General.*—The cargo body, which includes racks, roof bows, top paulin, rear curtain, tail light, marker light, reflectors, and splash shields, can be installed as an assembly. One or more chain falls are required to install body on chassis.

*b. Installation Procedures.*

(1) Place a wood body support sill on each frame side member, making sure cut-outs in sills match rivet heads on top of frame to permit sills to seat solidly on side members. Position cargo body on chassis and aline mounting bolt holes.

(2) At front flexible spring-type mountings, position $1\frac{1}{16}$-inch plain washer (E), inner compression spring (F), and outer compression spring (G) on each bolt (A). Position bolt (A) through frame and body brackets. Install $\frac{5}{8}$-18 nut (B) and tighten just enough to partly compress springs.

(3) At each rear rigid-type mounting, position $\frac{3}{4}$-16 x $2\frac{1}{2}$ cap screw (C) up through frame and body bracket, and install $\frac{3}{4}$-16 nut (D). Tighten nuts to torque of 165 to 220 pound-feet.

(4) Connect tail and marker light wiring harness to chassis wiring harness at bayonet type connectors.

### 80. Inspection

Perform a technical inspection as prescribed in AR 700–105, using DA Form 461–5, Limited Technical Inspection, and as outlined in TM 9–819A for the "6,000-mile" organizational maintenance services.

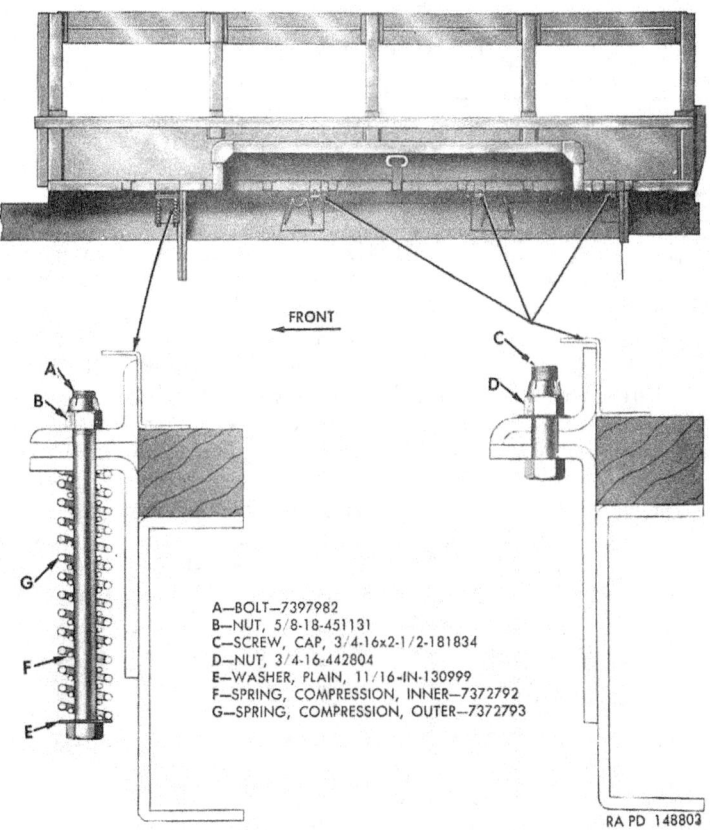

*Figure 46. Cargo body mountings.*

# CHAPTER 5

# TRANSFER ASSEMBLY

## Section I. DESCRIPTION AND DATA

### 81. Description

*Note.* Arrows and lines on views in figure 48 show power path through transfer assembly when vehicle is driven forward or backward and following text describes operation. Key letters in text refer to figure 49 unless otherwise indicated.

*a. General Description.*—Transfer assembly, which is mounted behind transmission assembly, provides a means for transmitting power to each of the rear axles and to the front axle when required. Input and output shafts are equipped with companion drive flanges to which propeller shafts are connected. Distribution of power to output shafts is by means of constantly meshed helical gears, supported on respective shafts which are mounted on nonadjustable bearing assemblies. All moving parts are splash-lubricated by lubricant contained in transfer case. Seals installed at drive flanges and shifter shafts prevent leakage of lubricant and entrance of dirt and water. Opening is provided at left side of transfer case to accommodate a power take-off assembly used to operate such equipment as winch, dump body hoist, and pumps on trucks with tank-type bodies. Parking brake mechanism is assembled at rear of transfer, and speedometer is driven by gears assembled at idler shaft front bearing retainer. The two shifter shafts mounted in support at front side of transfer case are operated by mechanical linkage. Lower shifter shaft is interconnected with transmission control and automatically positions front axle output shaft gear for forward and reverse driving. External views of transfer assembly are shown in figure 47.

*b. Operation.*

(1) *Forward.*—When transmission control is in neutral or any of the forward speeds, the front axle shifter shaft is in rearward position and shifter shaft front spring ((S), fig. 54), which is stronger than shifter shaft rear spring ((H), fig. 54), forces yoke toward rear and holds front axle output shaft gear in contact with clutch teeth on output shaft. Input shaft is driven from transmission, and since sliding gear (B) ex-

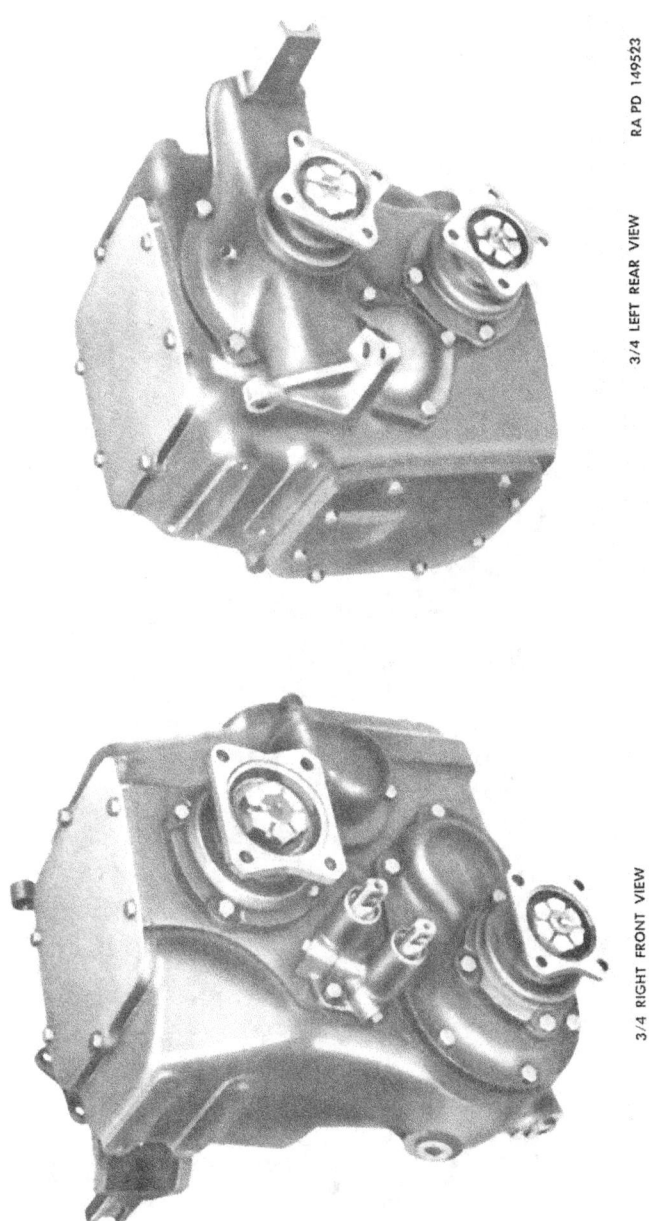

*Figure 47. Front and rear external views of transfer assembly.*

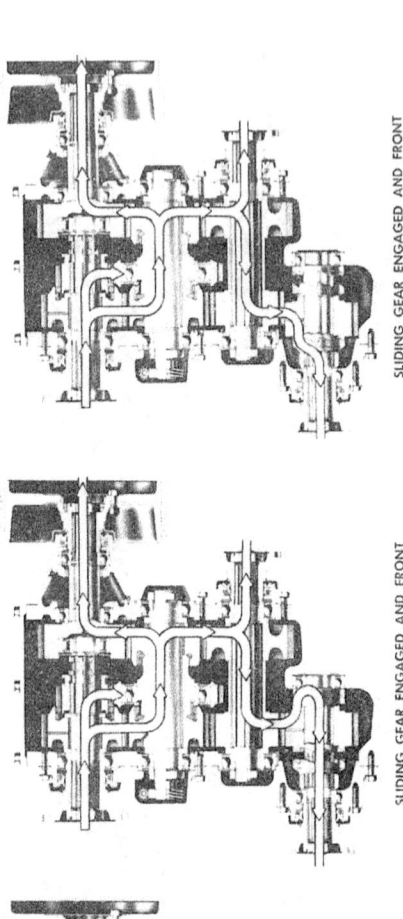

*Figure 48. Power flow through transfer.*

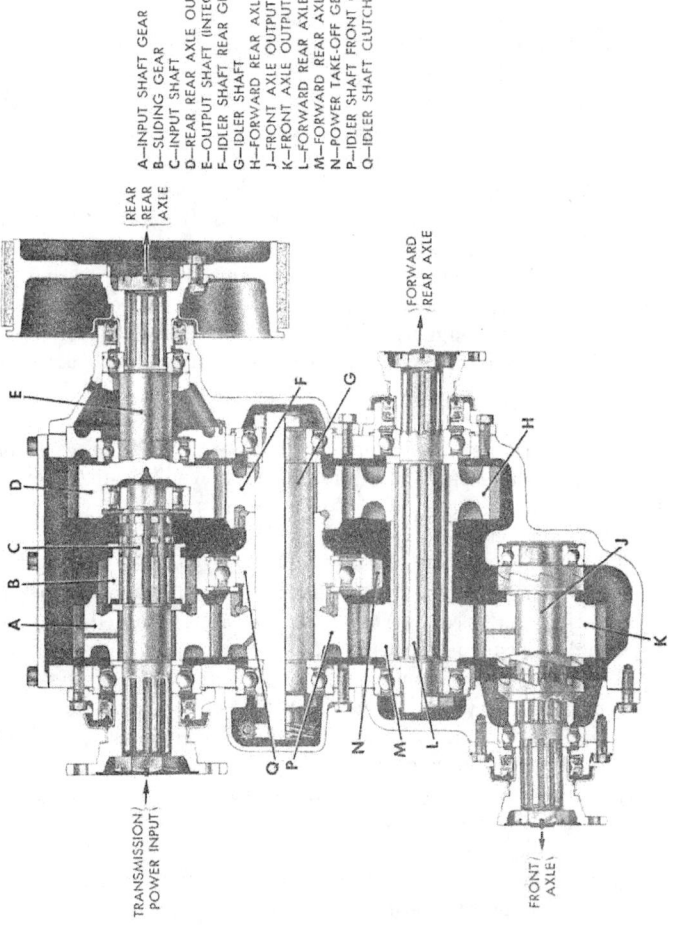

*Figure 49. Identification of transfer gears and shafts.*

ternal teeth are engaged with internal teeth at rear of input shaft gear, the input shaft (C), sliding gear (B), and input shaft gear (A) turn as an assembly. Input shaft gear drives idler shaft front gear (P) which causes idler shaft and idler shaft rear gear (F) to rotate also. Hence, power is transmitted through idler shaft and gears to rear rear axle output gear (D), and to forward rear axle output rear gear (H) at rear end of forward rear axle output shaft. This arrangement causes propeller shaft to each rear axle to turn at the same speed. Since forward rear axle output shaft front gear (M), splined to front end of forward rear axle output shaft, has one tooth less than the front axle output shaft gear (K) on front axle output shaft (J), front axle output shaft gear (K) turns slower than rear axle output shafts. However, when vehicle is operated on dry, hard-surfaced roads the front axle propeller shaft turns at same speed as rear axle propeller shafts, since the same ratio gears are used in all axles. Thus no power is transmitted to front axle unless sufficient slippage occurs at rear wheels to cause rear axle propeller shafts to turn faster than front axle propeller shaft, in which case, front axle output shaft gear (K) on front axle output shaft engages with mating clutch teeth on front axle output shaft (J) and effects power connection to drive the front axle assembly.

(2) *Reverse.*—Operation of transfer in reverse is same as for forward operation except for position of front axle output shaft gear, which is automatically moved ahead to reverse position by action of shifter mechanism linked with transmission control which moves shifter shaft ahead as transmission is shifted into reverse. With front axle shifter shaft ((V), fig. 54) in reverse position (toward front of transfer), shifter shaft front spring ((S), fig. 54) is rendered ineffective and rear spring moves front axle shifter fork ((R), fig. 54) and front axle output shaft gear (K) forward and in contact with clutch ((L), fig. 54). In this position, power connection for driving front axle will be effected should conditions arise when slippage at rear wheels occurs.

(3) *Power take-off gear.*—Power take-off gear (N), which is mounted on idler shaft clutch gear (Q), is constantly in mesh with sliding gear on input shaft. Sliding gear shifter fork ((E), fig. 54), installed on sliding gear shifter shaft ((X), fig. 54), moves sliding gear out of engagement with input shaft gear when necessary to operate power take-off with vehicle standing. Refer to TM 9–819A for arrangement of power take-off controls and for control operating instructions.

## 82. Data

| | |
|---|---|
| Type | Single-speed |
| Manufacturer | GMC Truck and Coach Div |
| Ordnance number | 7411327 |
| Ratio | 1.16 to 1 |

## Section II. DISASSEMBLY OF TRANSFER INTO SUBASSEMBLIES

## 83. General

*a. Scope of Procedures.*—Procedures for removal of transfer assembly from vehicle, and removal of power take-off assembly and parking brake mechanism from transfer assembly are covered in TM 9-819A. Procedures described herein are arranged in practical sequence covering the disassembly of transfer into component parts and subassemblies.

*b. Tools and Equipment.*—A suitable repair stand (fig. 50) should be available to support transfer assembly while various component parts and subassemblies are being removed. A suitable press is required to remove bearing from shafts. The required special tools are listed and illustrated with other special tools in chapter 2. Clean trays should be available to receive parts as they are removed during disassembly.

*c. Inspection During Disassembly.*—When overhauling transfer assembly, all oil seals, gaskets, lock washers, and cotter pins should be discarded and new parts should be used at assembly. Make visual

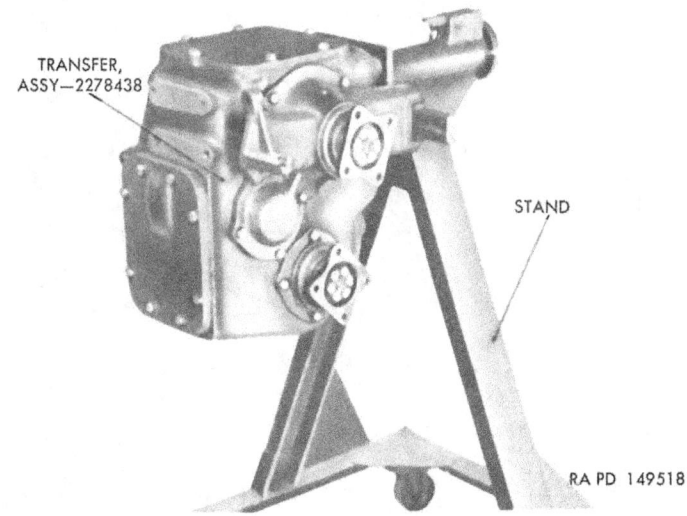

*Figure 50. Transfer assembly mounted in repair stand.*

inspection of parts as they are removed or disassembled. Discard all damaged or broken parts.

### 84. Removal of Covers, Plugs, and Flanges

*a.* Remove six cap screws and lock washers attaching cover to top of transfer case. Remove drain plug and filler plug and drain lubricant from case. If case is equipped with a cover at power take-off opening, remove eight cap screws and lock washers and remove power take-off opening cover from left side of transfer. Remove and discard gaskets.

*b.* Remove cotter pins used to lock retaining nuts at each of the four flanges to which propeller shafts are attached. Attach tool 41–T–3215–910 to companion flange on rear axle output gear as shown in figure 51. Remove flange nuts from input shaft and output shafts.

*Note.* From axle shifter shaft ((V), fig. 54) must be in forward position to lock input shaft to front gear to prevent shafts from turning while flange nuts are loosened.

*c.* Using soft metal hammer, drive flanges off input shaft, rear rear axle output shaft, and front axle output shaft. Flange should remain on forward rear axle output shaft until after lock nuts on

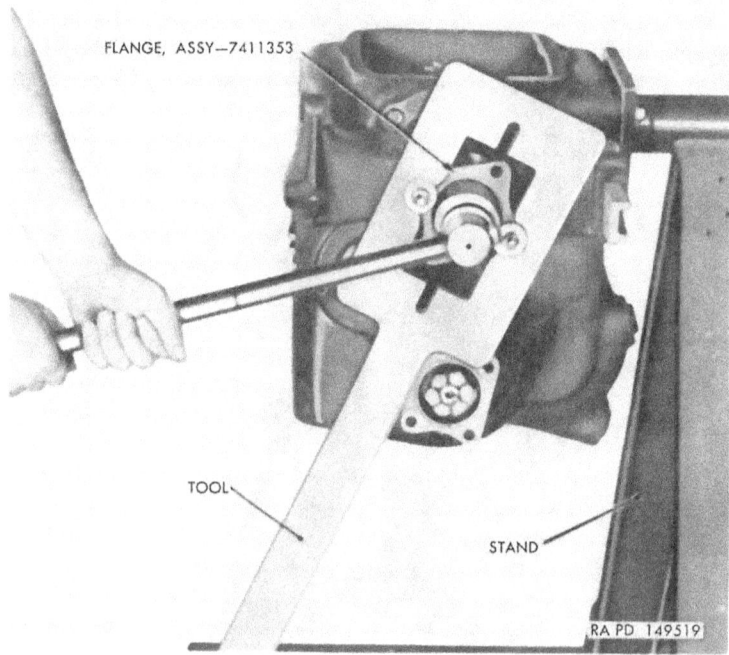

*Figure 51. Use of companion flange holding tool 41–T–3215–910.*

idler shaft and front end of forward rear axle output shaft have been loosened.

## 85. Removal of Output Gear Components

*Note.* Key letters in text refer to figure 52.

*a.* Remove seven cap screws and lock washers attaching output gear bearing retainer (K) transfer to case (ZZ).

*b.* Strike output gear bearing retainer (K) with lead hammer to free retainer from transfer case (ZZ); then withdraw output gear

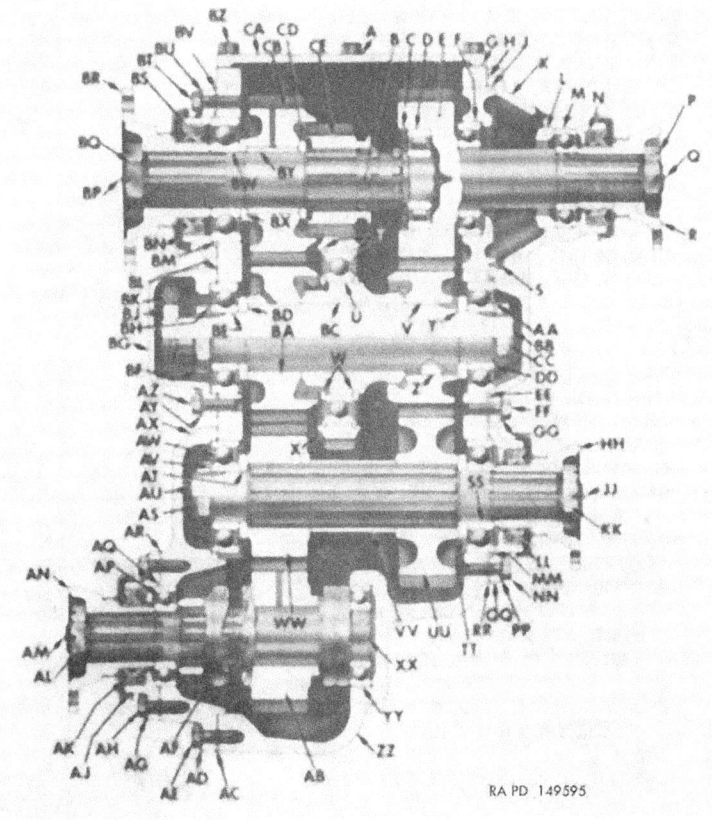

A—WASHER, LOCK, EXT-TEETH, ⅜-IN.—138489
B—SHAFT, INPUT—7412876
C—RING, LOCKING—7411340
D—BEARING, REAR, INPUT SHAFT—707655
E—GEAR, OUTPUT—7411356
F—BEARING, FRONT, OUTPUT GEAR—700539

*Figure 52. Sectional view of transfer assembly.*

G—GASKET, COVER—7411491
H—GASKET, RETAINER—7411398
J—RETAINER, BEARING, FRONT, ASSY—7002111
K—RETAINER, OUTPUT GEAR BEARING—7411361
L—SLEEVE, LOCK, REAR BEARING—7411409
M—BEARING, REAR, OUTPUT GEAR—7411351
N—SEAL, OIL—7411263
P—NUT, ⅞-14—7411259
Q—PIN, COTTER, ⅛ X 1¾—103388
R—FLANGE, ASSY—7411353
S—PIN, DOWEL, ⅛ X ⅜—141107
T—RING, SNAP, BEARING—7411400
U—BEARING, POWER TAKE-OFF GEAR—712866
V—GEAR, REAR, IDLER SHAFT—7411395
W—RING, SNAP, BEARING—7411401
X—GEAR, POWER TAKE-OFF—7411396
Y—SPACER, BEARING—7374703
Z—KEY, WOODRUFF, ¼ X ¾—127559
AA—LOCK, NUT—6245933
BB—SHAFT, IDLER—7411402
CC—NUT, LOCK, BEARING, 1½ X 16—6245935
DD—BEARING, REAR, IDLER SHAFT—710145
EE—RING, SNAP, BEARING—7411407
FF—SCREW, CAP, ⅜-16 X 1⅛—180123
GG—WASHER, LOCK, EXT-TEETH, ⅜-IN.—138489
HH—FLANGE, ASSY—7411353
JJ—PIN, COTTER, ⅛ X 1¾—103388
KK—NUT, ⅞-14—7411259
LL—SEAL, OIL—7411263
MM—RING, SNAP, BEARNG—7411407
NN—SCREW, CAP, ⅜-16 X 1⅛—180123
PP—WASHER, LOCK, EXT-TEETH, ⅜-IN.—138489
QQ—RETAINER FORWARD REAR AXLE OUTPUT SHAFT REAR BEARING—7411260
RR—GASKET, BEARING RETAINER—6244507
SS—BEARING, REAR, FORWARD REAR AXLE OUTPUT SHAFT—710145
TT—SPACER, BEARING—7374703
UU—GEAR, REAR, FORWARD REAR AXLE OUTPUT SHAFT—7411359
VV—SPACER, GEAR—7411389
WW—GEAR, FRONT, FORWARD REAR AXLE OUTPUT SHAFT—7411358
XX—SHAFT, OUTPUT, FRONT AXLE—7411388
YY—BEARING, REAR, FRONT AXLE OUTPUT SHAFT—6244507
ZZ—CASE, TRANSFER—7411331
AB—GEAR, FRONT AXLE OUTPUT SHAFT—7411360
AC—GASKET, BEARING SUPPORT—7411355
AD—WASHER, LOCK, EXT-TEETH, ⅜-IN.—138489
AE—SCREW, CAP, ⅜-16 X 1⅛—180123
AF—CLUTCH—7412879
AG—WASHER, LOCK, EXT-TEETH, ⅜-IN.—138489
AH—SCREW, CAP, ⅜-16 X 1⅛—180123
AJ—RETAINER, FRONT AXLE OUTPUT SHAFT FRONT BEARING—7411384
AK—SEAL, OIL—7411263
AL—NUT, ⅞-14—7411259

*Figure 52.*—Continued.

AM—PIN, COTTER, ⅛ X 1¾—103388
AN—FLANGE, ASSY—7411353
AP—BEARING, FRONT, FRONT AXLE OUTPUT SHAFT—6244507
AQ—RING, SNAP, BEARING—7411408
AR—GASKET, RETAINER—7411404
AS—NUT, LOCK, BEARING, 1½-16—6245935
AT—SPACER, BEARING—7374703
AU—SHAFT, OUTPUT, FORWARD REAR AXLE—7411386
AV—LOCK, NUT—6245933
AW—BEARING, FRONT, FORWARD REAR AXLE OUTPUT SHAFT—710145
AX—SUPPORT, BEARING—7411391
AY—WASHER, LOCK, EXT-TEETH, ⅜-IN.—138489
AZ—SCREW, CAP, ⅜-16 X 1⅛—180123
BA—GEAR, FRONT, IDLER SHAFT—6245934
BC—GEAR, CLUTCH, IDLER SHAFT—7411397
BD—SPACER, BEARING—7374703
BE—BEARING, FRONT, IDLER SHAFT—710145
BF—NUT, LOCK, BEARING, 1½-16—6245935
BG—RETAINER, IDLER SHAFT FRONT BEARING—7411399
BH—LOCK, NUT—6245933
BJ—GEAR, DRIVE, SPEEDOMETER—7412877
BK—GEAR, DRIVEN, SPEEDOMETER, ASSY—7412878
BL—GASKET, RETAINER—7411394
BM—RING, SNAP, BEARING—7411328
BN—RETAINER, INPUT SHAFT FRONT BEARING—7411329
BP—PIN, COTTER, ⅛ X 2—103389
BQ—NUT, 1⅛-12—7411336
BR—FLANGE, ASSY—7411324
BS—SEAL, OIL—7411330
BT—SCREW, CAP, ⅜-16 X 1⅛—180123
BU—WASHER, LOCK, EXT-TEETH, ⅜-IN.—138489
BV—GASKET, RETAINER—7411325
BW—BEARING, FRONT, INPUT SHAFT—700773
BX—SPACER, BEARING—7411339
BY—BEARING, BUSHING TYPE—7411230
BZ—SCREW, CAP, ⅜-16 X 1⅛—180123
CA—COVER, CASE, ASSY—7342349
CB—GEAR, INPUT SHAFT—6245938
CD—SPACER, SLIDING GEAR—7411339
CE—GEAR, SLIDING—7412875

*Figure 52.*—Continued.

(E) and retainer with output gear front and rear bearings (F and M) as an assembly.

*Note.* Inner race of input shaft rear bearing (D) is press fit and remains on input shaft (B) while rollers and outer race are removed with output gear (E).

*c.* Strike rear end of output gear shaft with soft metal hammer to force output gear (E) with output gear front bearing (**F**) **out of** output gear rear bearing (M) and front bearing retainer assembly (J). **R**emove front bearing retainer assembly (J).

*d.* Refer to paragraphs 95 through 98 for procedure covering further disassembly of these two subassemblies.

## 86. Removal of Shifting Mechanism

*Note.* Key letters noted in parentheses are in figure 54 unless otherwise indicated.

*a.* Clean exposed portion of shifter shafts (fig. 53). Remove poppet ball plugs from shifter shaft support.

*b.* Remove two bolts ((J), fig. 53) and lock washers ((K), fig. 53) holding shifter shaft support assembly (Z) on front of transfer case. While holding front axle shifter fork (R) and shifter shaft rear spring (H) inside transfer case, remove front axle shifter shaft (V), shifter shaft front spring (S), and support from front of case. Remove fork and rear spring from inside case. Remove support gasket.

A—CASE, TRANSFER—7411331
B—GASKET, SHIFTER SHAFT SUPPORT—7411236
C—PLUG, POPPET BALL, 7/16-14—6244661
D—SHAFT, SHIFTER, SLIDING GEAR—7411275
E—RETAINER, OIL SEAL—6244663
F—SHAFT, SHIFTER, FRONT AXLE—7411276
G—SUPPORT, SHIFTER SHAFT, ASSY—7411288
H—PLUG, POPPET BALL, 7/16-14—6244661
J—BOLT, 3/8-16 x 1-1/8—180123
K—WASHER, LOCK, EXT-TEETH, 3/8-IN—138489

RA PD 149563

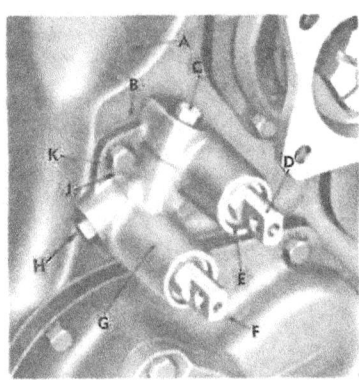

*Figure 53. Transfer shifter shafts installed.*

*c.* Allow poppet ball springs (U) and poppet balls (T) to fall out of shifter shaft support assembly (Z). Remove oil seals (W) and oil seal retainers (Y).

*d.* Through opening in rear of case, use wrench to remove 5/16–24 x 1½ bolt (F) from sliding gear shifter fork (E). Withdraw sliding gear shifter shaft (X) from sliding gear shifter fork (E) and pull out through front of case. Remove sliding gear shifter fork (E) from inside case.

*e.* Refer to paragraph 111 for inspection of shifter components and procedure for replacing front axle shifter shaft spacer (Q) on front axle shifter shaft (V), and shifter shaft support pilot (N) which is pressed into support.

## 87. Removal of Input Shaft and Gear Assembly

*Note.* Key letters noted in parentheses are in figure 52 unless otherwise indicated.

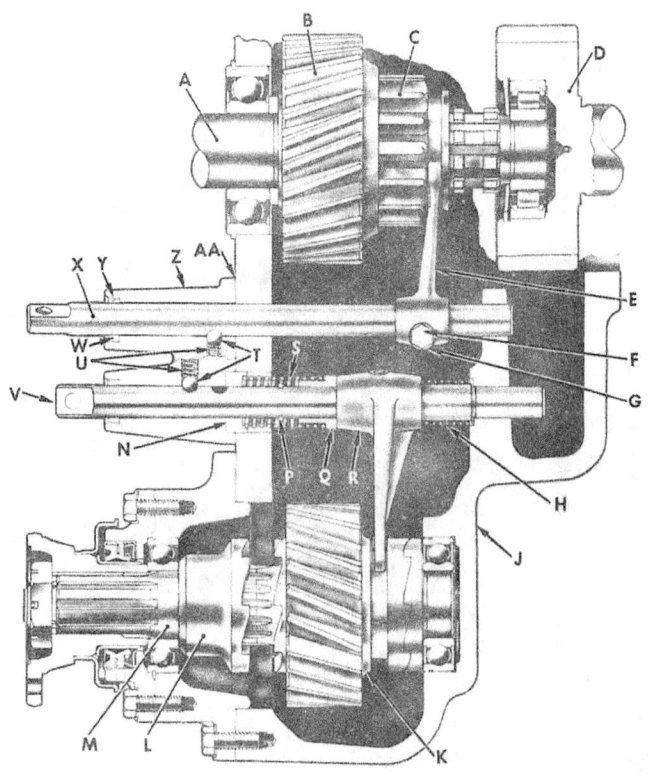

A—SHAFT, INPUT—7412876
B—GEAR, INPUT SHAFT—6245938
C—GEAR, SLIDING—7412875
D—GEAR, OUTPUT—7411356
E—FORK, SHIFTER, SLIDING GEAR—7411342
F—BOLT, 5/16-24x1-1/2—181614
G—WASHER, LOCK, 5/16-IN—120638
H—SPRING, SHIFTER SHAFT, REAR—7412874
J—CASE, TRANSFER—7411331
K—GEAR, FRONT AXLE OUTPUT SHAFT—7411360
L—CLUTCH—7412879
M—SHAFT, OUTPUT, FRONT AXLE—7411388
N—PILOT, SHIFTER SHAFT SUPPORT—7412857
P—RING, SNAP—7412858
Q—SPACER, FRONT AXLE SHIFTER SHAFT—7412859
R—FORK, SHIFTER, FRONT AXLE—YT-2290942
S—SPRING, SHIFTER SHAFT, FRONT—7412500
T—BALL, POPPET, 5/16-IN—147489
U—SPRING, POPPET BALL—7411347
V—SHAFT, SHIFTER, FRONT AXLE—7411276
W—SEAL, OIL—6244662
X—SHAFT, SHIFTER, SLIDING GEAR—7411275
Y—RETAINER, OIL SEAL—6244663
Z—SUPPORT, SHIFTER SHAFT, ASSY—7411288
AA—GASKET, SHIFTER SHAFT SUPPORT—7411236

RA PD 148790

*Figure 54. Sectional view of transfer shifting mechanism.*

95

*a.* Remove four cap screws and lock washers attaching input shaft front bearing retainer (BN) to transfer case (ZZ) and remove retainer assembly and gasket.

*b.* While holding input shaft gear (CB) inside case, drive input shaft assembly (fig. 55) toward rear and out of input shaft front bearing (BW).

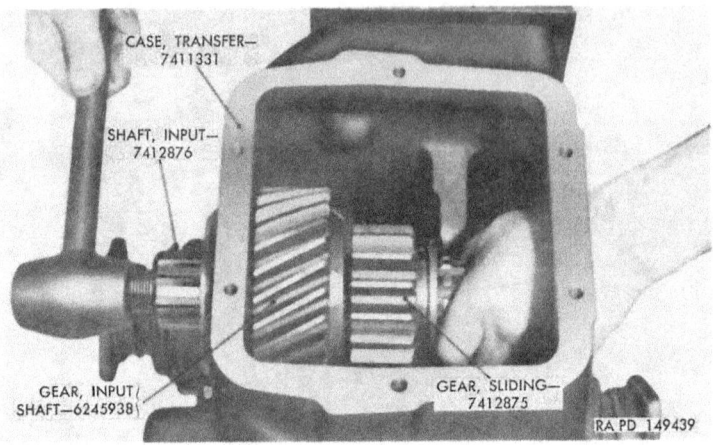

*Figure 55. Removing input shaft and sliding gear assembly.*

*c.* Remove input shaft with sliding gear (CE), sliding gear spacer (CD), bushing type bearing (BY), and inner race of input shaft rear bearing (D) as an assembly. Remove input shaft gear (CB) and bearing spacer (BX) from inside case.

*d.* Paragraph 92 gives instructions for further disassembly of components remaining on input shaft.

*e.* Remove input shaft front bearing (BW) from bore in front of transfer case.

*f.* Remove oil seal (BS) from input shaft front bearing retainer (BN).

## 88. Removal of Front Axle Output Shaft Components

*Note.* Key letters noted in parentheses are in figure 52.

*a.* Remove seven cap screws and lock washers which attach bearing support (AX) to front of transfer case.

*b.* Pull front axle output shaft with bearing support, gears, and front and rear bearings out of transfer case as an assembly. Remove bearing support gasket (AC).

*c.* Refer to paragraph 101 for instructions covering disassembly of front axle output shaft components.

## 89. Removal of Speedometer Gears

*Note.* Key letters noted in parentheses are in figure 52 unless otherwise indicated.

*a.* Using suitable wrench, remove speedometer driven gear and fitting assembly from idler shaft front bearing retainer (BG).

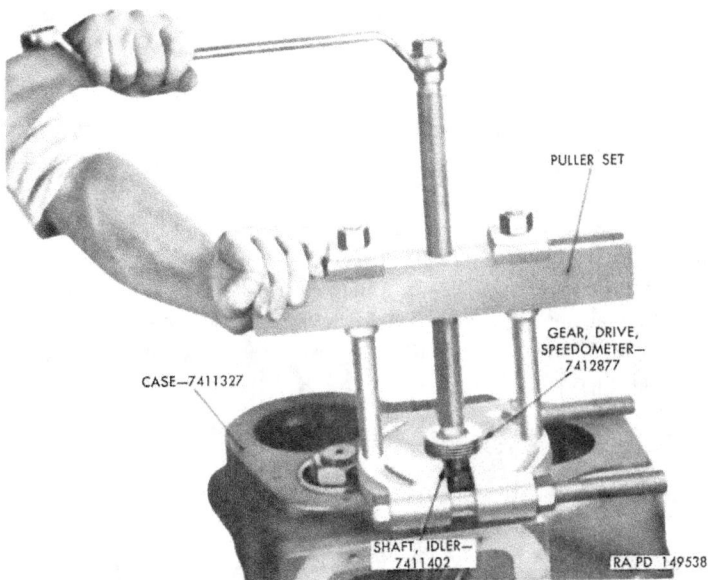

*Figure 56.—Removing speedometer drive gear using puller set 41–P–2905–60.*

*b.* Remove three bolts and lock washers which attach idler shaft front bearing retainer (BG) to transfer case; then remove retainer and retainer gasket (BL).

*c.* Remove speedometer drive gear (BJ) from idler shaft as shown in figure 56. Turn puller screw to remove gear from idler shaft.

## 90. Removal of Idler Shaft, Gears, and Bearings

*Note.* Key letters noted in parentheses are in figure 52 unless otherwise indicated.

*a.* Attach tool 41–T–3215–910 to flange on rear end of forward rear axle output shaft to hold shafts while loosening nuts as directed in *b* below. Use holding tool in manner illustrated in figure 51.

*b.* Bend nut locks (AA and BH) away from lock nut at each end of idler shaft (BB) and bend nut lock (AV) away from lock nut at front end of forward rear axle output shaft (AU). Remove three bearing lock nuts and nut locks.

97

c. Using soft metal hammer, drive on front end of idler shaft to force idler shaft rearward. As idler shaft is moved rearward, idler shaft rear bearing (DD) is forced out of bore in case and Woodruff key (Z), which fits into keyway in idler shaft rear gear (V), is removed with idler shaft (BB).

d. As idler shaft is pulled out of transfer case, remove bearing spacers (Y and BD), idler shaft front gear (BA), power take-off gear (X), and idler shaft rear gear (V) from inside transfer case.

e. Remove idler shaft front bearing (BE) from bore in front of case.

## 91. Removal of Forward Rear Axle Output Shaft and Gears

*Note.* Key letters noted in parentheses are in figure 52 unless otherwise indicated.

a. Using soft metal hammer, drive flange assembly (HH) off rear end of forward rear axle output shaft (AU).

b. Remove four cap screws and lock washers attaching forward rear axle output shaft rear bearing retainer (QQ) to transfer case. Remove retainer and oil seal (LL) from case; then remove oil seal from retainer.

*Figure 57. Removing forward rear axle output shaft and gears.*

c. Drive forward rear axle output shaft (AU) rearward with soft metal hammer. As shaft moves out of case, rear bearing assembly and spacer (fig. 57) will be carried with shaft. When rear bearing assembly is free from bore in case, support gears and pull shaft out of gears. Remove gears and spacers (fig. 57) from inside transfer case.

d. Remove forward rear axle output shaft front bearing (AW) from bore in case.

## Section III. REBUILD OF INPUT SHAFT AND COMPONENTS

### 92. Disassembly of Input Shaft Components

*Note.* Key letters noted in parentheses are in figure 52 unless otherwise indicated.

*a.* Place input shaft with sliding gear in arbor press supported on attachment 41–A–345–328 (part of puller set 41–P–2905–60) as shown in figure 59.

*b.* Using suitable driver under press ram, apply pressure at input shaft (M) to remove bearing inner race (N) from rear end of shaft.

*Note.* The preceding operation may be deferred until after inspection (par. 93*b*).

*c.* Change position of input shaft assembly in press with sliding gear spacer (K) supported on attachment (fig. 59).

*d.* Press on front end of input shaft to force bushing type bearing (J) and sliding gear spacer (K) off input shaft.

### 93. Cleaning, Inspection, and Repair of Input Shaft Components

*Note.* Key letters noted in parentheses are in figure 52 unless otherwise indicated.

*a. Cleaning.*—Wash all input shaft components shown in figure 58 in dry-cleaning solvent or volatile mineral spirits and wipe or blow dry. Handle parts with care to prevent damage to ground surfaces and to prevent damage to splines.

*b. Inspection.*

*Note.* Refer to paragraph 343 for dimensions and fits.

  (1) Examine surface of bearing inner race (N) for evidence of pitting, grooving, and roughness. If race is not in good condition, replace input shaft rear bearing ((D), fig. 52), including inner race.
  (2) Note condition of splines at input shaft (M) and sliding gear (L). Also inspect teeth at sliding gear. If teeth are broken, chipped, or otherwise damaged, discard gear and use new part when assembling.
  (3) Measure outside diameter of bushing type bearing (J) over bearing area. If diameter is less than shown in paragraph 343*a*, install new bearing.
  (4) Inspect bearing spacer (G) and sliding gear spacer (K) for wear and other damage; discard if outer edge is worn.
  (5) Inspect teeth on input shaft gear (H). Measure diameter of bore through gear (par. 343*a*). Replace gear if damaged or excessively worn.
  (6) Examine input shaft front bearing (F). Rotate bearing races by hand to detect roughness. If wear, roughness, or

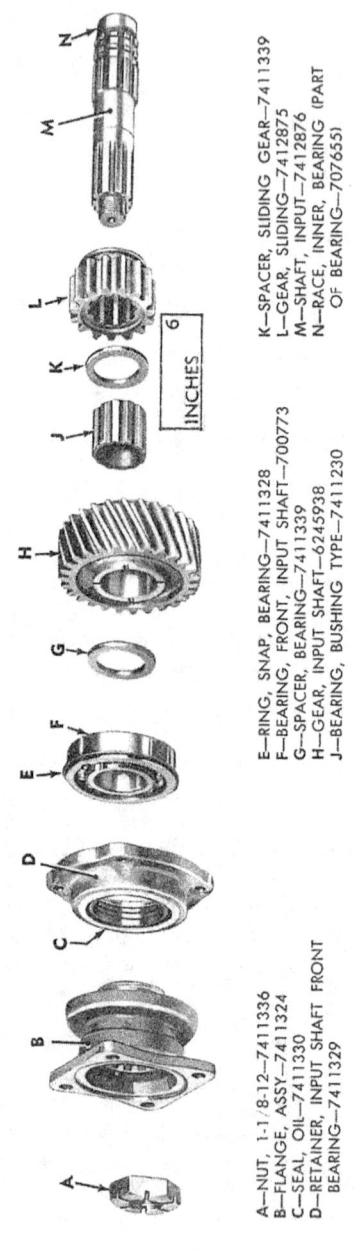

A—NUT, 1-1/8-12—7411336
B—FLANGE, ASSY—7411324
C—SEAL, OIL—7411330
D—RETAINER, INPUT SHAFT FRONT BEARING—7411329
E—RING, SNAP, BEARING—7411328
F—BEARING, FRONT, INPUT SHAFT—700773
G—SPACER, BEARING—7411339
H—GEAR, INPUT SHAFT—6245938
J—BEARING, BUSHING TYPE—7411230
K—SPACER, SLIDING GEAR—7411339
L—GEAR, SLIDING—7412875
M—SHAFT, INPUT—7412876
N—RACE, INNER, BEARING (PART OF BEARING—707655)

RA PD 148818

*Figure 58. Transfer input shaft, sliding gear, and related components.*

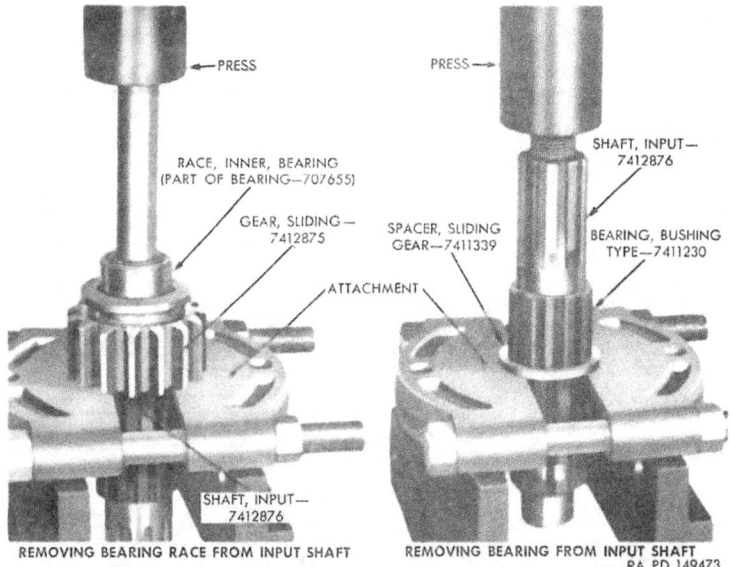

*Figure 59. Removing bearing race and input shaft gear bearing from input shaft with attachment 41-A-345-328 (part of puller set 41-P-2905-60).*

damage is evident, replace bearing. Bearing snap ring (E) should be checked for distortion and for fit in groove in bearing.

(7) Examine input shaft front bearing retainer (D). If evidence of excessive wear appears at surfaces contacted by bearing and snap ring, or if retainer is distorted or otherwise damaged, the retainer should be replaced.

(8) Inspect deflector and oil seal sleeve on input shaft flange assembly (fig. 60). If surface of sleeve is scored or grooved at point of contact with oil seal lip, or if deflector is bent or otherwise damaged, these parts should be replaced. Refer to *c* below for procedure necessary to replace sleeve and deflector.

(9) Check width of fork groove on sliding gear (L). Replace gear if groove is worn (par. 343*a*).

*c. Repair of Input shaft Components.*

(1) *Replacing input shaft flange deflector.*

(*a*) Using a hack saw, make a diagonal cut through outer edge of deflector, then with a sharp cold chisel, cut through remaining portion of deflector. If deflector is welded to flange, use suitable chisel to break the weld and remove deflector.

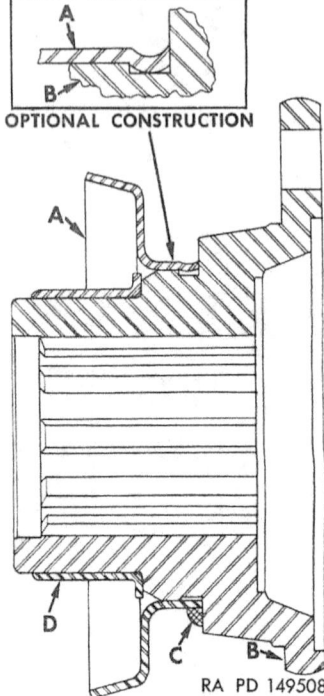

A—DEFLECTOR—YT-2278618
B—FLANGE, ASSY—7411324
C—ARC WELD
D—SLEEVE—7411338

*Figure 60. Sectional view of input shaft flange assembly.*

    (*b*) File off any weld metal at flange shoulder which would prevent new deflector from seating against shoulder; then locate new deflector on flange as shown in figure 60.

    (*c*) Either of the methods shown in figure 60 may be used to attach new deflector to flange. When peening deflector into groove as shown in inset (fig. 60), use peening tool and peen in four equally spaced places. If welding is employed, tack weld deflector to flange in three places.

(2) *Replacing oil seal sleeve on input shaft flange assembly.*

    (*a*) Using a light-weight hammer, tap sleeve (fig. 60) over entire exposed area to stretch metal and loosen sleeve on flange; then use chisel to cut through sleeve flange. Pry sleeve off flange assembly.

    (*b*) Inspect surface of flange assembly for burs or roughness and smooth with file if necessary.

    (*c*) Install new oil seal sleeve on flange, using replacer 41–R–2395–535 as illustrated in figure 61 to drive sleeve flange firmly against shoulder at input shaft flange assembly.

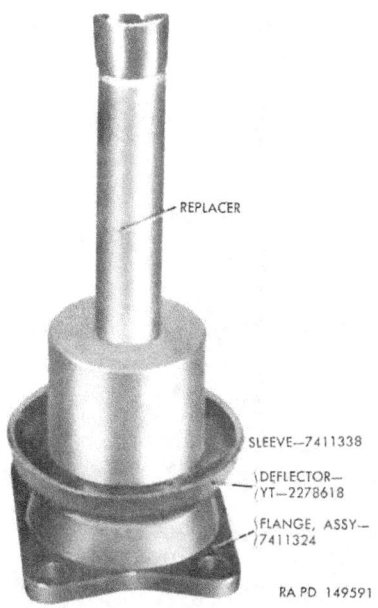

*Figure 61. Use of replacer 41-R-2395-535 to install oil seal sleeve on input shaft flange.*

## 94. Assembly of Input Shaft Components

*Note.* Key letters noted in parentheses are in figure 58.

*a.* Place sliding gear spacer (K) against shoulder on input shaft (M); then install bushing type bearing (J) on shaft and press or drive bearing firmly against sliding gear spacer (K).

*Note.* If bearing inner race (N) is on shaft, place sliding gear (L) on shaft before bushing type bearing (J) is installed.

*b.* If bearing inner race (N) has been removed from input shaft (M), install sliding gear (L) on shaft with shifter fork groove toward rear end of input shaft; then support bearing inner race on press plate and press input shaft into inner bearing race (fig. 62). Figure 63 shows input shaft and components assembled and ready for installation in transfer case.

## Section IV. REBUILD OF OUTPUT GEAR AND COMPONENTS

## 95. Disassembly of Rear Rear Axle Output Gear Components

*Note.* Key letters noted in parentheses are in figure 52.

*a.* Pry locking ring (C) out of groove at front of output gear (E). Using suitable puller, remove input shaft rear bearing (D) from output gear (E).

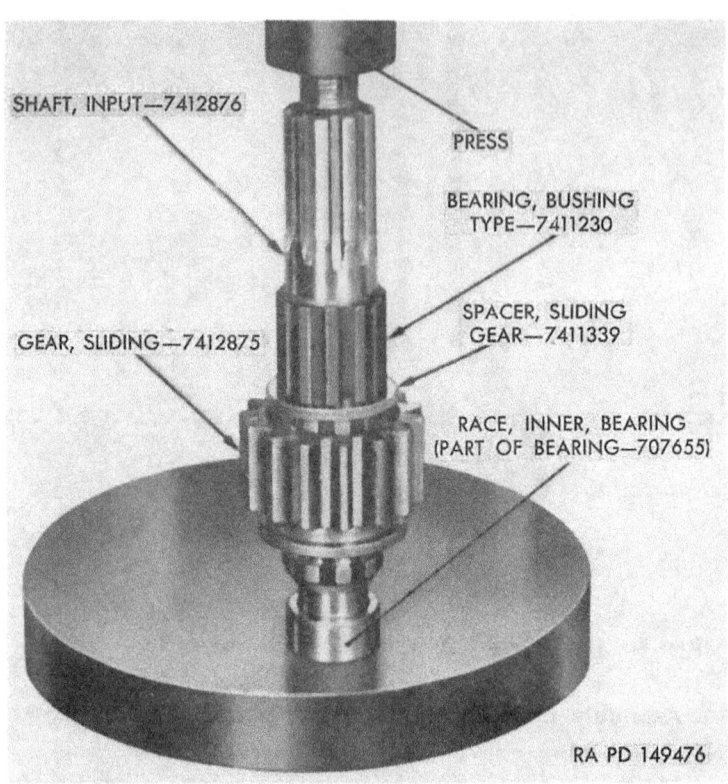

*Figure 62. Use of arbor press to install input shaft bearing inner race.*

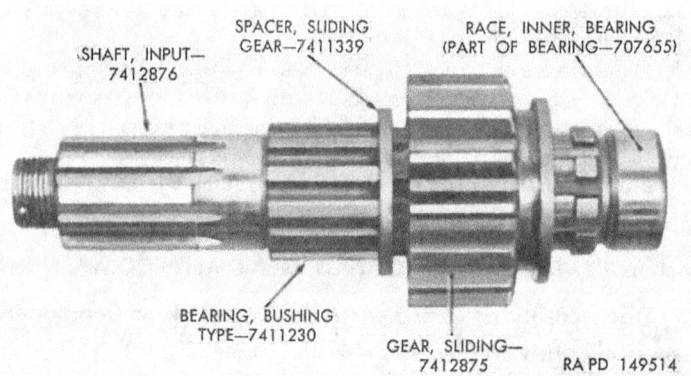

*Figure 63. Input shaft and components assembled for installation in transfer case.*

*b.* Assemble attachment 41–A–345–328 (part of puller set 41–P–2905–60) at output gear front bearing (F). Position output gear and bearing assembly with attachment in arbor press as shown in figure 64 and remove bearing from gear.

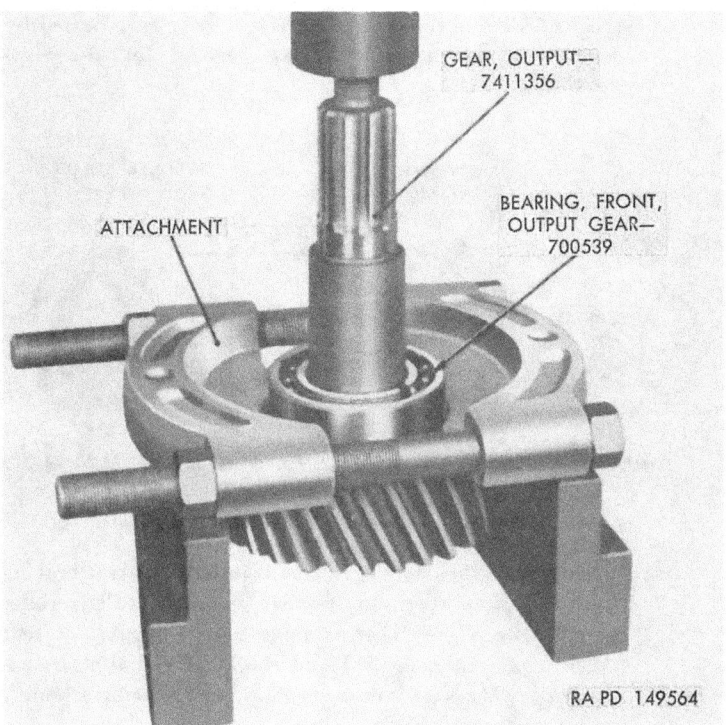

*Figure 64. Using arbor press and attachment 41–A–345–328 (part of puller set 41–P–2905–60) to remove output gear front bearing.*

## 96. Cleaning and Inspection of Output Gear Components.

*a. Cleaning.* Wash all output gear components shown in figure 65 in dry-cleaning solvent or volatile mineral spirits, and wipe or blow dry.

**Caution:** Do not spin bearings with air as damage may result.

*b. Inspection.*
  (1) Examine gear teeth on output gear for wear, nicks, and chipped or broken teeth. Also inspect splines and threads at rear end of shaft which is forged integral with output gear. If any damage is found, discard gear as no repair is recommended on this part.

(2) Inspect front bearing retainer assembly (fig. 65) for damage and check 1/8 x 3/8 dowel pin ((S), fig. 52) which should be tight in hole in retainer.

(3) Examine output gear front bearing, which is single-row ball type. Rotate bearing races by hand to detect roughness. If races or balls are rough or if excessive wear is indicated by looseness of parts, obtain a new bearing for use when assembling.

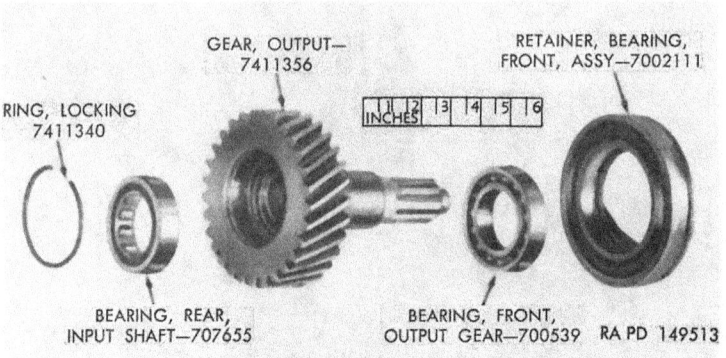

*Figure 65. Output gear, bearings, and related components.*

(4) Visually inspect surface of rollers in input shaft rear bearing (fig. 65). Note condition of roller separator. Place roller bearing assembly on bearing inner race (installed at rear end of input shaft, fig. 63) and check for radial clearance. If excessive looseness is evident, obtain new bearing assembly for use when assembling.

(5) Check bearing locking ring for distortion.

## 97. Assembly of Output Gear Components

*Note.* Key letters noted in parentheses are in figure 52 unless otherwise indicated.

*a.* Set output gear (E) on flat solid support; then use replacer 41–R–2390–415 to drive output gear front bearing (F) into place at shoulder on output gear (fig. 66).

*b.* Install input shaft rear bearing (D) (except inner race) in recess at front side of output gear. Drive bearing outer race squarely into place. Install locking ring (C) to retain bearing.

*Note.* Inner race of input shaft rear bearing (D) must be installed on rear end of input shaft (B). Refer to paragraphs 92a and b, and 94b for instructions covering replacement of input shaft rear bearing inner race on input shaft.

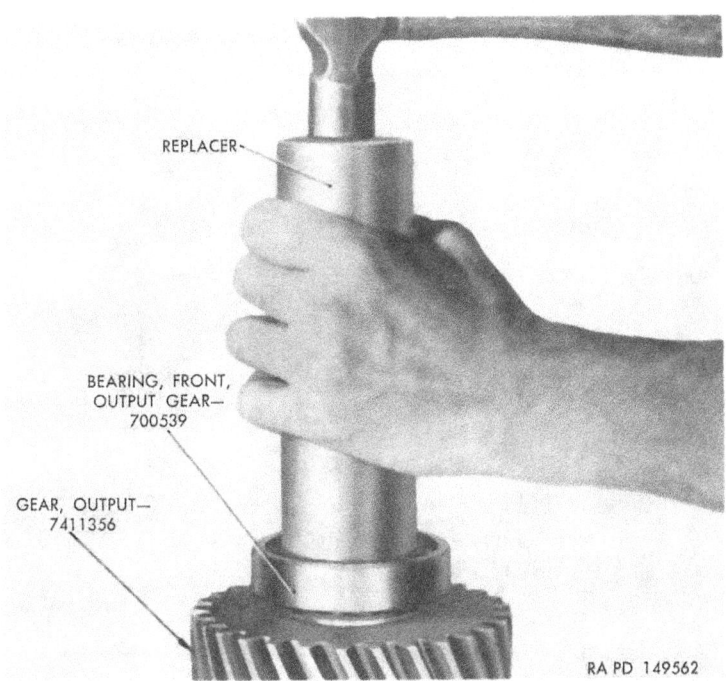

*Figure 66. Installing front bearing on output gear with replacer 41-R-2390-415.*

## Section V. REBUILD OF OUTPUT GEAR BEARING RETAINER AND RELATED COMPONENTS

### 98. Disassembly of Output Gear Bearing Retainer Components

*Note.* Key letters noted in parentheses are in figure 52 unless otherwise indicated.

  *a.* Loosen jam nuts on $1/2$–20 x $1\frac{7}{32}$ special screws (fig. 67); then remove two special screws which hold rear bearing lock sleeve (L) in contact with output gear rear bearing (M). Remove rear bearing lock sleeve (L) from output gear bearing retainer (K).

  *b.* Remove and discard oil seal (N) installed at rear of output gear bearing retainer; then using a suitable driver and light-weight hammer, remove output gear rear bearing (M), driving bearing out toward front side of retainer.

### 99. Cleaning, Inspection, and Repair of Output Gear Bearing Retainer Components

  *a. Cleaning.*—Wash all components (fig. 67) in dry-cleaning solvent or volatile mineral spirits. Wipe parts or dry with air.

107

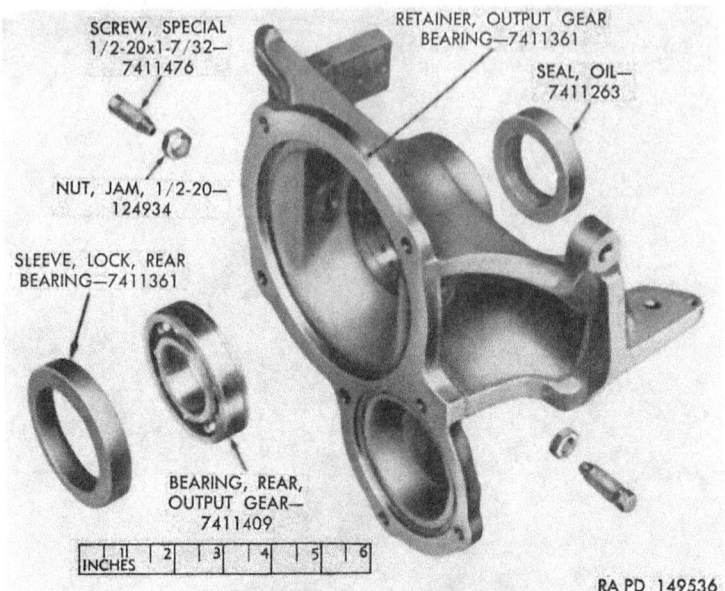

*Figure 67. Output gear bearing retainer and related components.*

b. *Inspection* (fig. 67).
  (1) Visually inspect output gear bearing retainer for damage, cracks, and warpage. Front machined surface which contacts gasket must be flat. Inspect threads in tapped holes.
  (2) Check threads on special screws and jam nuts, and replace any damaged parts.
  (3) Rotate races of output gear rear bearing by hand to detect roughness. Examine bearing races for cracks. If bearing is damaged or excessively worn, replace with new bearing.
  (4) Examine rear bearing lock sleeve. Obtain new sleeve if old sleeve is broken or distorted.

c. *Repair.* Sleeve and deflector on flange assembly ((R), fig. 52) must be replaced if inspection shows parts to be grooved or worn. Flange assembly is identical with flange assembly used at pillow block. Refer to paragraph 224 for inspection and repair procedure for flange assembly.

## 100. Assembly of Output Gear Bearing Retainer Components

*Note.* Key letters noted in parentheses are in figure 52 unless otherwise indicated.

a. Install output gear rear bearing (M) in recess in output gear bearing retainer (K). Place rear bearing lock sleeve (L) against bearing; then thread one 1/2–20 jam nut on each of two 1/2–20 x 1 7/32 special screws (fig. 67), coat screw threads with plastic type gasket cement, and install screws in tapped holes in output gear bearing retainer (K). Conical points on screws must engage tapered side of rear bearing lock sleeve (L). Tighten screws to torque of 20 to 25 pound-feet to force sleeve firmly against outer race of bearing, then tighten jam nuts.

*b.* Place output gear bearing retainer on flat surface and coat oil seal recess with plastic type gasket cement. Drive new oil seal (N) into recess with spring-loaded lip pointing inward. Use wood block or suitable driver to apply pressure at outer circumference of seal assembly.

*c.* Locate front bearing retainer assembly (J) with 1/8 x 3/8 dowel pin (S) alined with notch in output gear bearing retainer (K); then drive front bearing retainer assembly (J) into counterbore in output gear bearing retainer. Insert splined end of output gear and bearing assembly shown in figure 65 through front bearing retainer assembly (J) and out through output gear rear bearing (M). Coat surface of oil seal sleeve on flange assembly (R) with universal gear lubricant; then install flange assembly on splines at rear end of output gear (E). Drive flange assembly onto splines far enough to install 7/8–14 nut and partially tighten nut to pull gear and bearing assembly into place in output gear bearing retainer (K). Final tightening of nut is done

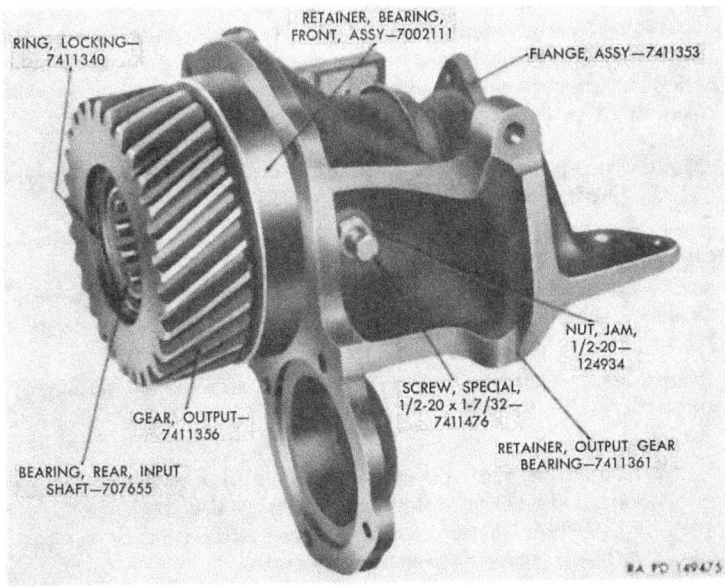

*Figure 68. Output gear and bearing retainer assembled.*

after subassemblies are installed in transfer case (par. 119). Figure 68 shows output gear and bearing retainer components assembled prior to installation.

## Section VI. REBUILD OF FRONT AXLE OUTPUT SHAFT COMPONENTS

### 101. Disassembly of Front Axle Output Shaft Components

*Note.* Key letters noted in parentheses are in figure 52 unless otherwise indicated.

*a.* Remove four 3/8–18 x 1 1/8 cap screws (AH) and 3/8-inch external-teeth lock washers (AG) which attach front axle output shaft front bearing retainer (AJ) to bearing support (AX); then remove retainer assembly and retainer gasket (AR) from bearing support.

*b.* Remove oil seal (AK) from bearing retainer.

*Note.* Discard oil seal, as new seals should be installed when assembling transfer.

*c.* Position output shaft and bearing support assembly in arbor press and press front axle output shaft assembly out of front bearing (fig. 69).

*d.* Remove clutch (AF) from front axle output shaft; then remove clutch snap ring ((J), fig. 71) from groove in front axle output shaft. Remove front axle output shaft gear (AB) from output shaft.

*e.* Assemble attachment 41–A–345–328 (part of puller set 41–P–2905–60) on output shaft and bearing assembly, and position this assembly in arbor press as shown in figure 70. Press against end of output shaft to remove front axle output shaft rear bearing (YY).

### 102. Cleaning, Inspection, and Repair of Front Axle Output Shaft Components

*Note.* Key letters noted in parentheses are in figure 71 unless otherwise indicated.

*a. Cleaning.*—Wash all components shown in figure 71 in dry-cleaning solvent or volatile mineral spirits. Dry parts with clean cloth or with air.

**Caution:** Do not spin ball-type bearings with air as damage to balls and races will result.

*b. Inspection.*

(1) Inspect deflector, oil seal sleeve, and splines on flange assembly (B). If deflector or sleeve are damaged or if sleeve is worn or grooved at seal surface, these parts may be replaced. Refer to *c* below for repair procedure.

(2) Oil seal (C) should always be replaced when transfer is overhauled.

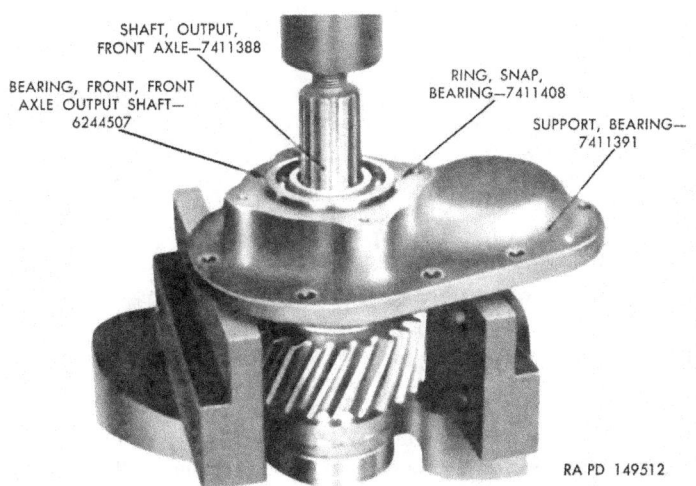

*Figure 69. Pressing front axle output shaft out of front bearing.*

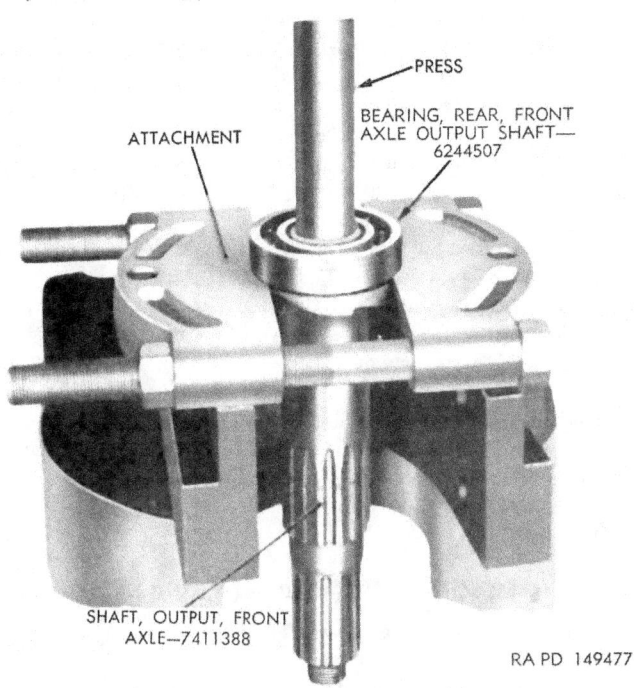

*Figure 70. Removing rear bearing from front axle output shaft using attachment 41-A-345-328 (part of puller set 41-P-2905-60) and arbor press.*

111

(3) Inspect front axle output shaft front bearing retainer (D) and bearing support (G). Gasket surfaces on each part must be flat. Check flatness with straightedge. Examine threads at tapped holes in bearing support (G), also inspect 3/8–16 x 1 1/8 cap screws ((AH), and (AE), fig. 52). Discard parts if threads are stripped.

(4) Inspect front axle output shaft front and rear bearings (F) and (M) for wear and for damage. Rotate bearing races by hand while exerting pressure to determine if balls or races are rough. If roughness is felt or if excessive wear is evident as indicated by looseness of balls, obtain new bearings for use when assembling.

*Note.* Bearing snap ring (E) is used at front axle output shaft front bearing (F). Front axle output shaft rear bearing (M) has a snap ring groove, but no snap ring is used.

(5) Examine teeth on clutch (H) and mating teeth on front axle output shaft gear (K). Also check clutch teeth at rear side of gear and mating teeth on front axle output shaft (L). If any of the teeth are chipped or broken, parts having damaged teeth must be replaced.

(6) Inspect splines at hub of clutch (H) and splines on front axle output shaft (L). Discard parts having splines worn or otherwise damaged.

(7) Measure diameter of bore through front axle output shaft gear (K) and compare with limits listed in paragraph 343*b*. Also measure diameter of front axle output shaft (L) at surface on which gear operates. Compare diameter with limits listed for this part in paragraph 343*b*. Desired radial clearance between front axle output shaft gear (K) and front axle output shaft (L) is also listed in paragraph 343*b*. Measure width of fork groove in gear (par. 343*b*). Replace gear if groove is worn.

*c. Repair.*—The only repair recommended at front axle output shaft and related components is the replacement of deflector and/or oil seal sleeve on output shaft flange assembly (B). This flange assembly is identical to flange assembly used at pillow block. Refer to paragraph 224 for inspection and repair procedures for flange assembly.

## 103. Assembly of Front Axle Output Shaft Components

*Note.* Key letters noted in parentheses are in figure 71 unless otherwise indicated.

*a.* Lay front axle output shaft rear bearing (M) on press plate, position front axle output shaft (L) on bearing inner race and press shaft squarely into bearing. Inner race must seat solidly on shaft.

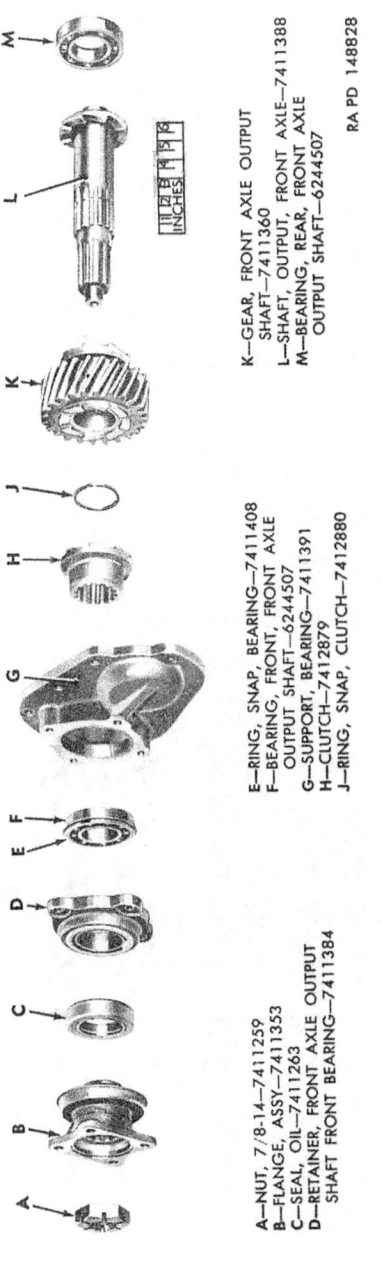

*Figure 71. Front axle output shaft and related components.*

A—NUT, 7/8-14—7411259
B—FLANGE, ASSY—7411353
C—SEAL, OIL—7411263
D—RETAINER, FRONT AXLE OUTPUT SHAFT FRONT BEARING—7411384
E—RING, SNAP, BEARING—7411408
F—BEARING, FRONT, FRONT AXLE OUTPUT SHAFT—6244507
G—SUPPORT, BEARING—7411391
H—CLUTCH—7412879
J—RING, SNAP, CLUTCH—7412880
K—GEAR, FRONT AXLE OUTPUT SHAFT—7411360
L—SHAFT, OUTPUT, FRONT AXLE—7411388
M—BEARING, REAR, FRONT AXLE OUTPUT SHAFT—6244507

RA PD 148828

113

*b.* Coat surface of front axle output shaft with universal gear lubricant (GO), then install front axle output shaft gear (K) on shaft with shift fork groove toward rear end of shaft. Install clutch snap ring (J) in groove in output shaft.

*c.* Install clutch (H) on splines and drive the clutch tightly against snap ring (J).

*d.* Coat oil seal recess in front axle output shaft front bearing retainer (D) with plastic type gasket cement; then drive new oil seal (C) into recess with spring-loaded lip toward rear side of retainer. Use wood block or suitable driver to apply pressure at outer circumference of seal assembly.

*e.* Install front axle output shaft front bearing (F) in bearing support (G), with bearing snap ring (E) against surface of support. Place new retainer gasket ((AR), fig. 52) at bearing support (G) and install retainer and oil seal assembly using four $\frac{3}{8}$–16 x $1\frac{1}{8}$ cap screws with $\frac{3}{8}$-inch external-teeth lock washers ((AH) and (AG), fig. 52). Tighten cap screws to torque of 20 to 25 pound-feet.

*f.* Place support assembly over front end of front axle output shaft (L) and push front axle output shaft front bearing (F) onto shaft as far as possible. Apply coat of universal gear lubricant on surface of oil seal sleeve; then start flange assembly (B) onto splines at front

A—FLANGE, ASSY—7411353
B—RETAINER, FRONT AXLE OUTPUT SHAFT FRONT BEARING—7411384
C—SUPPORT, BEARING—7411391
D—BEARING, REAR, FRONT AXLE OUTPUT SHAFT—6244507
E—SHAFT, OUTPUT, FRONT AXLE—7411388
F—GEAR, FRONT AXLE OUTPUT SHAFT—7411360
G—CLUTCH—7412879
H—GASKET, RETAINER—7411404

RA PD 149481

*Figure 72. Front axle output shaft components assembled.*

axle output shaft and through oil seal in front axle output shaft front bearing retainer (D). Drive flange assembly onto shaft with lead hammer until flange hub forces inner race of bearing into contact with clutch (H).

*g.* Install 7/8–14 nut (A) on shaft threads and tighten lightly. Final tightening is done after subassembly is installed in transfer case (par. 115). Front axle output shaft components appear as shown in figure 72 when properly assembled.

## Section VII. REBUILD OF FORWARD REAR AXLE OUTPUT SHAFT COMPONENTS

### 104. Disassembly of Forward Rear Axle Output Shaft Components

*a. General.*—Forward rear axle output shaft components are disassembled as they are removed from transfer case (par. 91), except forward rear axle output shaft rear bearing (K, fig. 74) and bearing spacer (J, fig. 74) which must be removed as directed in *b* below.

*b. Removal of Forward Rear Axle Output Shaft Rear Bearing.*—Assemble attachment 41–A–345–328 (part of puller set 41–P–2905–60)

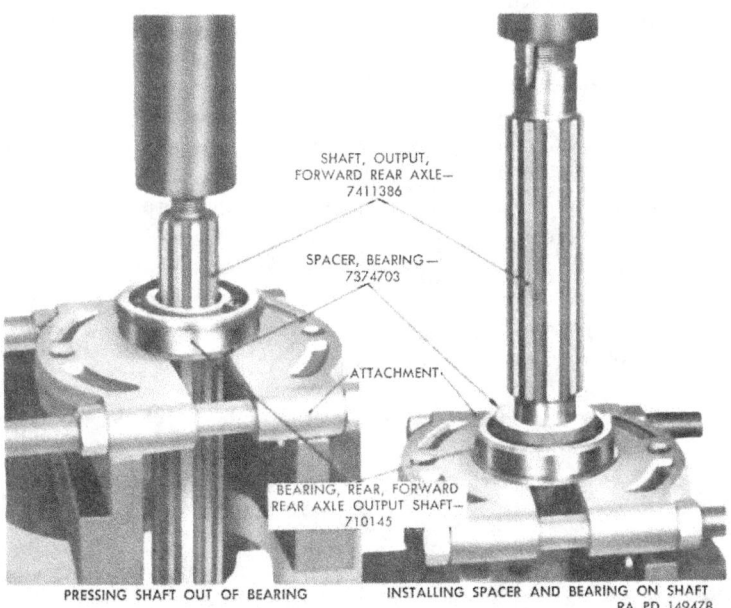

*Figure 73. Use of attachment 41–A–345–328 (part of puller set 41–P–2905–60) for removing and installing rear bearing on forward rear axle output shaft.*

on output shaft and support shaft and bearing assembly in arbor press as shown in figure 73. Press shaft out of bearing and remove spacer from shaft.

## 105. Cleaning, Inspection, and Repair of Forward Rear Axle Output Shaft Components

*Note.* Key letters noted in parentheses are in figure 74.

*a. General.*—Gaskets, oil seal, and bearing nut lock should be discarded and new parts obtained for use when assembling. The only one of this group of components which may be repaired is the flange assembly (P).

*b. Cleaning.*—Wash all forward rear axle output shaft components in dry-cleaning solvent or volatile mineral spirits and wipe or blow dry.

**Caution:** Do not spin ball bearing assemblies with air as damage to bearings may result.

*c. Inspection.*
 (1) Inspect threads in bearing lock nut (A) and 7/8–14 nut (Q) and threads on each end of forward rear axle output shaft (G). Obtain new parts for assembly if threads are stripped or damaged.
 (2) Rotate races of bearings by hand and check for roughness at races and balls. Also check bearings for wear and examine for cracked or broken balls and races.
 (3) Bearing spacers (D and J) and gear spacer (F) must be inspected for damage. Wear will not normally occur at these parts since the shaft, gears, and spacers rotate as an assembly.
 (4) Examine splines on forward rear axle output shaft (G) for evidence of twisting and other damage. Replace shaft if any damage is evident.
 (5) Examine teeth and splines on forward rear axle front (E) and rear (H) gears. Replace gears having chipped, broken, or worn teeth or splines.
 (6) Inspect forward rear axle output shaft rear bearing retainer (N) for damage. Check gasket surface for flatness with straightedge. Replace retainer if damaged.
 (7) Inspect splines, deflector, and oil seal sleeve on flange assembly (P). If deflector or sleeve require replacement, refer to *d* below for procedure.
 (8) Inspect rear bearing snap ring (L) for distortion and wear. Replace snap ring if not in good condition.

*d. Repair.* Refer to paragraph 224 for instructions for replacing deflector and sleeve on flange assembly (P).

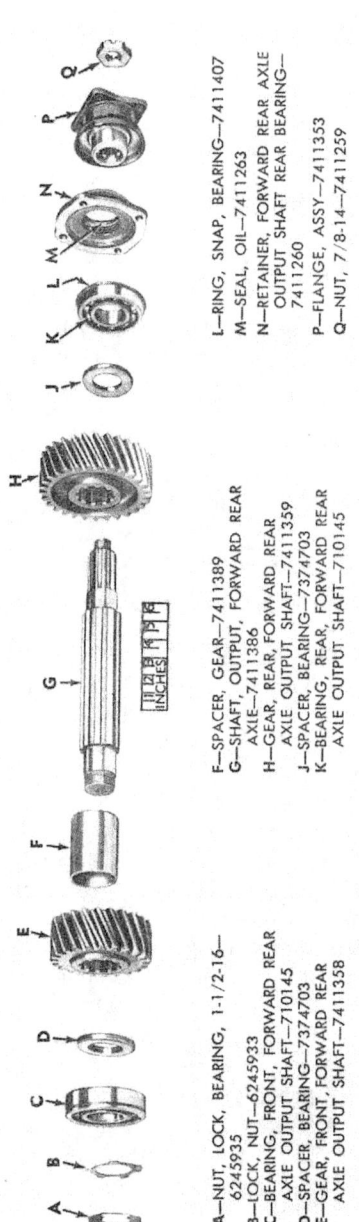

A—NUT, LOCK, BEARING, 1-1/2-16—6245935
B—LOCK, NUT—6245933
C—BEARING, FRONT, FORWARD REAR AXLE OUTPUT SHAFT—7101145
D—SPACER, BEARING—7374703
E—GEAR, FRONT, FORWARD REAR AXLE OUTPUT SHAFT—7411358
F—SPACER, GEAR—7411389
G—SHAFT, OUTPUT, FORWARD REAR AXLE—7411386
H—GEAR, REAR, FORWARD REAR AXLE OUTPUT SHAFT—7411359
J—SPACER, BEARING—7374703
K—BEARING, REAR, FORWARD REAR AXLE OUTPUT SHAFT—710145
L—RING, SNAP, BEARING—7411407
M—SEAL, OIL—7411263
N—RETAINER, FORWARD REAR AXLE OUTPUT SHAFT REAR BEARING—7411260
P—FLANGE, ASSY—7411353
Q—NUT, 7/8-14—7411259

RA PD 148819

*Figure 74. Forward rear axle output shaft components.*

## 106. Assembly of Forward Rear Axle Output Shaft Components

*a. General.*—All components of forward rear axle output shaft and gears except forward rear axle output shaft rear bearing (K, fig. 74) and bearing spacer (J, fig. 74) are assembled during installation of components in transfer case (par. 114). Procedure for installing output shaft rear bearing and spacer is given in *b* below.

*b. Installing Rear Bearing and Bearing Spacer on Forward Rear Axle Output Shaft.*

(1) Adjust opening in attachment 41–A–345–328 (part of puller set 41–P–2905–60) to support output shaft rear bearing at inner race; then place bearing and spacer and attachment on arbor press in position shown in right view of figure 73.

(2) Insert splined end of output shaft downward through spacer and bearing; then press shaft through bearing until shaft shoulder bottoms on spacer.

## Section VIII. REBUILD OF IDLER SHAFT COMPONENTS

### 107. Disassembly of Idler Shaft Components

*Note.* Key letters noted in parentheses are in figure 77 unless otherwise indicated.

*a. General.* When idler shaft is removed from transfer case during disassembly of transfer into subassemblies, all components are removed from idler shaft except rear bearing, bearing spacer, and Woodruff key. The instructions for further disassembly of power take-off gear assembly (J, fig. 76) and removal of bearing and spacer from idler shaft are covered below.

*b. Removing Rear Bearing and Bearing Spacer From Idler Shaft.*

(1) Remove Woodruff key from idler shaft; then position idler shaft and bearing assembly in arbor press with bearing spacer supported on attachment 41–A–345–328 (part of puller set 41–P–2905–60) in same manner as shown in figure 73.

(2) Press idler shaft out of bearing and remove spacer from idler shaft.

*c. Disassembly of Power Take-Off Gear Components.*

(1) Remove bearing snap ring (C) from groove in power take-off gear (E) at front side of power take-off gear bearing (D).

(2) Support power take-off gear in arbor press and apply pressure at idler shaft clutch gear (B) to force clutch gear and power take-off gear bearing (D) out of power take-off gear (E). Refer to figure 75.

(3) Remove bearing snap ring (A) from groove in idler shaft clutch gear (B) at front side of power take-off gear bearing (D). Place clutch gear and bearing assembly in arbor press

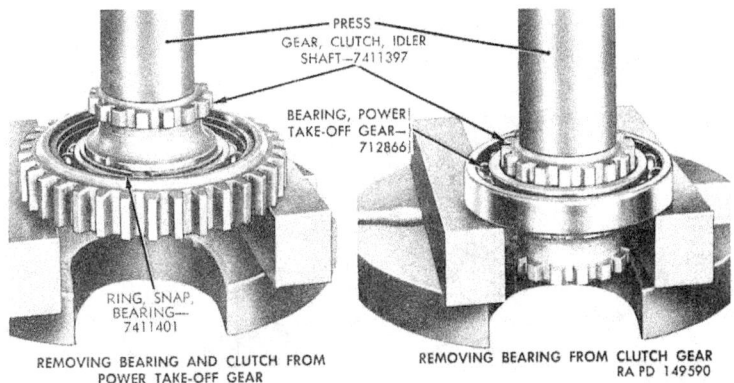

*Figure 75. Power take-off gear and bearing removal.*

as shown in right view of figure 75 and press on clutch gear to separate clutch gear from bearing.

### 108. Cleaning and Inspection of Idler Shaft Components

*Note.* Key letters noted in parentheses are in figure 76 unless otherwise indicated.

*a. Cleaning.* Wash idler shaft, gears, bearings, and other components (figs. 77 and 76) in dry-cleaning solvent or volatile mineral spirits. Wipe parts or dry with compressed air.

**Caution:** Do not spin bearings with air as balls and races may be damaged. Be sure all oil holes through gears are open.

*b. Inspection of Idler Shaft Components.*

(1) Inspect idler shaft front bearing retainer (A) for damage. Also check gasket surface for flatness using straightedge. Threads in which speedometer driven gear assembly (B) is installed must be in good condition.

(2) Inspect speedometer driven gear assembly (B) and speedometer drive gear (C) for damage and wear at teeth. Replace parts if damaged or worn.

(3) Inspect idler shaft (K) for damage at threads on either end and inspect threads in bearing lock nuts (D). If threads are stripped or damaged, obtain new parts for assembly. Inspect Woodruff key (L) and keyway in idler shaft. Key must fit tightly in keyway.

(4) Examine helical teeth on idler shaft front and rear gears (H and M) and the internal clutch teeth on each gear. Gears must be replaced if any of the teeth are found broken or chipped, or if drive surfaces are pitted.

(5) Inspect bearing spacers (G) which must be flat and have no evidence of wear.

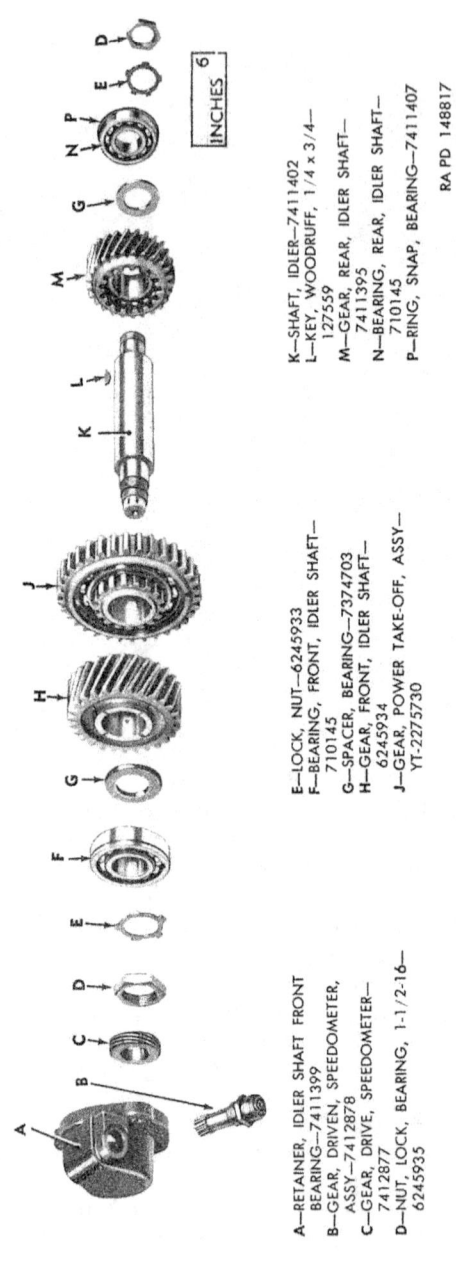

A—RETAINER, IDLER SHAFT FRONT BEARING—7411399
B—GEAR, DRIVEN, SPEEDOMETER, ASSY—7412878
C—GEAR, DRIVE, SPEEDOMETER—7412877
D—NUT, LOCK, BEARING, 1-1/2-16—6245935
E—LOCK, NUT—6245933
F—BEARING, FRONT, IDLER SHAFT—710145
G—SPACER, BEARING—7374703
H—GEAR, FRONT, IDLER SHAFT—6245934
J—GEAR, POWER TAKE-OFF, ASSY—YT-2275730
K—SHAFT, IDLER—7411402
L—KEY, WOODRUFF, 1/4 x 3/4—127559
M—GEAR, REAR, IDLER SHAFT—7411395
N—BEARING, REAR, IDLER SHAFT—710145
P—RING, SNAP, BEARING—7411407

RA PD 148817

*Figure 76. Transfer idler shaft and related components.*

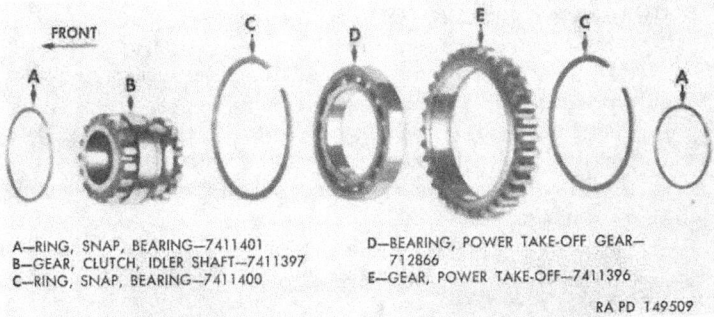

A—RING, SNAP, BEARING—7411401
B—GEAR, CLUTCH, IDLER SHAFT—7411397
C—RING, SNAP, BEARING—7411400
D—BEARING, POWER TAKE-OFF GEAR—712866
E—GEAR, POWER TAKE-OFF—7411396

RA PD 149509

*Figure 77. Power take-off gear and related components.*

(6) Power take-off gear assembly (J) is composed of the components shown in figure 77. Refer to *c* below for inspection procedure for these components.

*c. Inspection of Power Take-Off Gear Components.*

(1) Inspect bearing snap rings (A and C, fig. 77) for distortion and for wear. Rings must be flat and round. Discard rings if not in good condition.

(2) Inspect idler shaft clutch gear (B, fig. 77) which has clutch teeth at both ends. If any of the teeth are chipped, broken, or worn, clutch gear must be replaced with new part.

(3) Inspect power take-off gear bearing (D, fig. 77) for roughness of races and balls by turning bearing races by hand. If rough spots are felt or if races are cracked or broken, replace bearing. Excessive looseness of balls allowing radial movement of inner race indicates worn bearing which must be replaced.

(4) Examine teeth on power take-off gear (E, fig. 77). If any teeth are found to be broken, chipped, or pitted, the gear must be replaced.

## 109. Assembly of Idler Shaft Components

*Note.* Key letters noted in parentheses are in figure 77 unless otherwise indicated.

*a. General.*—Components of power take-off gear (fig. 77) must be assembled into a subassembly before installing idler shaft in transfer case. Power take-off gear assembly (J, fig. 76) and balance of idler shaft components except Woodruff key (L, fig. 76), bearing spacer (G, fig. 76), idler shaft rear bearing (N, fig. 76), nut lock (E, fig. 76), and bearing lock nut (D, fig. 76) are assembled in transfer case as described in procedure for installing idler shaft (par. 116). Refer to *b* and *c* below for procedure for assembling power take-off gear assembly and for installing idler shaft rear bearing and bearing spacer.

*b. Assembling Power Take-Off Gear Components.*
  (1) Install one bearing snap ring (A) on idler shaft clutch gear (B) in groove toward rear end of clutch gear. Support power take-off gear bearing (D) on arbor press plate, position clutch gear on bearing, and press clutch gear into bearing (fig. 78) until snap ring contacts bearing inner race. Install other bearing snap ring (A) in groove at front side of bearing.

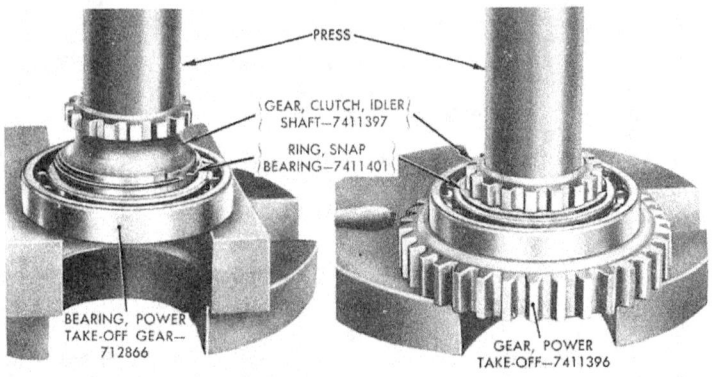

INSTALLING BEARING ON CLUTCH GEAR    PRESSING BEARING INTO POWER TAKE-OFF GEAR

*Figure 78. Installing power take-off gear and bearing on clutch gear.*

  (2) Install one bearing snap ring (C) in groove in power take-off gear (E); then lay gear on press plate with clearance below for clutch gear hub. Position clutch gear and bearing assembly on power take-off gear; then press on clutch gear to force bearing outer race into power take-off gear (fig. 78) until bearing outer race contacts large bearing snap ring (C). Install remaining large bearing snap ring (C) in groove in power take-off gear (E).

*c. Installing Idler Shaft Rear Bearing and Spacer on Idler Shaft.*
  (1) Support idler shaft rear bearing (N, fig. 76) on attachment 41-A-345-328 (part of puller set 41-P-2905-60) in arbor press. Lay bearing spacer (G, fig. 76) on bearing and insert rear end of idler shaft (K, fig. 76) through spacer and into bearing in same manner as illustrated in right view of figure 73.
  (2) Press idler shaft into bearing until shoulder on shaft contacts bearing spacer (G, fig. 76).
  (3) Remove idler shaft (K, fig. 76) from press and drive 1/4 x 3/4 Woodruff key (L, fig. 76) into place in keyway in idler shaft.
  (4) Grip idler shaft (K, fig. 76) in vise equipped with soft jaws.

Install new bearing nut lock (E, fig. 76) and bearing lock nut (D, fig. 76) on rear end of idler shaft. Final tightening of nut is deferred until idler shaft has been installed in transfer case (par. 116).

## Section IX. REBUILD OF SHIFTING MECHANISM

### 110. General

*Note.* Key letters noted in parentheses are in figure 79 unless otherwise indicated.

Shifting mechanism, consisting of the components shown in figure 79, is disassembled to the extent shown. Front axle shifter shaft

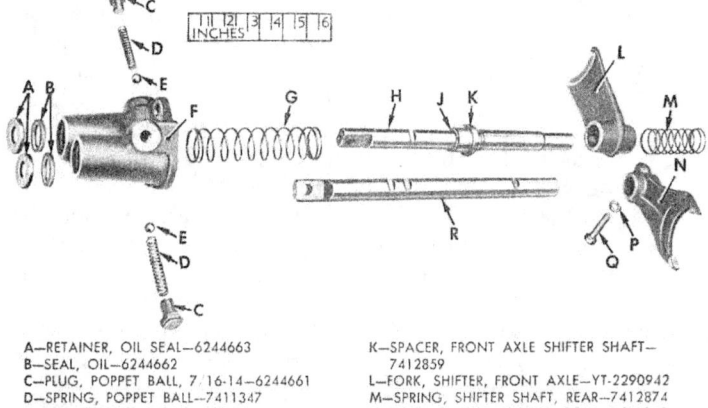

A—RETAINER, OIL SEAL—6244663
B—SEAL, OIL—6244662
C—PLUG, POPPET BALL, 7/16-14—6244661
D—SPRING, POPPET BALL—7411347
E—BALL, POPPET, 3/8-IN—147489
F—SUPPORT, SHIFTER SHAFT, ASSY—7411288
G—SPRING, SHIFTER SHAFT, FRONT—7412500
H—SHAFT, SHIFTER, FRONT AXLE—7411276
J—RING, SNAP—7412858
K—SPACER, FRONT AXLE SHIFTER SHAFT—7412859
L—FORK, SHIFTER, FRONT AXLE—YT-2290942
M—SPRING, SHIFTER SHAFT, REAR—7412874
N—FORK, SHIFTER, SLIDING GEAR—7411342
P—WASHER, LOCK, 5/16-IN—120638
Q—BOLT, 5/16-24x1-1/2—181614
R—SHAFT, SHIFTER, SLIDING GEAR—7411275

RA PD 148816

*Figure 79. Transfer shifting mechanism components.*

spacer (K) need not be removed from front axle shifter shaft (H) unless inspection indicates necessity for replacement of parts. Shifter shaft support pilot (N, fig. 54) is pressed into shifter shaft support assembly (F) and is replaceable. Specifications for various parts are given in paragraph 343c.

### 111. Cleaning, Inspection, and Repair of Shifting Mechanism

*Note.* Key letters noted in parentheses are in figure 79 unless otherwise indicated.

a. *Cleaning.*—Wash all parts of shifting mechanism in dry-cleaning solvent or volatile mineral spirits. Be sure all paint or other deposits are removed from portion of shifter shafts which extend on outside of support. When parts are clean, wipe dry or blow dry with air.

*b. Inspection.*

(1) Inspect two poppet ball springs (D) for distortion. If springs appear in good condition, test for pressure and free length (par. 343c).

(2) Inspect shifter shaft front and rear springs (G and M) for distortion and if springs appear to be in good condition, measure free length and test spring pressure (par. 343c).

(3) Inspect poppet ball plugs (C) and note condition of threads. Also inspect mating threads in shifter shaft support assembly (F). Replace parts if threads are stripped or otherwise damaged.

(4) Examine shifter shaft support assembly (F). Shifter shaft support pilot (N, fig. 54) must be tight fit in support.

(5) Visually inspect front axle shifter shaft (H) and sliding gear shifter shaft (R). If wear is evident at poppet ball notches and at pin holes in front end of shafts, replace shafts. Measure diameter of shifter shafts at points indicated on figure 243. If worn beyond limits (par. 343c), new shafts must be installed when assembling transfer. Examine front axle shifter shaft spacer (K) and snap ring (J). Spacer must be free fit on shifter shaft and snap ring must be in good condition.

(6) Place front axle shifter fork (L) and sliding gear shifter fork (N) in groove in respective gears and check fork-to-groove clearance with feeler gage. If clearance is excessive (par. 343c), replace parts as necessary. Measure bore through front axle shifter fork and compare with dimension listed in paragraph 343c. Forks must be replaced if worn beyond limits.

*c. Repair.*

(1) *Shifter shaft spacer.*—If inspection indicates necessity for replacing front axle shifter shaft spacer (K, fig. 79), remove snap ring (J, fig. 79) with snap ring pliers and slide spacer off front end of front axle shifter shaft (H, fig. 79). Slide new spacer on shifter shaft with flange toward rear end of shifter shaft. Install new snap ring (J, fig. 79) in groove in shifter shaft.

(2) *Shifter shaft support pilot.*—If inspection indicates that shifter shaft support pilot (N, fig. 54) requires replacement, remove pilot from shifter shaft support assembly (F, fig. 79) and press new pilot into recess, with counterbore in pilot facing outward.

## Section X. CLEANING AND INSPECTION OF TRANSFER CASE AND COVERS

### 112. Cleaning

*Note.* Key letters noted in parentheses are in figure 80.

*a.* Thoroughly wash transfer case (K), both inside and out, using dry-cleaning solvent or volatile mineral spirits. If necessary, use a stiff brush or other cleaning tool to dislodge accumulations of dirt.

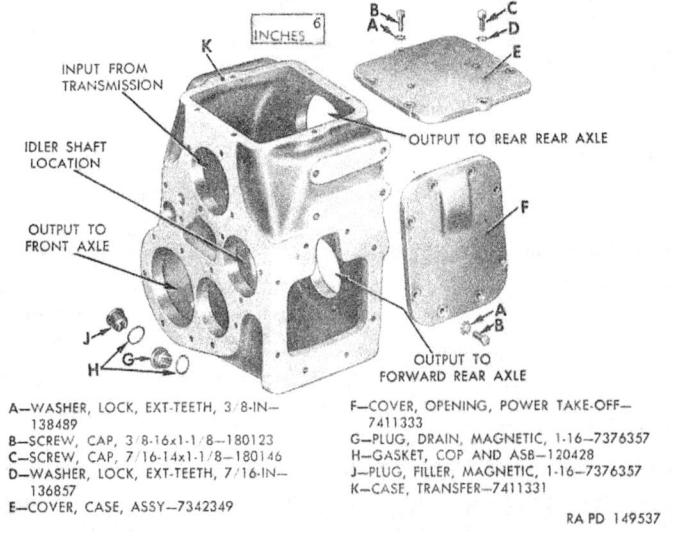

A—WASHER, LOCK, EXT-TEETH, 3/8-IN—138489
B—SCREW, CAP, 3/8-16x1-1/8—180123
C—SCREW, CAP, 7/16-14x1-1/8—180146
D—WASHER, LOCK, EXT-TEETH, 7/16-IN—136857
E—COVER, CASE, ASSY—7342349
F—COVER, OPENING, POWER TAKE-OFF—7411333
G—PLUG, DRAIN, MAGNETIC, 1-16—7376357
H—GASKET, COP AND ASB—120428
J—PLUG, FILLER, MAGNETIC, 1-16—7376357
K—CASE, TRANSFER—7411331

RA PD 149537

*Figure 80. Transfer case and covers.*

*b.* Remove metal particles from magnetic drain and filler plugs (G and J).

*c.* Clean case cover assembly (E) and power take-off opening cover (F), using same method described in *a* above.

*d.* Blow out dirt from tapped holes in case. Dry parts by wiping or with compressed air.

### 113. Inspection

*Note.* Key letters noted in parentheses are in figure 80.

*a.* Examine transfer case (K) carefully for evidence of cracks, using approved method for locating defects in iron castings. If case is cracked or broken, obtain new case for use when assembling.

*b.* Inspect case cover assembly (E) and note condition of oil trough riveted to under side of cover. Also inspect power take-off opening cover (F) (if used). If covers are broken or otherwise damaged, obtain new parts to be used when assembling transfer.

  c. Inspect threads on magnetic drain and filler plugs (G and J). If plug threads are damaged, use new plugs when assembling.

## Section XI. ASSEMBLY OF TRANSFER FROM SUBASSEMBLIES AND COMPONENTS

### 114. Installation of Forward Rear Axle Output Shaft and Gears

*Note.* Key letters noted in parentheses are in figure 52 unless otherwise indicated.

  a. Using plastic type gasket cement, attach bearing retainer gasket (RR) to transfer case at opening labelled "OUTPUT TO FORWARD REAR AXLE" on figure 80. Insert front end of output shaft assembly through opening in transfer case (ZZ) and install forward rear axle output shaft rear gear (UU), gear spacer (VV), and forward rear axle output shaft front gear (WW) as shaft is pushed toward front of case. Figure 57 shows gears and spacer in place inside case. Move shaft assembly forward until bearing snap ring (MM) in rear bearing outer race contacts transfer case.

  b. Coat bore in forward rear axle output shaft rear bearing retainer (QQ) with plastic type gasket cement; then use wood block or suitable driver to install oil seal (LL) in retainer. Seal must bottom against shoulder in retainer. Position forward rear axle output shaft rear bearing retainer (QQ) over end of shaft against case; then install four $3/8$-16 x $1\frac{1}{8}$ cap screws (NN), and new $3/8$-inch external teeth lock washers (PP) to attach bearing retainer to case. Tighten cap screws to torque of 20 to 25 pound-feet.

  c. Place bearing spacer (AT) on front end of forward rear axle output shaft (AU); then install forward rear axle output shaft front bearing (AW) on shaft, using replacer 41–R–2390–415 in manner illustrated in figure 82. Bearing inner race must seat firmly against bearing spacer (AT).

*Note.* Outer race of forward rear axle output shaft front bearing (AW) has a snap ring groove but no snap ring is used.

  d. Place a new nut lock (AV) on front end of forward rear axle output shaft (AU) and install bearing lock nut (AS) loosely.

  e. Install flange assembly (HH) on forward rear axle output shaft (AU) and retain with $7/8$-14 nut (KK).

  f. Assemble companion flange holding tool 41–T–3215–910 on forward rear axle output shaft flange in manner shown in figure 51; then tighten bearing lock nut (AS) to minimum torque of 100 pound-feet.

Bend nut lock (AV) against flat on nut. Tighten 7/8–14 nut (KK) to minimum torque of 130 pound-feet and install 1/8 x 1 3/4 cotter pin (JJ).

*Note.* Leave holding tool attached to flange until balance of bearing lock nuts and flange retaining nuts have been tightened.

### 115. Installation of Front Axle Output Shaft Components

*Note.* Key letters noted in parentheses are in figure 52 unless otherwise indicated.

*a.* Using plastic type gasket cement, attach new bearing support gasket (AC) to front of transfer case with holes in gasket alined with holes in case.

*b.* Install front axle output shaft and components assembly (shown in figure 72) in opening in front of transfer case labelled "OUTPUT TO FRONT AXLE" on figure 80. Front axle output shaft rear bearing (YY) must be guided into recess in transfer case as the assembly is moved into position.

*c.* Aline holes through front bearing support with holes in gasket and transfer case and install eight 3/8–16 x 1 1/8 cap screws (AE) with new 3/8-inch external-teeth lock washers (AD).

*Note.* Bolts installed in five holes tapped through case must be coated with plastic type gasket cement.

Tighten cap screws to torque of 20 to 25 pound-feet.

*d.* Move front axle output shaft gear (AB) into engagement with clutch teeth at rear end of front axle output shaft (XX); then hold shaft from turning with holding tool and tighten 7/8–14 nut (AL) to minimum torque of 130 pound-feet.

### 116. Installation of Idler Shaft, Gears, and Bearings

*Note.* Key letters noted in parentheses are in figure 52 unless otherwise indicated.

*a.* Make up an assembly composed of idler shaft rear gear (V) and idler shaft front gear (BA) mated with power take-off gear assembly ((J), fig. 76) and hold these parts in position inside transfer case.

*Note.* Sectional view of transfer (fig. 52) shows correct position of parts. Larger (27-tooth) gear is located at front.

*b.* Insert front end of idler shaft assembly through opening at rear of transfer case and through idler shaft rear gear (V), idler shaft clutch gear (BC), and idler shaft front gear (BA). Aline 1/4 x 3/4 Woodruff key (Z) in idler shaft with keyway in rear gear as idler shaft (fig. 81) is moved forward. Use lead hammer to drive on rear end of idler shaft to force idler shaft into place with bearing snap ring (EE) against rear face of transfer case.

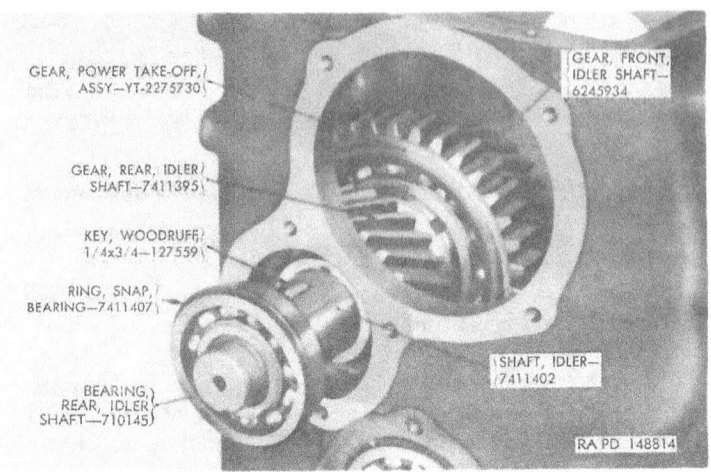

*Figure 81. Installing idler shaft and gears.*

c. Place bearing spacer (BD) on front end of idler shaft, then install idler shaft front bearing (BE), using replacer 41–R–2390–415 as shown in figure 82 to drive bearing onto idler shaft.

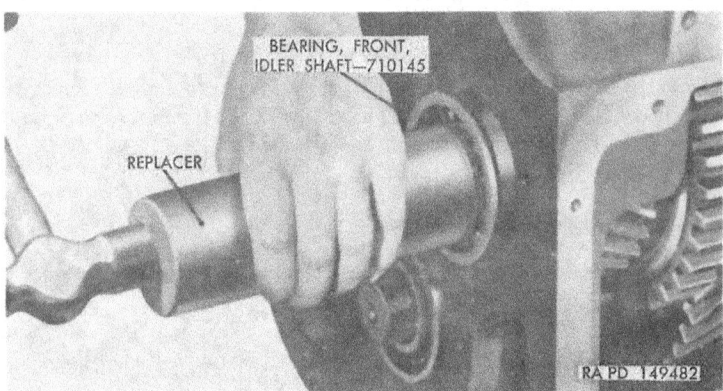

*Figure 82. Use of replacer 41–R–2390–415 for installing idler shaft front bearing.*

d. Install nut lock (BH); then install bearing lock nut (BF). Tighten bearing lock nuts (BF and CC) to minimum torque of 100 pound-feet, using holding tool 41–T–3215–910 on forward rear axle output shaft flange to prevent idler shaft from turning. Bend nut locks (AA and BH) against nuts to prevent loosening.

e. Install speedometer drive gear (BJ) on front end of idler shaft, using replacer 41–R–2390–415 shown in figure 82 to drive gear into place on idler shaft.

*f.* Place a new retainer gasket (BL) at transfer case; then install idler shaft front bearing retainer (BG), using three 3/8–16 x 1 1/8 cap screws (AZ), with new 3/8-inch external-teeth lock washers; coat cap screw threads with plastic type gasket cement to prevent leaks. Tighten cap screws to torque of 20 to 25 pound-feet.

*g.* Coat threads with plastic type gasket cement; then install speedometer driven gear assembly (BK) in idler shaft front bearing retainer (BG). Tighten gear assembly into retainer with torque wrench to torque 30 to 35 pound-feet.

### 117. Installation of Input Shaft and Gear Assembly

*Note.* Key letters noted in parentheses are in figure 52 unless otherwise indicated.

*a.* Coat bearing surface of bushing type bearing (BY) and surface of splines on input shaft (B) with universal gear lubricant (GO). Also apply same lubricant on surface of oil seal sleeve on flange assembly.

*b.* While holding input shaft gear (CB) inside transfer case, insert input shaft (B) through opening in rear of case, through input shaft gear (CB), and out front of case through opening marked "INPUT FROM TRANSMISSION" on figure 80.

*c.* Place bearing spacer (BX) on input shaft; then install input shaft front bearing (BW), driving bearing into place until bearing snap ring (BM) contacts transfer case.

*d.* Coat bore in input shaft front bearing retainer (BN) with plastic type gasket cement; then use wood block or suitable driver to install oil seal (BS) in retainer. Front face of seal assembly must be flush with front of retainer.

*e.* Place retainer gasket (BV) around opening at front of transfer case. Install bearing retainer and oil seal assembly on transfer case, using four 3/8–16 x 1 1/8 cap screws (BT) and new 3/8-inch external-teeth lock washers (BU). Tighten cap screws to torque of 20 to 25 pound-feet.

*f.* Install flange assembly (BR) on front end of input shaft and retain with 1 1/8–12 nut (BQ). Tighten 1 1/8–12 nut (BQ) to minimum torque of 130 pound-feet. Install 1/8 x 2 cotter pin (BP) to secure nut.

### 118. Installation of Shifting Mechanism

*Note.* Key letters noted in parentheses are in figure 54 unless otherwise indicated.

*a.* Hold sliding gear shifter fork (E) in groove at sliding gear (C) with threaded side of fork upward. Insert sliding gear shifter shaft (X) through hole in front of transfer case and move shaft through fork to a point where notch in shifter shaft is alined with bolt hole in fork. Install one 5/16–24 x 1½ bolt (F) with new 5/16-inch lock washer (G) and tighten bolt to torque of 15 to 20 pound-feet.

*b.* Locate front axle shifter fork (R) in groove in front axle output shaft gear (K). Insert front axle shifter shaft (V) through hole in front of transfer case. Pass shifter shaft through front axle shifter fork (R), then hold shifter shaft rear spring (H) between fork and boss in transfer case. Push shifter shaft through spring and into boss in case.

*c.* Place shifter shaft front spring (S) on front axle shifter shaft (V). Using plastic type gasket cement, attach shifter shaft support gasket (AA) to front of transfer case.

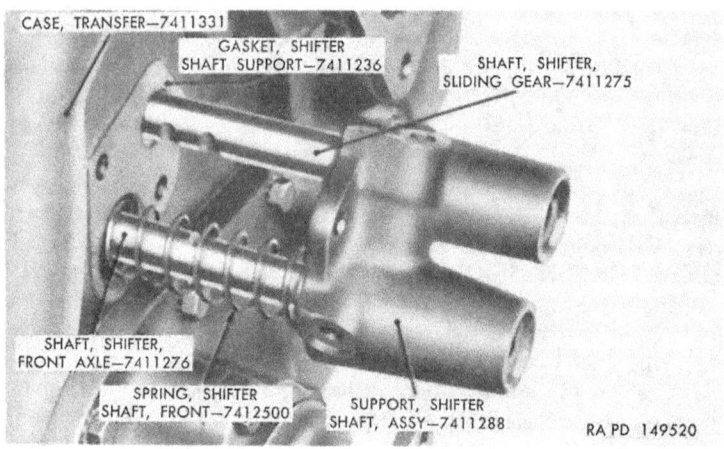

*Figure 83. Installing shifter shaft support.*

*d.* Slide shifter shaft support assembly (Z) onto shifter shafts as shown in figure 83. Push support into place at case. Coat threads of two 3/8–16 x 1⅛ bolts with plastic type gasket cement, and attach support to transfer case with bolts and new lock washers. Tighten bolts to torque of 20 to 25 pound-feet.

*e.* Install oil seals (W) and oil seal retainers (Y) in recesses in shifter shaft support assembly (Z). Use shifter shaft oil seal replacer 41–R–2394–115 as shown in figure 84 to drive oil seal retainers into place.

*f.* Insert one 5/16-inch poppet ball (T) and poppet ball spring (U) in each poppet well in shifter shaft support assembly (Z). Apply plastic type gasket cement on threads of 7/16–14 poppet ball plugs (C and H, fig. 53); then install plugs in shifter shaft support and tighten plugs firmly.

*Note.* If necessary, turn front axle shifter shaft so that poppet ball engages detent in shaft.

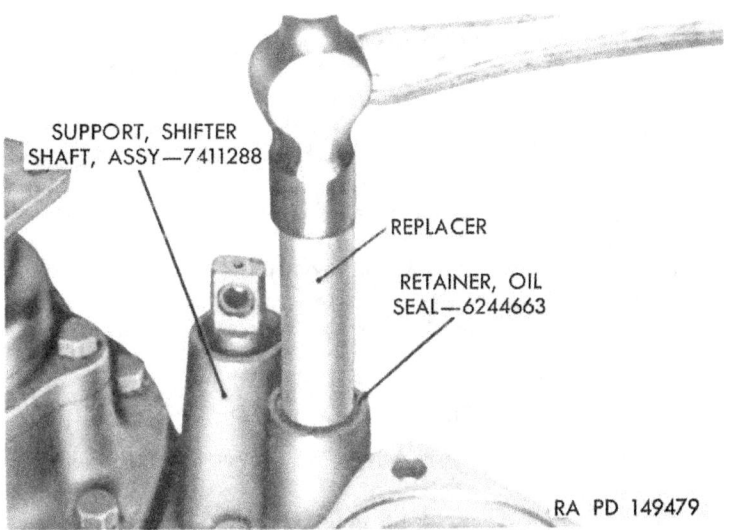

*Figure 84. Use of shifter shaft oil seal replacer 41–R–2394–115 to install oil seal retainers.*

### 119. Installation of Output Gear Components

*Note.* Key letters noted in parenthesis are in figure 52 unless otherwise indicated.

*a. General.*—Figure 68 shows output gear, bearing retainer, and related components assembled preparatory to installation. Prior to installation of assembly shown in figure 68, forward rear axle output shaft must be rotated so that a line drawn through flange upper mounting bolt holes is parallel to top surface of transfer case.

*b. Installation.*

   (1) Using plastic type gasket cement sparingly, attach retainer gasket (H) to output gear bearing retainer (K), with holes in gasket and retainer alined.

   (2) Hold output gear assembly at opening in transfer case, and before engaging output gear (E) with idler shaft rear gear (V), turn flange assembly (R) so that a line through flange upper mounting holes is parallel with top of transfer case. Move output gear and retainer assembly toward transfer case.

   (3) Before installing 3/8–16 x 1 1/8 cap screws (FF), check relationship of flange assemblies (R and HH) with top of transfer case (ZZ). With holes in flange assembly (HH) parallel with top of case, corresponding holes in flange assembly (R) must also be parallel within six degrees. If flanges are not alined within six degrees, remove one of the flanges and shift

131

position of flange one tooth from original position. Repeat procedure described in (2) above to bring about the desired conditions.

(4) Using plastic type gasket cement on cap screw threads, install seven 3/8–16 x 1 1/8 cap screws (FF) with new 3/8-inch external-teeth lock washers (GG) to attach output gear bearing retainer (K) to transfer case. Tighten cap screws to torque of 20 to 25 pound-feet.

(5) Tighten 7/8–14 nut (P) on rear end of output gear to minimum torque of 100 pound-feet and install 1/8 x 1 3/4 cotter pin (Q). Remove holding tool from forward rear axle output shaft flange.

### 120. Installation of Plugs and Covers

*Note.* Key letters noted in parentheses are in figure 80 unless otherwise indicated.

*a.* Install 1–16 magnetic drain plug (G) and 1–16 magnetic filler plug (J) using new copper and asbestos gaskets (H).

*b.* Using eight 3/8–16 x 1 1/8 cap screws (B) and new 3/8-inch external-teeth lock washers (A), install power take-off opening cover (F) with new gasket. Cap screws used at holes tapped through case must be coated with plastic type gasket cement to prevent leaks. Tighten all cap screws to torque of 20 to 25 pound-feet.

*c.* Place a new cover gasket (G, fig. 52) on top of transfer case (K) with holes alined in transfer case. Place case cover assembly (E) on case and install one 7/16–14 x 1 1/8 cap screw (C) with new 7/16-inch external-teeth lock washer through largest hole in cover. Install 3/8–16 x 1 1/8 cap screws (B) with new 3/8-inch external-teeth lock washers in five remaining holes. Cap screws used in holes tapped through case must be coated with plastic type gasket cement before installation. Tighten transfer case cover cap screws to torque of 20 to 25 pound-feet.

*d.* Test Transfer Case for Leaks as Follows.—
(1) Connect air supply with gage and shut-off valve to vent line fitting in output gear bearing retainer (K, fig. 52).
(2) Fill transfer case with air at 15 psi and close shut-off valve.
(3) Air pressure must not drop at a rate in excess of 5 pounds in 45 seconds.
(4) If leakage is excessive, cause must be determined and corrected.

# CHAPTER 6

# POWER TAKE-OFF

## Section I. DESCRIPTION AND DATA

### 121. Description of Power Take-Off

*a. General.*—Power take-off equipment is provided for driving winch only, winch and dump body hoist, or for operating pump on tank type trucks. The power take-off assembly is installed at left side of transfer and is driven by power take-off gear in transfer. Power take-off can be operated either with truck standing or in motion. Power take-off receives lubrication from the transfer assembly.

*b. Power Take-Off for Operating Winch Only.*—The power take-off used on vehicles equipped with winch only is shown in figure 85. This power take-off has single output shaft which can be operated in either forward or reverse direction. Winch drive shaft and control cable both connect at forward side of power take-off.

*c. Power Take-Off for Operating Winch and Dump Body Hoist.*—Trucks with dump bodies are equipped with a power take-off on which is installed an accessory drive unit (fig. 86). On this installation the winch (when used) is driven in same manner as described in *b* above. The accessory drive unit has an output shaft toward rear and a shifter shaft at the rear to which accessory control linkage is attached. The separate controls permit independent operation of winch and dump body. Figure 98 shows arrangement of accessory drive components for above installation. Refer to paragraph 142*b* for detailed description of accessory drive unit.

*d. Power Take-Off for Operating Pump On Tank Trucks.*—Trucks with tank bodies do not have winch but a drive is provided for operating pump. On these vehicles the installation is similar to that described under *c* above except that winch drive shaft and control parts are omitted from the power take-off. A plain cap is installed to close winch drive shaft opening in front of housing and plug (fig. 87) with washer and seal is installed in winch control cable opening. Front and rear views of power take-off used to operate pump are shown in figure 87. Figure 97 shows arrangement of accessory drive unit com-

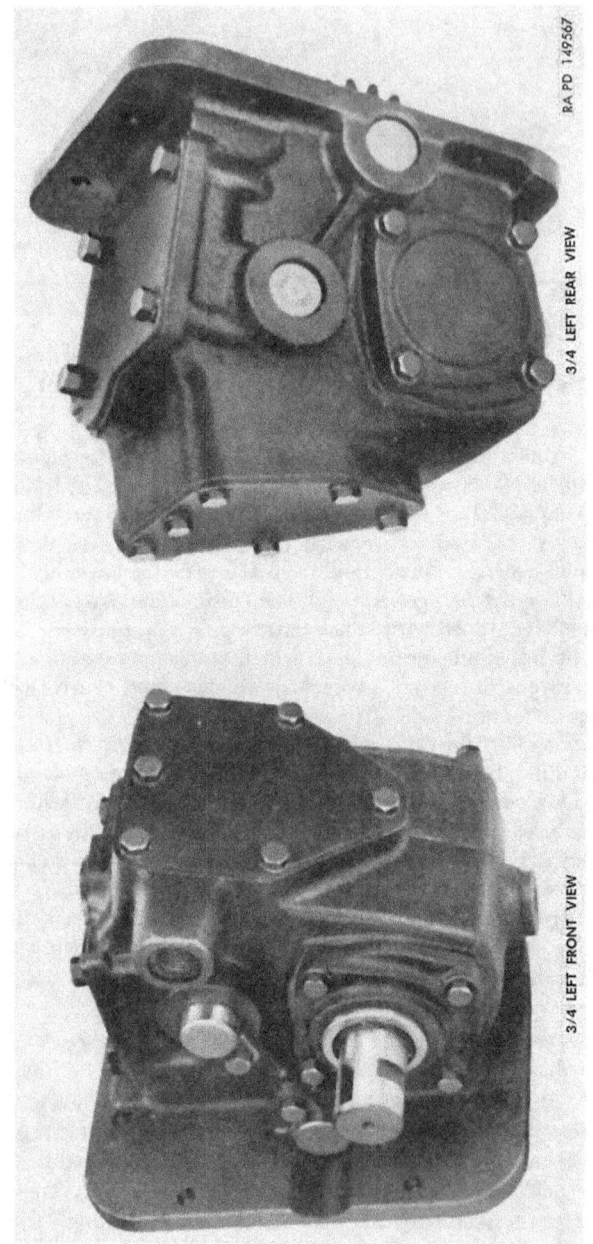

*Figure 85. Front and rear views of power take-off used to drive winch only.*

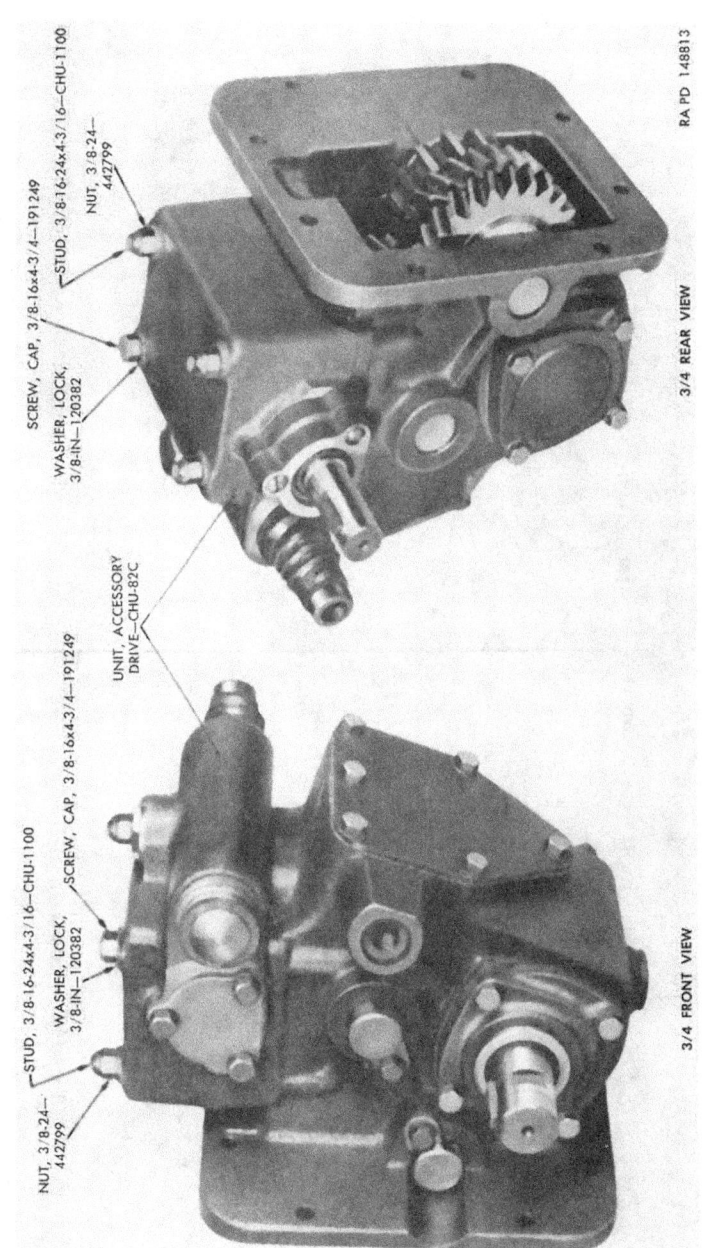

Figure 86. Front and rear views of power take-off used to operate winch and dump body hoist.

135

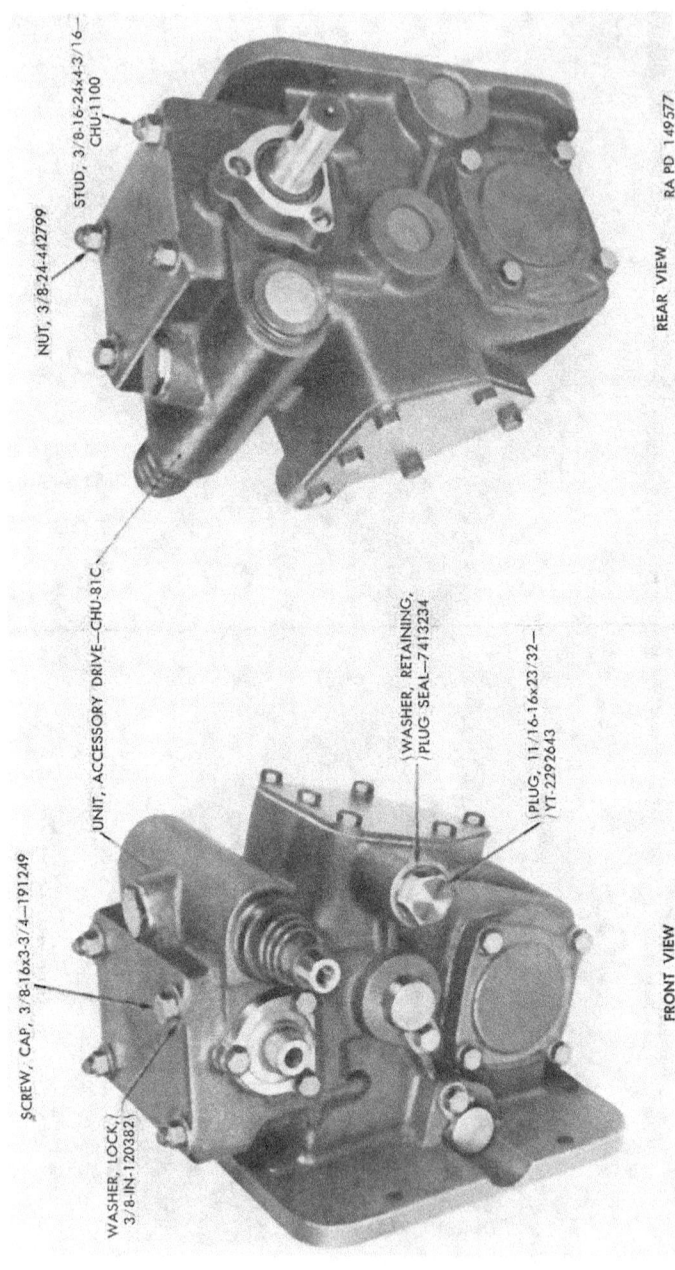

Figure 87. Front and rear views of power take-off equipped with accessory drive unit for operating tank truck pump.

136

ponents for above installation. Note that shifter shaft extends toward front and a special cap (drive shaft front bearing cap (A, fig. 97)) is installed on accessory drive housing. Special cap serves as an adapter for connecting drive cable used to govern the truck engine speed while operating tank pump. Refer to paragraph 142*a* for detailed description of accessory drive unit.

## 122. Data

```
Manufacturer_____ Chelsea Products
Power take-off model:
    Trucks w/winch only_____ 87C1
    Trucks w/winch and dump body_____ 85C
    Trucks w/winch and tank body_____ 85C
Type _____ reversible
Speed _____ single
Accessory drive model:
    Trucks w/dump body_____ 82C
    Trucks w/tank body_____ 81C
```

## Section II. DISASSEMBLY OF POWER TAKE-OFF

## 123. General

Power take-off can be disassembled without the use of special tools. Procedure given in this section covers removal from housing of all component parts in logical sequence. Replacement of accessory drive unit is covered in paragraphs 143 and 144. Rebuild of tank truck accessory drive unit is covered in paragraphs 145 through 148. Rebuild of dump truck accessory drive unit is covered in paragraphs 149 through 152. During disassembly, all gaskets and oil seals should be removed and discarded as these parts should be replaced with new ones when overhauling power take-off. Arrangement of component parts is shown in sectional view of power take-off (fig. 88).

## 124. Removal of Plugs and Covers

*Note.* Key letters noted in parentheses are in figure 89 unless otherwise indicated.

*a. General.*—If power take-off is equipped with an accessory drive unit, remove (par. 143) and disassemble unit as instructed in sections VI or VII of this chapter. All power take-off housings have a magnetic-type drain plug and if power take-off is the type used on tank truck (fig. 87), a plug is installed in front of housing and serves to close hole at which shifter control is installed when used on other vehicles.

*b. Removing Plugs.*—Remove housing drain plug (BB) and drain plug gasket (AA). Remove plug (fig 87), washer, and seal used to close shifter hole in power take-off used on tank trucks.

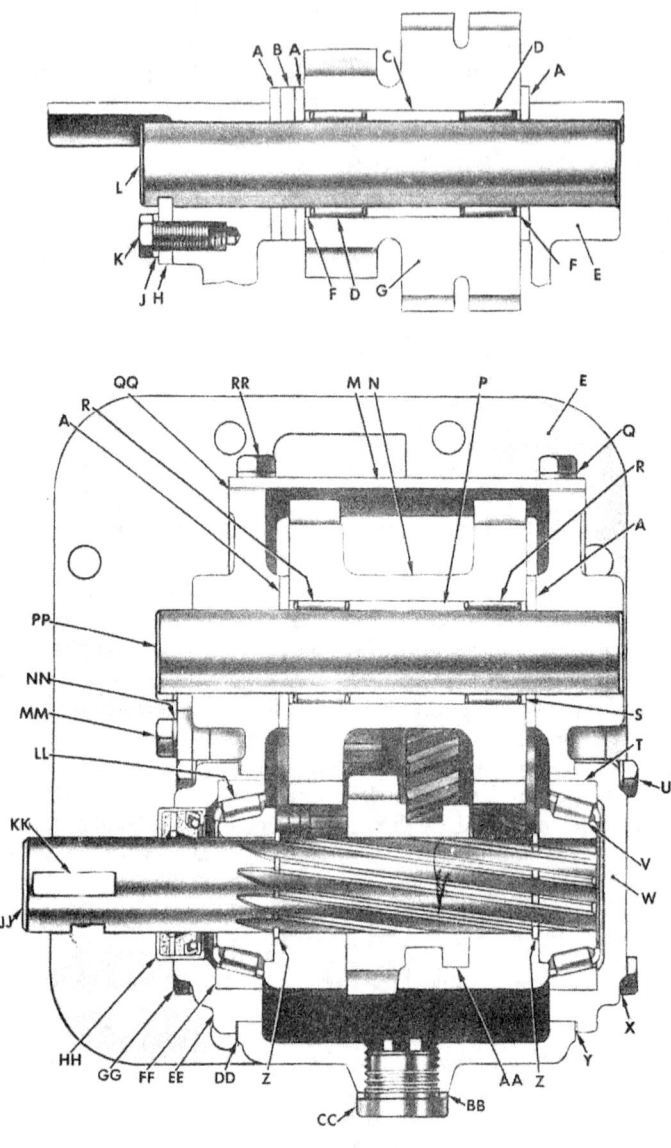

*Figure 88. Sectional view of winch power take-off.*

A—WASHER, THRUST, IDLER GEAR—7412905
B—SPACER, IDLER GEAR—7412902
C—SPACER, BEARING—7412903
D—BEARING, ROLLER, IDLER GEAR—7412832
E—HOUSING—7412891
F—WASHER, THRUST, IDLER GEAR BEARING—CHU-1415
G—GEAR, IDLER—7412889
H—PLATE, RETAINING, IDLER SHAFT—7412829
J—WASHER, LOCK, 5/16-IN.—120214
K—SCREW, CAP, 5/16-18 X 3/4—180077
L—SHAFT, IDLER—7412901
M—PLATE, COVER, TOP—7412893
N—GEAR, REVERSE—7412890
P—SPACER, BEARING, REVERSE GEAR—7412904
Q—GASKET—105451
R—BEARING, ROLLER, REVERSE GEAR—7412832
S—WASHER, THRUST, REVERSE GEAR BEARING—CHU-1415
T—CUP, BEARING, REAR—706820
U—SCREW, CAP, 5/16-18 X 1—180079
V—CONE, BEARING, REAR—705804
W—{ CAP, DRIVE SHAFT REAR BEARING—7412882
      CAP, FRONT (WITHOUT WINCH DRIVE)—7412882
X—WASHER, COP—7412883
Y—{ GASKET, CAP, REAR BEARING 0.010-IN.—7412886
      GASKET, CAP, REAR BEARING 0.020-IN.—7412887
Z—RING, LOCKING, REAR BEARING—7735382
AA—GEAR, SLIDING—7412888
BB—GASKET, DRAIN PLUG—120428
CC—PLUG, DRAIN, HOUSING—7376357
DD—{ GASKET, CAP, FRONT BEARING 0.010-IN.—7412886
       GASKET, CAP, FRONT BEARING 0.020-IN.—7412887
EE—CAP, FRONT BEARING—7412881
FF—CUP, BEARING, FRONT—706820
GG—SCREW, CAP, 5/16-18 X 1—180079
HH—SEAL, OIL, DRIVE SHAFT—7412899
JJ—SHAFT, DRIVE—7412900
KK—KEY, WOODRUFF, 5/16 X 1 1/8—117979
LL—CONE, BEARING, FRONT—705395
MM—SCREW, CAP, 5/16-18 X 3/4—180077
NN—WASHER, LOCK, 5/16-IN.—120214
PP—SHAFT, REVERSE GEAR—7412901
QQ—GASKET, TOP COVER PLATE—7412884
RR—SCREW, CAP, 3/8-16 X 7/8—180121

*Figure 88.*—Continued.

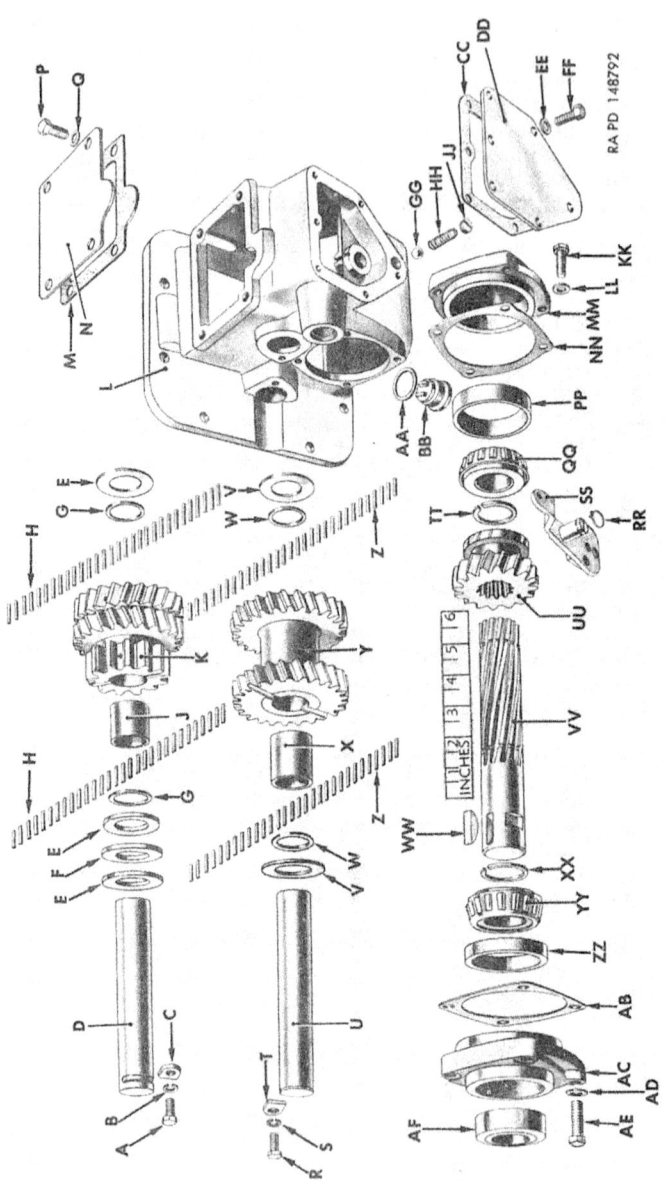

A—SCREW, CAP, 5/16-18 X 3/4—180077
B—WASHER, LOCK, 5/16-IN.—120214
C—PLATE, RETAINING, IDLER SHAFT—7412829
D—SHAFT, IDLER—7412901
E—WASHER, THRUST, IDLER GEAR—7412905
F—SPACER, IDLER GEAR—7412902
G—WASHER, THRUST, IDLER GEAR BEARING—CHU-1415
H—BEARING, ROLLER, IDLER GEAR—7412882
J—SPACER, BEARING—7412903
K—GEAR, IDLER—7412889
L—HOUSING—7412891
M—GASKET, TOP COVER PLATE—7412884
N—PLATE, COVER, TOP—7412893
P—SCREW, CAP, 3/8-16 X 7/8—180121
Q—GASKET—105451
R—SCREW, CAP, 5/16-18 X 3/4—180077
S—WASHER, LOCK, 5/16-IN.—120214
T—PLATE, RETAINING, REVERSE SHAFT—7412829
U—SHAFT, REVERSE GEAR—7412901
V—WASHER, THRUST, REVERSE GEAR—7412905
W—WASHER, THRUST, REVERSE GEAR BEARING—CHU-1415
X—SPACER, BEARING, REVERSE GEAR—7412904
Y—GEAR, REVERSE—7412890
Z—BEARING, ROLLER, REVERSE GEAR—7412882
AA—GASKET, DRAIN PLUG—120428
BB—PLUG, DRAIN, HOUSING—736357
CC—GASKET, SIDE COVER PLATE—7412885
DD—PLATE, COVER, SIDE—7412894
EE—WASHER, 7412883
FF—SCREW, CAP, 5/16-18 X 3/4—180077
GG—BALL, DETENT, 7/16-IN.—104921
HH—SPRING, DETENT BALL—7412835
JJ—PLUG, DETENT BALL, SPRING—7412830
KK—SCREW, CAP, 5/16-18 X 1—180079
LL—WASHER, COPPER—7412883
MM—{CAP, DRIVE SHAFT REAR BEARING—7412882
     {CAP, FRONT (WITHOUT WINCH DRIVE)—7412882
NN—{GASKET, CAP, REAR BEARING, 0.010-IN.—7412886
     {GASKET, CAP, REAR BEARING, 0.020-IN.—7412887
PP—CUP, BEARING, REAR—706820
QQ—CONE, BEARING, REAR—705804
RR—RING, LOCKING, SHIFTER PLATE—7412831
SS—PLATE, SHIFTER, ASSY—7412892
TT—RING, LOCKING, REAR BEARING—735382
UU—GEAR, SLIDING—7412888
VV—SHAFT, DRIVE—7412900
WW—KEY, WOODRUFF, 5/16 X 1 1/8—117979
XX—RING, LOCKING, FRONT BEARING—735382
YY—CONE, BEARING, FRONT—705395
ZZ—CUP, BEARING, FRONT—706820
AB—{GASKET, CAP, FRONT BEARING, 0.010-IN.—7412886
     {GASKET, CAP, FRONT BEARING, 0.020-IN.—7412887
AC—CAP, FRONT BEARING—7412881
AD—WASHER, COPPER—7412883
AE—SCREW, CAP, 5/16-18 X 1—180079
AF—SEAL, OIL, DRIVE SHAFT—7412899

*Figure 89. Power take-off components with winch drive.*

*c. Removing Covers.*—Remove four top cover plate 3/8–16 x 7/8 cap screws (**P**) and gaskets (**Q**); then remove top cover plate (**N**) and top cover plate gasket (**M**). Remove five cap screws and gaskets used to attach side cover plate (**DD**) and remove side cover plate and side cover plate gasket (**CC**).

*d. Removing Caps (Power Take-Off Without Winch Drive Shaft Only).*—Remove four 5/16–18 x 1 cap screws (**KK**) and copper washers (**LL**) used at each end of power take-off housing to attach caps which cover winch drive shaft and bearing openings. Remove caps and gaskets.

### 125. Removal of Idler Gear, Shaft, and Bearings

*Note.* Key letters noted in parentheses are in figure 89 unless otherwise indicated.

*a.* Remove 5/16–18 x 3/4 cap screw (**A**) and 5/16-inch lock washer (**B**) attaching idler shaft retaining plate (**C**). Remove plate from power take-off housing; then, using hammer and brass drift, drive idler shaft (**D**) forward and remove from housing.

*b.* Remove parts shown in figure 96 together with idler gear roller bearings (**H**), idler gear bearing thrust washers (**G**), and bearing spacer (**J**). Remove bearings, washers, and spacer from idler gear.

### 126. Removal of Reverse Gear, Shaft, and Bearings

*Note.* Key letters noted in parentheses are in figure 89 unless otherwise indicated.

*a.* Remove reverse shaft retaining plate (**T**) attached to power take-off housing with cap screw and lock washer.

*b.* Using hammer and brass drift, drive reverse gear shaft (**U**) forward, and remove from housing. Remove reverse gear and reverse gear thrust washers (fig. 95) together with reverse gear roller bearings (**Z**), reverse gear bearing thrust washers (**W**), and reverse gear bearing spacer (**X**).

### 127. Removal and Disassembly of Drive Shaft and Bearing Components

*Note.* Key letters noted in parentheses are in figure 89.

*a. General.*—If power take-off is for use on truck equipped with tank body, winch drive shaft, bearings, and related components are not installed in power take-off housing and the instructions in this paragraph do not apply; however, the operations described in *b* and *c* below will apply to power take-off used on all other types of trucks.

*b. Removal of Drive Shaft and Bearings.*

(1) If 5/16 x 1 1/8 Woodruff key (**WW**) has not been removed from drive shaft, remove key; then remove four 5/16–18 x 1

cap screws (AE) and copper washers (AD) which attach drive shaft front bearing cap (AC) to power take-off housing. Remove front bearing cap (AC) with drive shaft oil seal (AF) and front bearing cup (ZZ). Remove front bearing gasket cap. Remove oil seal from cap. Remove front bearing cup from cap.

(2) Remove four $5/16$–18 x 1 cap screws (**KK**) and copper washers (**LL**) attaching drive shaft rear bearing cap (**MM**)) to power take-off housing. Strike front end of drive shaft lightly with lead hammer to remove rear bearing cap (**MM**). Remove rear bearing gasket cap (**NN**). Remove rear bearing cup (**PP**) from cap.

(3) Remove drive shaft (**VV**) with sliding gear (**UU**) and drive shaft front and rear bearing cones (**YY** and **QQ**) as an assembly.

*c. Removal of Sliding Gear and Bearings From Drive Shaft.*
(1) Using arbor press, remove front and rear bearing cones from drive shaft.
(2) Remove front and read bearing locking rings (**XX** and **TT**), using snap ring pliers. Remove sliding gear (**UU**) from rear end of drive shaft.

## 128. Removal of Shifter Plate and Shifter Detent Components

*a. General.*—If power take-off is used on truck equipped with a tank body, no shifter mechanism is used since these power take-offs do not have a drive shaft for operating winch. The procedure given in *b* below is applicable to all trucks equipped with winch.

*b. Removing Shifting Mechanism.*
(1) Remove detent ball spring plug, detent ball spring, and detent ball (fig. 90).
(2) Using needle-nose snap ring pliers, remove shifter plate locking ring, then remove shifter plate assembly from inside housing.

## Section III. REBUILD OF POWER TAKE-OFF COMPONENTS

## 129. General

The procedures described in this section cover all the component parts of power take-off. On some applications, some of the parts mentioned are not used. Gaskets and drive shaft oil seal should be discarded and new parts obtained for use at assembly.

## 130. Cleaning

Wash all components in dry-cleaning solvent or volatile mineral spirits, using a stiff brush if necessary to dislodge accumulations of grease and dirt. Scrape off all portions of gaskets and sealing compound. Use a wire to clean out oil holes through gears. Immerse roller bearing assemblies in cleaning solution until all lubricant is dissolved and bearings are clean. Use necessary care to remove all metallic particles from magnetic drain plug.

### 131. Inspection of Gears

*Note.* Key letters noted in parentheses are in figure 89.

*a.* Inspect all gear teeth for evidence of wear. Also look for chipped or broken teeth. If any of the teeth are damaged, discard gear and obtain new part. Compare diameter of bore of idler gear (K) with dimensions given in paragraph 344. If bore is worn beyond limits, new gear must be used when assembling power take-off. Make same inspection of bore of through reverse gear (Y).

*b.* Inspect sliding gear (UU) for wear at splines which mate with helical splines on drive shaft (VV). If excessive wear is evident select new parts for use when assembling. Refer to paragraph 135 for instructions covering inspection of groove in sliding gear and clearance of shifter plate assembly in groove.

### 132. Inspection of Shafts

*Note.* Key letters noted in parentheses are in figure 89.

*a.* Idler shaft (D) and reverse gear shaft (U) are identical and should be inspected for wear at surfaces contacted by bearing rollers. Diameter of new shafts is given in paragraph 344.

*b.* Inspect drive shaft (VV) for condition at splines and at keyway. If drive shaft is not in good condition, it must be replaced by new drive shaft when assembling power take-off.

### 133. Inspection of Bearings and Thrust Washers

*Note.* Key letters noted in parentheses are in figure 89.

*a.* Inspect idler gear and reverse gear roller bearings (H and Z). If any of the rollers are broken or worn excessively, obtain a complete set of 60 rollers for use in each gear at assembly. Also note condition of bearing spacer (J) and reverse gear bearing spacer (X). Spacers must not show evidence of excessive wear and must be free fit in respective gears and slide freely over idler shaft (D) and reverse gear shaft (U).

*b.* Check condition of idler gear bearing thrust washers (G) and reverse gear bearing thrust washers (W). Replace thrust washers if worn or damaged.

*c.* Examine each roller in front bearing cone (YY) and rear bearing cone (QQ) for chipping, pitting, or other damage. Also inspect rear bearing cup (PP) and front bearing cup (ZZ) for damage and wear at surface contacted by rollers. Replace cups if damaged.

*d.* Measure thickness of three idler gear thrust washers (E) and two reverse gear thrust washers (V). Thickness must be within limits listed in paragraph 344. Inspect idler gear spacer (F) for scoring and for wear. If any of the thrust washers or the spacer is worn, the damaged parts must be replaced.

### 134. Inspection of Power Take-Off Housing and Covers

*Note.* Key letters noted in parentheses are in figure 89.

*a.* Inspect power take-off housing (L) for cracked or broken condition. Also examine threads in tapped holes. If housing is damaged, replace with new housing when assembling.

*b.* Examine top cover plate (N) and side cover plate (DD). If covers are bent or otherwise damaged, replace with new parts.

*c.* Inspect drive shaft rear bearing cap (MM) for cracks and for distortion. Replace cap if damaged.

### 135. Inspection of Miscellaneous Parts

*Note.* Key letters noted in parentheses are in figure 89.

*a.* Inspect front bearing locking ring (XX) and rear bearing locking ring (TT) for bent condition. If rings are bent or broken, obtain new rings for installation at assembly.

*b.* Inspect threads on housing drain plug (BB). Replace plug if threads are not in good condition.

*c.* Inspect $5/16$ x $1\frac{1}{8}$ Woodruff key (WW), and idler shaft and reverse shaft retaining plates (C) and (T). Replace if these parts are not in good condition.

*d.* Examine shifter plate assembly (SS) for fit in housing and position shifter shoe (part of plate assembly) in groove in sliding gear (UU) and check fit of shoe in groove. If clearance exceeds maximum clearance listed in paragraph 344, replace worn parts.

*e.* Check shifter detent ball spring for free length. If spring is not within specifications (par. 344), obtain new spring for use at assembly.

## Section IV. ASSEMBLY OF POWER TAKE-OFF

### 136. General

Procedures for assembling power take-off differ, depending on the type of vehicle on which power take-off is installed. On trucks with tank type bodies, no drive shaft is used in power take-off, since these vehicles have no winch. Shifting mechanism is also omitted from power take-off used on these vehicles. When an accessory drive unit is used, the top cover plate ((N), fig. 89) is omitted and studs with self-locking nuts are used to mount the accessory drive unit. Refer to figures 85, 86, and 87 for external views of various power take-off assemblies used. Refer to paragraphs 142 through 152 for procedures for removing, rebuilding, and installing accessory drive units.

### 137. Installation of Shifter Plate Assembly and Components

*Note.* Information in this paragraph is not applicable to power take-off used on vehicles without winch. Key letters noted in parentheses are in figure 89 unless otherwise indicated.

*a.* From inside housing (**L**), position shifter plate assembly (**SS**) with plate shaft through hole in housing. Using snap ring pliers, install shifter plate locking ring (**RR**) in groove in shaft (fig. 90).

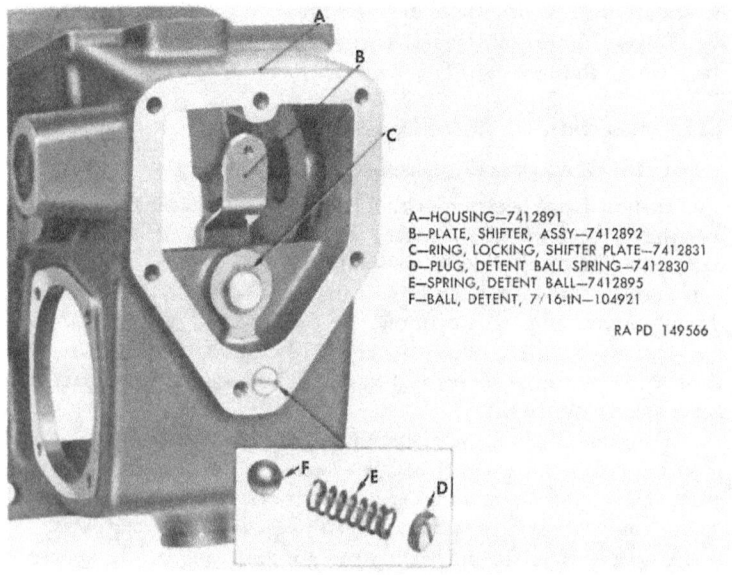

*Figure 90. Shifter plate assembly installed in housing.*

*b.* Assemble $7/16$-inch detent ball (fig. 90) detent ball spring, and detent ball spring plug in poppet well in housing. Tighten plug firmly with screwdriver.

### 138. Assembly and Installation of Drive Shaft and Related Components

*Note.* Key letters noted in parentheses are in figure 89 unless otherwise indicated.

*a.* Using snap ring pliers, install front bearing locking ring (XX) in groove at front end of splines in drive shaft (VV). Support front side of front bearing cone (YY) on arbor press; then press drive shaft through bearing until locking ring seats firmly against rear of front bearing cone.

*b.* Install sliding gear (UU) on drive shaft splines with shifter groove toward rear end of drive shaft. Install rear bearing locking ring (TT) in groove near rear end of drive shaft; then press rear bearing cone (QQ) on drive shaft with locking ring contacting bearing cone.

*c.* Apply universal gear lubricant (GO) on front and rear bearing cones and on surface of drive shaft which is contacted by oil seal.

*d.* Press front bearing cup (ZZ) into place in front bearing cap (AC). Coat bore in front bearing cap (AC) with light coat of plastic type gasket cement; then press drive shaft oil seal into bore with spring-loaded lip pointing inward. Press oil seal in until inner edge is just flush with inner edge of bore. Refer to sectional view of winch power take-off (fig. 88). Wipe off any excess cement after oil seal is installed.

*e.* Wrap shim stock around front end of drive shaft (VV) to protect oil seal from damage during installation; then install front bearing cap on front end of drive shaft. Figure 91 shows drive shaft components assembled, ready for installation in housing.

*f.* Using light coat of plastic type gasket cement, mount two front bearing cap gaskets (AB) at housing; then place drive shaft assembly through bore in front of housing and move into place with bearing

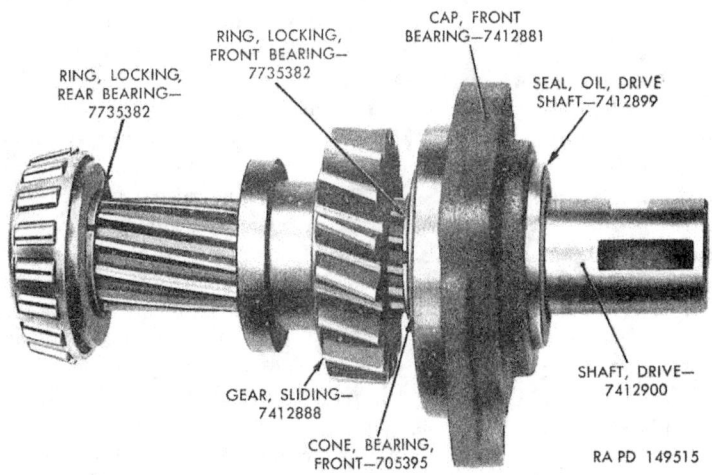

*Figure 91. Drive shaft, sliding gear, and bearings assembled.*

cap alined with holes in housing and shoe on shifter plate assembly engaged in groove in sliding gear (UU). Install four 5/16–18 x 1 cap screws (AE) and four copper washers (AD) to attach cap to housing. Tighten cap screws firmly.

*g.* Press rear bearing cup (PP) into drive shaft rear bearing cap (MM); then, using two rear bearing cap gaskets (NN), install drive shaft rear bearing cap on housing and retain with four 5/16–18 x 1 cap screws (KK) and copper washers (LL). Tighten cap screws gradually, meanwhile turning drive shaft by hand to detect any tightness at bearings. If tightening cap screws causes tightness in bearings, remove rear cap and use thicker gaskets or add one 0.010-inch gasket. When sufficient gaskets have been selected to prevent tightness at bearings, tighten rear bearing cap attaching cap screws firmly.

*Note.* Shoe on shifter plate assembly (SS) must engage groove in sliding gear. Figure 92 shows drive shaft and sliding gear installed.

*h.* Check bearing adjustment with dial indicator mounted as shown in figure 93. Adjustment is correct when there is 0.005-inch end play in drive shaft (VV).

### 139. Installation of Reverse Gear, Shaft, and Bearings

*Note.* Key letters noted in parentheses are in figure 89.

*a.* Coat reverse gear thrust washers and bore of reverse gear (Y) with universal gear lubricant (GO).

*b.* Set reverse gear (Y) on end on work bench. Using a tool made of metal or wood approximately 1.100-inch in diameter and long enough to extend through gear, assemble 30 reverse gear roller bearings (Z) around tool. Use reverse gear bearing spacer (X) to push bearings down to lower side of gear; then assemble 30 more bearings around tool as shown in figure 94.

*c.* Start reverse gear shaft (U) through boss at front side of housing with end of shaft slightly beyond inside surface of case. Locate one reverse gear thrust washer (V) on end of reverse gear shaft (U).

*d.* Lay reverse gear (Y) on side and place one reverse gear bearing thrust washer (W) against rollers at each end of reverse gear. Position reverse gear in housing as shown in figure 95. Note that end of gear with rounded teeth must be toward front of housing.

*e.* Hold reverse gear thrust washer between rear end of gear and housing; then drive lightly on reverse gear shaft to push tool out through bore in rear of housing and force reverse gear shaft into place in housing. Be sure slot in reverse gear shaft is toward retaining plate bolt hole in housing.

*f.* Position reverse shaft retaining plate in slot in shaft and install one 5/16–18 x 3/4 cap screw (R) and 5/16-inch lock washer (S) to hold retaining plate in place.

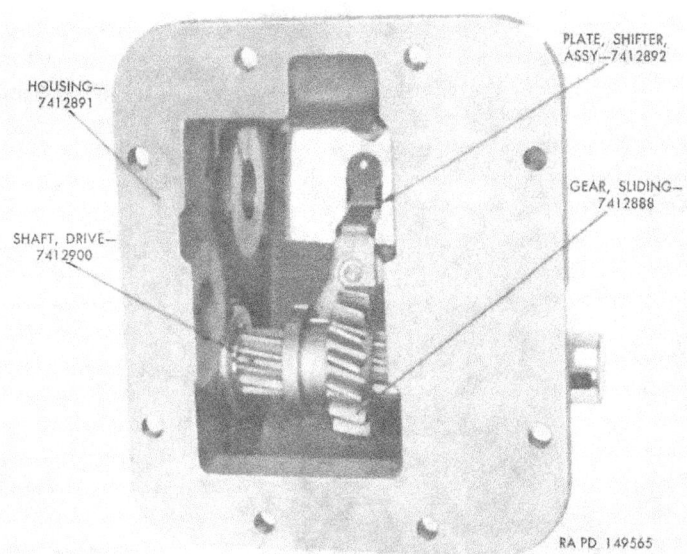

*Figure 92. Power take-off drive shaft and sliding gear installed.*

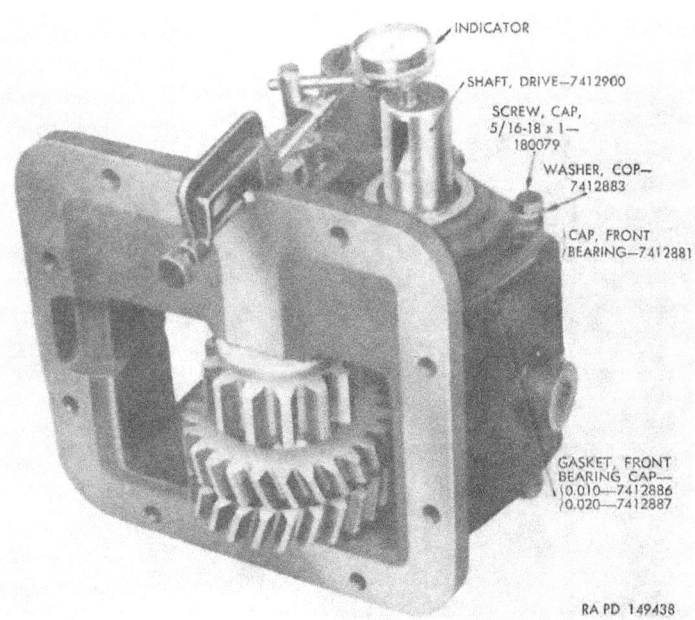

*Figure 93. Checking drive shaft bearing adjustment.*

149

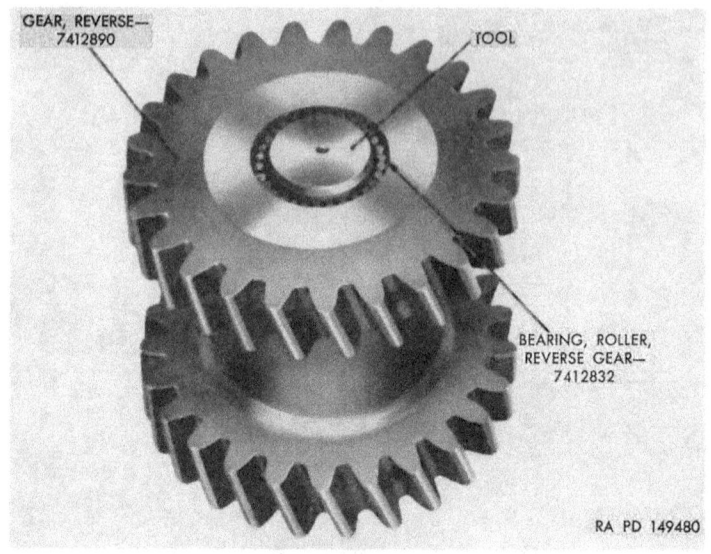

*Figure 94. Assembling rollers in reverse gear.*

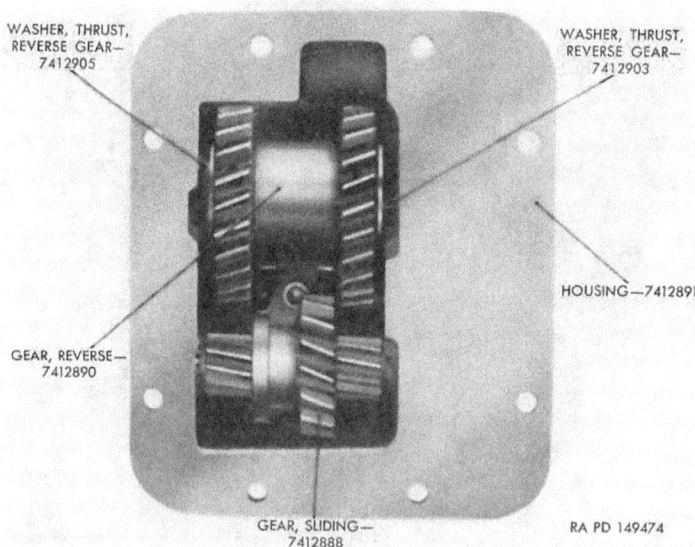

*Figure 95. Power take-off reverse gear installed.*

## 140. Installation of Idler Gear, Shaft, and Bearings

*Note.* Key letters noted in parentheses are in figure 89 unless otherwise indicated.

*a.* Coat three idler gear thrust washers (E) and idler gear spacer (F) with universal gear lubricant (GO). Also apply same lubricant on bore of idler gear (K).

*b.* Set idler gear on work bench and assemble 30 idler gear roller bearings in each end of bore with bearing spacer (J) between sets of bearings. Use tool in same manner as described in paragraph 139*b* to hold bearings in place while assembling.

*c.* Start idler shaft (D) through boss at front side of housing (L) with end of shaft extending slightly beyond inside of case. Place one idler gear thrust washer (E) on end of idler shaft.

*d.* Lay idler gear on side and place one idler gear bearing thrust washer (G) against bearings at each end of idler gear. Place one thrust washer (E) and idler gear spacer (F) at front side of idler gear; then locate idler gear (fig. 96) and bearings in housing. Place idler gear rear thrust washer between gear and housing; then drive lightly on front end of idler shaft to push out assembly tool through rear of housing and force idler shaft into place in housing. Refer to figure 96 and check relative position of gear and thrust washers.

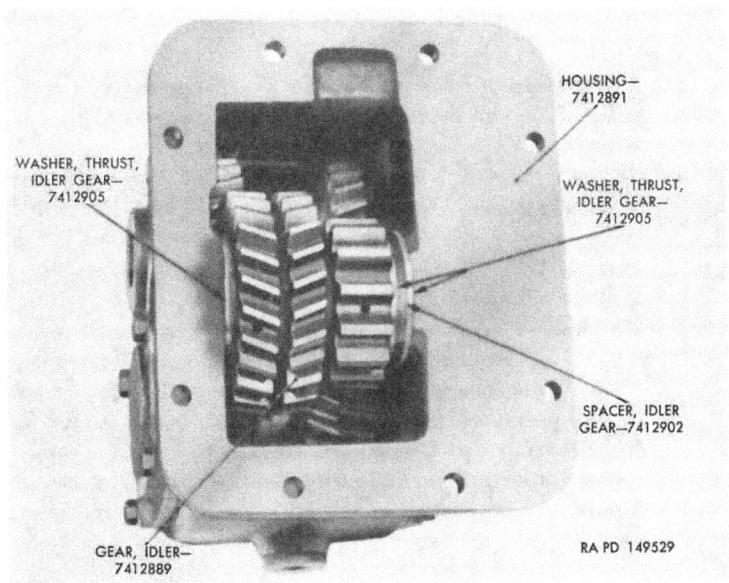

*Figure 96. Power take-off idler gear installed.*

*e*. Locate idler shaft retaining plate (C) in slot in idler shaft and attach plate with $\frac{5}{16}$–18 x $\frac{3}{4}$ cap screw (A) and $\frac{5}{16}$-inch lock washer (B).

### 141. Installation of Plugs and Covers on Power Take-Off Housing

*Note.* Key letters noted in parentheses are in figure 89 unless otherwise indicated.

*a. General.* If power take-off is used on truck which requires an accessory drive unit, no top cover plate (N) is used; however, a side cover plate (DD) must be installed, and housing drain plug (BB) is always used. The hex-head plug shown in figure 87 is installed in tapped hole at front side of housing on power take-off used on tank-type trucks. Refer to paragraphs 143 and 144 for procedure for installation of accessory drive units on power take-off housing.

*b. Installation of Plugs.*
  (1) Install housing drain plug (BB), using new drain plug gasket (AA).
  (2) Using a new plug seal (same as item (J), fig. 104) install $1\frac{1}{16}$–16 x $2\frac{3}{32}$ plug (fig. 87, front view) and new seal retaining washer in threaded hole at front of housing.

  *Note.* Installation of this plug applies to power take-off used on trucks with tank bodies only.

*c. Installation of Covers.*
  (1) Using new side cover plate gasket (CC), position side cover plate (DD) and install six $\frac{5}{16}$–18 x $\frac{3}{4}$ cap screws (FF) and washers (EE). Tighten cap screws firmly.
  (2) If power take-off is used on truck equipped with winch but without accessory drive, install top cover plate (N) using new top cover plate gasket (M). Attach cover plate, using four $\frac{3}{8}$–16 x $\frac{7}{8}$ cap screws (P) and new gaskets (Q). Tighten cap screws firmly.
  (3) On power take-off which does not have a winch drive shaft (fig. 87), install a drive shaft rear bearing cap and front cap (MM), using one new rear bearing cap gasket (N) under rear cap and one new front bearing cap gasket (AB) at front bearing cap. Use four $\frac{5}{16}$–18 x 1 cap screws and copper washers to attach each bearing cap. Tighten cap screws firmly.

## Section V. DESCRIPTION AND REPLACEMENT OF ACCESSORY DRIVE UNIT

### 142. Description

*Note.* Key letters noted in parentheses are in figure 97 unless otherwise indicated.

*a. Tank Truck Accessory Drive.*—The accessory drive unit (fig. 97) used with water and gasoline tank trucks is mounted on power take-off as illustrated in figure 87. The assembly consists of a drive shaft (H) upon which a sliding gear (G) operates on spiral spline on shaft. A shifter fork (U), mounted on shifter shaft (Z), positions sliding gear (G) to engage power take-off reverse gear when shifter shaft (Z) is manually positioned through linkage. The governor drive engages front end of drive shaft to operate limiting speed auxiliary governor.

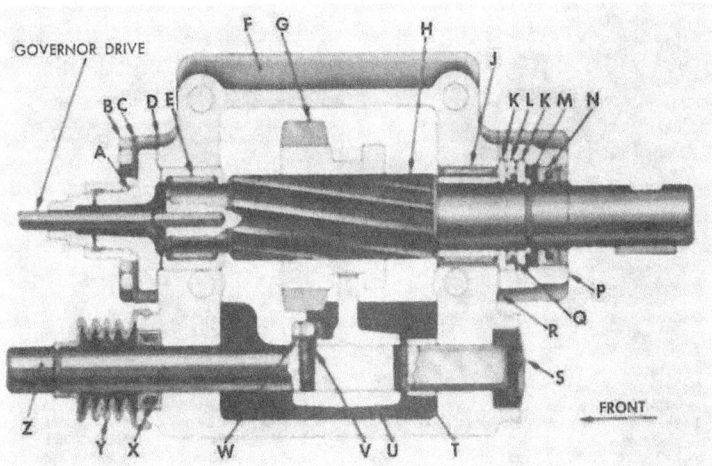

A—CAP, FRONT BEARING, DRIVE SHAFT—CHU-1452
B—SCREW, CAP, 5/16-18x3/4—180077
C—WASHER, LOCK, 5/16-IN—120214
D—GASKET, CAP, FRONT BEARING—CHU-407
E—BEARING, FRONT, DRIVE SHAFT, ASSY—709460
F—HOUSING—CHU-1450
G—GEAR, SLIDING—7412888
H—SHAFT, DRIVE—CHU-1460
J—BEARING, REAR, DRIVE SHAFT, ASSY—TR-BR1616
K—WASHER, THRUST—CHU-1455
L—RING, RETAINER, BEARING BALL—CHU-1457
M—RING, LOCKING—CHU-403A
N—SEAL, OIL, DRIVE SHAFT, ASSY—CHU-1464
P—CAP, BEARING, REAR, DRIVE SHAFT—CHU-1451
Q—BALL, BEARING, DRIVE SHAFT—104916
R—GASKET, CAP, DRIVE SHAFT REAR BEARING—CHU-407
S—PLUG, HOUSING—CHU-1467
T—SPACER, SHIFTER FORK—CHU-1470
U—FORK, SHIFTER—CHU-1456
V—SCREW, CAP, 1/4-20x5/8—180018
W—WASHER, LOCK, 1/4-IN—120380
X—SEAL, OIL, SHIFTER SHAFT, ASSY—CHU-1465
Y—BOOT, SHIFTER SHAFT—CHU-1462
Z—SHAFT, SHIFTER—CHU-1463

RA PD 148795

*Figure 97. Sectional view of accessory drive unit used with tank truck.*

b. *Dump Truck Accessory Drive.*—The accessory drive unit (fig. 98) used with dump trucks is mounted on power take-off as illustrated in figure 86. The assembly consists of the same parts as described for tank truck accessory drive (*a* above) except the drive shaft front bearing cover ((A), fig. 98) is plain while the drive shaft front bearing cap (A) on the tank accessory drive provides for limiting speed auxiliary governor drive. The shifter shaft ((Z), fig. 98) is installed

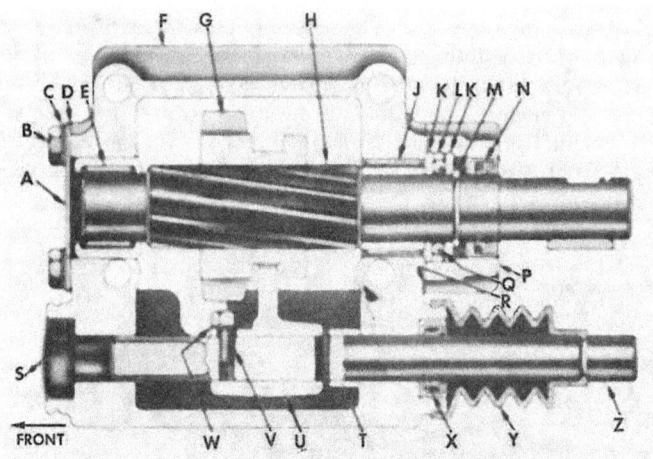

A—COVER, FRONT BEARING, DRIVE SHAFT—CHU-1458
B—SCREW, CAP, 5/16-18x1/2—180073
C—WASHER, LOCK, 5/16-IN—120214
D—GASKET, COVER, FRONT BEARING—CHU-407
E—BEARING, FRONT, DRIVE SHAFT, ASSY—709460
F—HOUSING—CHU-1450
G—GEAR, SLIDING—7412888
H—SHAFT, DRIVE—CHU-1460
J—BEARING, REAR, DRIVE SHAFT, ASSY—TR-BR1616
K—WASHER, THRUST—CHU-1455
L—RING, RETAINER, BEARING BALL—CHU-1457
M—RING, LOCKING—CHU-403A
N—SEAL, OIL, DRIVE SHAFT, ASSY—CHU-1464
P—CAP, BEARING, REAR, DRIVE SHAFT—CHU-1451
Q—BALL, BEARING, DRIVE SHAFT—104916
R—GASKET, CAP, DRIVE SHAFT REAR BEARING—CHU-407
S—PLUG, HOUSING—CHU-1467
T—SPACER, SHIFTER FORK—CHU-1470
U—FORK, SHIFTER—CHU-1456
V—SCREW CAP, 1/4-20x5/8—180018
W—WASHER, LOCK, 1/4-IN—120380
X—SEAL, OIL, SHIFTER SHAFT, ASSY—CHU-1465
Y—BOOT, SHIFTER SHAFT—CHU-1462
Z—SHAFT, SHIFTER—CHU-1463

RA PD 148794

*Figure 98. Sectional view of accessory drive unit used with dump truck.*

to operate from the rear instead of the front as on tank accessory drive. When shifter shaft is manually shifted, shifter fork ((U), fig. 98) meshes sliding gear ((G), fig. 98) with power take-off reverse gear.

## 143. Removal of Accessory Drive Unit

*a.* Remove three stud nuts and one cap screw which attach accessory drive to power take-off.

*b.* Lift accessory drive assembly straight up from power take-off assembly. Remove accessory drive-to-power-take-off gasket (CC), fig. 99 or (JJ), fig. 104).

## 144. Installation of Accessory Drive Unit

*a.* Position accessory drive-to-power-take-off gasket on power take-off.

*b.* Position accessory drive unit over the three studs. Install one 3/8–16 x 3 3/4 cap screw and three 3/8–24 stud nuts. Tighten nuts and cap screw firmly.

## Section VI. REBUILD OF TANK TRUCK ACCESSORY DRIVE UNIT

### 145. Disassembly of Tank Truck Accessory Drive

*Note.* Key letters in text refer to figure 99 unless otherwise indicated. The sectional view (fig. 97) shows the various components in their correct positions.

*a.* Remove three cross-recess 5/16–18 x 1 1/2 fillister head screws (Z) and 5/16-inch special lock washers (AA) which attach drive shaft rear bearing cap (X) to housing. Remove bearing cap and drive shaft rear bearing cap gasket (W).

*b.* Remove 5/16–18 x 3/4 cap screws (C) and 5/16-inch lock washers (D) which attach drive shaft front bearing cap (E) to housing (H).

*c.* With brass drift applied to front end of drive shaft (Q), force drive shaft through housing. Drive shaft front bearing assembly (G) will remain in housing. The inner race (R) of the drive shaft rear bearing assembly (P) will remain on drive shaft. Remove sliding gear (H) from housing.

*d.* Remove locking ring (V) from drive shaft. Remove two thrust washers (S), bearing ball retainer ring (U), and 21 drive shaft bearing balls (T).

*e.* Remove shifter shaft boot (A).

*f.* Remove ball spring retainer (J), ball spring retainer washer (K), ball spring retainer seal (L), shifter poppet ball spring (M), and shifter poppet ball (N).

*g.* Remove 1/4–20 x 5/8 cap screw (GG) and 1/4-inch lock washer (FF) which attach shifter fork (DD) to shifter shaft (B).

*h.* With soft hammer applied at front end of shifter shaft (B), drive shaft toward rear of housing until housing plug (BB) is forced from housing. Shaft can then be removed from housing. Remove shifter fork (DD) and shifter fork spacer (EE) from housing.

*i.* Drive shaft front and rear bearing assemblies (G) and (P) may remain in housing until inspected (par. 146*b*). Drive shaft oil seal assembly (Y) and shifter shaft oil seal assembly (JJ) may remain in place until after inspection (par. 146*b*).

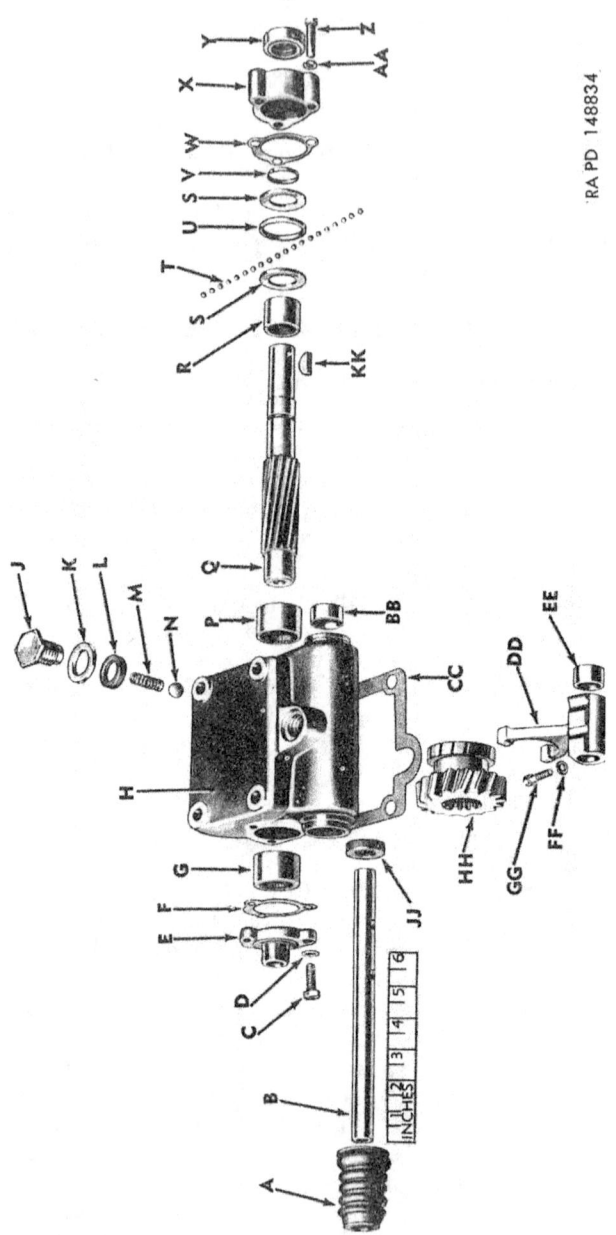

A—BOOT, SHIFTER SHAFT—CHU-1462
B—SHAFT, SHIFTER—CHU-1463
C—SCREW, CAP, 5/16–18 X 3/4—180077
D—WASHER, LOCK, 5/16-IN.—120214
E—CAP, FRONT BEARING, DRIVE SHAFT—CHU-1458
F—GASKET, CAP, FRONT BEARING—CHU-407
G—BEARING, FRONT, DRIVE SHAFT, ASSY—709460
H—HOUSING—CHU-1450
J—RETAINER, SPRING, BALL—CHU-1468
K—WASHER, BALL SPRING RETAINER—743234
L—SEAL, RETAINER, BALL SPRING—743233
M—SPRING, BALL, SHIFTER POPPET—737S128
N—BALL, SHIFTER POPPET—104921
P—BEARING, REAR, DRIVE SHAFT, ASSY—TR-BR1616
Q—SHAFT, DRIVE—CHU-1460
R—RACE, INNER (PART OF REAR BEARING ASSY—TR-BR1616)
S—WASHER, THRUST—CHU-1455
T—BALL, BEARING, DRIVE SHAFT—104916
U—RING, RETAINER, BEARING BALL—CHU-1457
V—RING, LOCKING—CHU-403A
W—GASKET, CAP, DRIVE SHAFT REAR BEARING—CHU-407
X—CAP, BEARING, REAR, DRIVE SHAFT—CHU-1451
Y—SEAL, OIL, DRIVE SHAFT, ASSY—CHU-1464
Z—SCREW, FIL-HD, CROSS-RECESS, 5/16–18 X 1 1/2—154000
AA—WASHER, LOCK, SPECIAL, 5/16-IN.—CHU-1087
BB—PLUG, HOUSING—CHU-1467
CC—GASKET, ACCESSORY DRIVE-TO-POWER TAKE-OFF—741288.4
DD—FORK, SHIFTER—CHU-1456
EE—SPACER, SHIFTER FORK—CHU-1470
FF—WASHER, LOCK, 1/4-IN.—126S80
GG—SCREW, CAP, 1/4–20 X 5/8—180018
HH—GEAR, SLIDING—741288S
JJ—SEAL, OIL, SHIFTER SHAFT, ASSY—CHU-1465
KK—KEY, WOODRUFF—12139

*Figure 99. Components of accessory drive unit for tank trucks.*

## 146. Cleaning and Inspection of Tank Truck Accessory Drive

*Note.* Key letters in text refer to figure 99 unless otherwise indicated.

*a. Cleaning.*—Clean all parts thoroughly with dry-cleaning solvent or volatile mineral spirits. Thoroughly scrub drive shaft front bearing assembly (G) and drive shaft rear bearing assembly (P) so that needle bearings can be inspected. Do not clean shifter shaft boot (A) with solvent.

*b. Inspection.*
  (1) *Housing.*—Thoroughly inspect housing (H) for cracks and damaged threads. Examine mounting surface for scores or roughness. Small nicks may be honed out. Check clearance of shifter shaft (B) in passages in housing. Replace shifter shaft or housing if clearance is excessive (par. 345).
  (2) *Drive shaft front and rear bearings.*—Examine needle bearings in drive shaft front bearing assembly (G) and drive shaft rear bearing assembly (P). If needle bearings are checked, bent, or otherwise damaged, replace with new parts (par. 147). The inner race (R) which is a part of drive shaft rear bearing assembly is pressed on drive shaft (Q). Examine outer surface of race for roughness or scores. If race is damaged, the rear bearing assembly (with inner race) must be replaced (par. 147).
  (3) *Oil seals.*—Examine shifter shaft oil seal assembly (JJ) for damage or looseness in housing. Examine drive shaft oil seal assembly (Y) in drive shaft rear bearing cap (X) for similar damage. Replace if necessary as described in paragraph 147.
  (4) *Shifter shaft.*—Inspect shifter shaft (B) for damage. Check outside diameter and clearance as described in (1) above. Refer to paragraph 345 for dimensions. Inspect shifter shaft boot (A) for stretched or damaged condition. The boot must fit tightly in shifter shaft groove and groove on housing (refer to fig. 97). Replace boot if damaged.
  (5) *Sliding gear.*—Check inner splines of sliding gear (HH) for roughness or damage. Check teeth for chipped, rough, or damaged condition. Small nicks or burrs on teeth may be honed out. Check condition of shifter fork groove on sliding gear. Check clearance of shifter fork pads to groove ((6) below). If groove in gear is excessively worn (par. 345), replace gear.
  (6) *Shifter fork.*—Inspect shifter fork (DD) for damaged or sprung fork legs and worn fork pads. Check clearance of fork pads in sliding gear groove ((5) above). If clearance is excessive (par. 345), check width of pads for excessive wear (par. 345).

(7) *Drive shaft.*—Check spiral splines on drive shaft (Q) for damaged, chipped, or worn condition. Replace shaft if splines are damaged. Check outer surface of shaft at front end. The drive shaft front bearing assembly (G) operates directly on shaft. Check surface for excessive wear (par. 345). Check keyway for damage. Check outer surface of rear end of drive shaft where drive shaft bearings balls (T) contact. Surface must not be grooved, rough, or worn (par. 345). Shaft must be replaced if damaged as described.

(8) *Drive shaft thrust washers and bearing balls.*—Inspect thrust washers (S) for excessive wear (par. 345). Inspect drive shaft bearing balls (T) for roughness or wear. There are 21 bearing balls. Replace balls if any damage is evident. Inspect bearing ball retainer ring (U) for damage or excessively worn inside surface.

(9) *Drive shaft rear bearing cap.*—Check drive shaft rear bearing cap (X) for damage. Replace drive shaft oil seal assembly (Y) if necessary as described in paragraph 147. Drive shaft rear bearing cap gasket (W) must be replaced with new part at assembly.

(10) *Drive shaft front bearing cap.*—Inspect threads on drive shaft front bearing cap (E) for stripped or damaged condition. Front bearing cap gasket (F) must be replaced at assembly.

(11) *Shifter poppet ball and retainer.*—Inspect ball spring retainer (J) for stripped or damaged threads. Ball spring retainer seal (L) must be replaced with new part at assembly. Check free length of shifter poppet ball spring (M). Replace if not to standard (par. 345). Examine shifter poppet ball (N) for roughness, excessive wear, or checks. Replace if damaged.

(12) *Housing plug.*—Housing plug (BB) may be sprung or damaged when shifter shaft is removed. Replace plug with new part at assembly.

### 147. Repair of Tank Truck Accessory Drive

*Note.* Key letters in text refer to figure 99 unless otherwise indicated.

*a. Drive Shaft Front Bearing Assembly Replacement.*
  (1) With a suitable tool, remove drive shaft front bearing assembly (G) from housing. Use care not to damage housing bore.
  (2) Drive new bearing into case, using plastic hammer (fig. 100). Drive bearing assembly into housing one-sixteenth of an-inch from outer surface housing.

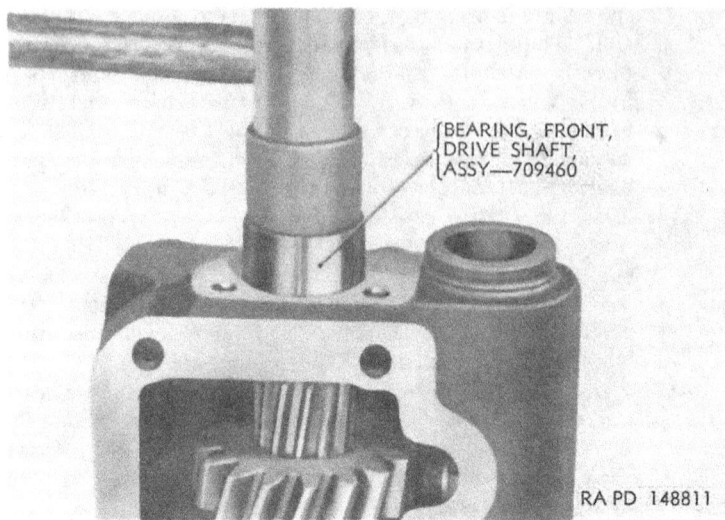

*Figure 100. Installing drive shaft front bearing assembly.*

b. *Drive Shaft Rear Bearing Assembly Replacement.*
   (1) Inner race (R) is pressed onto drive shaft (Q). Remove old inner race from shaft, using care not to damage shaft.
   (2) Press new inner race (R) onto shaft until it bottoms on shoulder of shaft. Refer to figure 103 for location of inner race.
   (3) With suitable tool, remove drive shaft rear bearing assembly (P) from housing.
   (4) Press new bearing into housing until it is flush with outer surface of housing. Make certain that tool used to install bearing assembly contacts entire edge surface of bearing to prevent damage to bearing assembly.

c. *Shifter Shaft Oil Seal Assembly Replacement.*
   (1) With a suitable tool, drive out old shifter shaft oil seal assembly (JJ) from housing.
   (2) Coat outer surface of seal assembly with plastic type gasket cement. Do not coat inner surface.
   (3) Apply small quantity of universal gear lubricant (GO) on lip of seal.
   (4) Install seal in housing with lip of seal, toward outer edge until seal bottoms in housing. Refer to item X, figure 97 for correct position of oil seal assembly. After installation,

wipe surplus gasket cement from exposed surface of seal assembly.

*d. Drive Shaft Oil Seal Assembly Replacement.*
   (1) Drive shaft oil seal assembly (Y) is pressed into drive shaft rear bearing cap (X). With suitable tool, remove old seal sembly.
   (2) Coat outer surface of seal assembly with plastic type gasket cement. Do not coat inner surface.
   (3) Apply small quantity of universal gear lubricant (GO) on lips of seal.
   (4) Press seal assembly into drive shaft rear bearing cap (X) with spring-loaded seal lip toward inside. Press seal flush with outer surface of cap. Refer to item N, figure 97 for correct position of oil seal assembly. After installation, wipe surplus gasket cement from exposed surface of seal assembly.

### 148. Assembly of Tank Truck Accessory Drive

*Note.* Key letters in text refer to figure 99 unless otherwise indicated. The sectional view (fig. 97) shows the various components in their correct positions.

   *a.* With shifter fork (DD) in position in housing, insert shifter shaft (B) through front hole in housing and through shifter fork. Install shifter fork spacer (EE) on shaft; then insert shifter shaft into rear hole in housing. Shaft must be installed in housing with threaded holes in shaft in position to attach shifter fork.

   *b.* Install 1/4–20 x 5/8 cap screw and 1/4-inch lock washer which attach fork to shaft. Note items V and W, figure 97 for correct in-

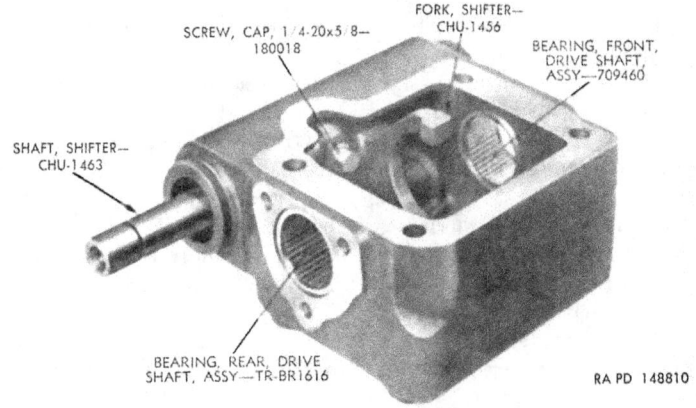

*Figure 101. Shifter fork and shaft installed.*

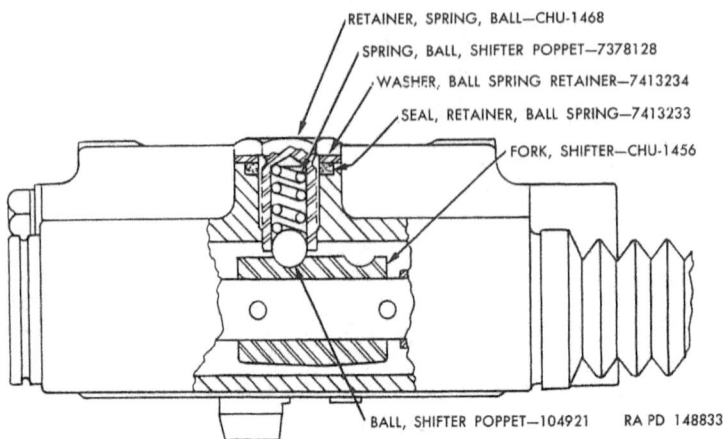

*Figure 102. Sectional view showing shifter poppet ball installed.*

stallation of cap screw and washer. Tighten cap screw firmly. Refer also to figure 101 which shows shifter fork installed.

  c. Insert shifter poppet ball (N) into housing to engage notch on shifter fork (DD). Refer to figure 102 for correct installation of poppet ball and retaining parts. Insert shifter poppet ball spring (M), new ball spring retainer seal (L), ball spring retainer washer (K), and ball spring retainer (J). Tighten ball spring retainer firmly.

  d. Install new housing plug (BB) into housing, bottoming plug on housing shoulder.

  e. With drive shaft (Q) positioned as shown in figure 103, install a thrust washer (S), then install bearing ball retainer ring (U).

  f. Lubricate drive shaft bearing balls (T); then place 21 bearing balls into retainer ring.

  g. Place second thrust washer (S) over bearing balls. Install new locking ring (V) onto shaft groove of shaft.

  h. Thoroughly lubricate drive shaft front (G) and rear (P) bearing assemblies. With sliding gear (HH) in place on shifter fork (groove toward rear of housing), insert assembled drive shaft in housing through drive shaft rear bearing assembly (P), then through sliding gear and drive shaft front bearing assembly (G). Make certain that thrust washer (S) at front bottoms against drive shaft rear bearing assembly.

  i. With new drive shaft rear bearing cap gasket (W) in place, install drive shaft rear bearing cap (X) with drive shaft oil seal assembly (Y) in place. Use three $5/16$-18 x $1\frac{1}{4}$ fillister head cross-recess screws and $5/16$-inch special lock washers. Tighten screws firmly.

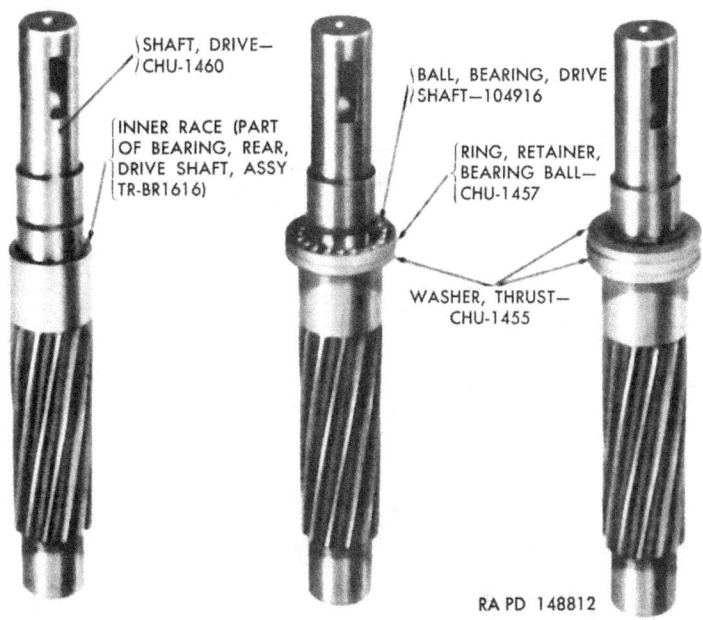

*Figure 103. Method of installing drive shaft bearing balls.*

*j.* With new drive shaft front cap gasket (F) in place, install drive shaft front bearing cap (E) with three $\frac{5}{16}$–18 x $\frac{3}{4}$ cap screws and $\frac{5}{16}$-inch lock washers. Tighten cap screws firmly.

*k.* Install shifter shaft boot (A) over drive shaft and on housing. Clamp both ends of boot with wire if necessary.

## Section VII. REBUILD OF DUMP TRUCK ACCESSORY DRIVE UNIT

### 149. Disassembly of Dump Truck Accessory Drive

*Note.* Key letters noted in parentheses are in figure 104 unless otherwise indicated. The sectional view (fig. 98) shows the various components in their correct positions.

*a.* Remove three $\frac{5}{16}$–18 x $1\frac{1}{4}$ cross-recess head screws (X) and $\frac{5}{16}$-inch special lock washers (Y) which attach drive shaft rear bearing cap (V). Remove bearing cap and drive shaft rear bearing cap gasket (U).

*b.* Remove $\frac{5}{16}$–18 x $\frac{1}{2}$ cap screws (A) and $\frac{5}{16}$-inch lock washers (B) which attach drive shaft front bearing cap (C) to housing (F).

*c.* With brass drift applied to front end of drive shaft (N), force drive shaft through housing. Drive shaft front bearing assembly

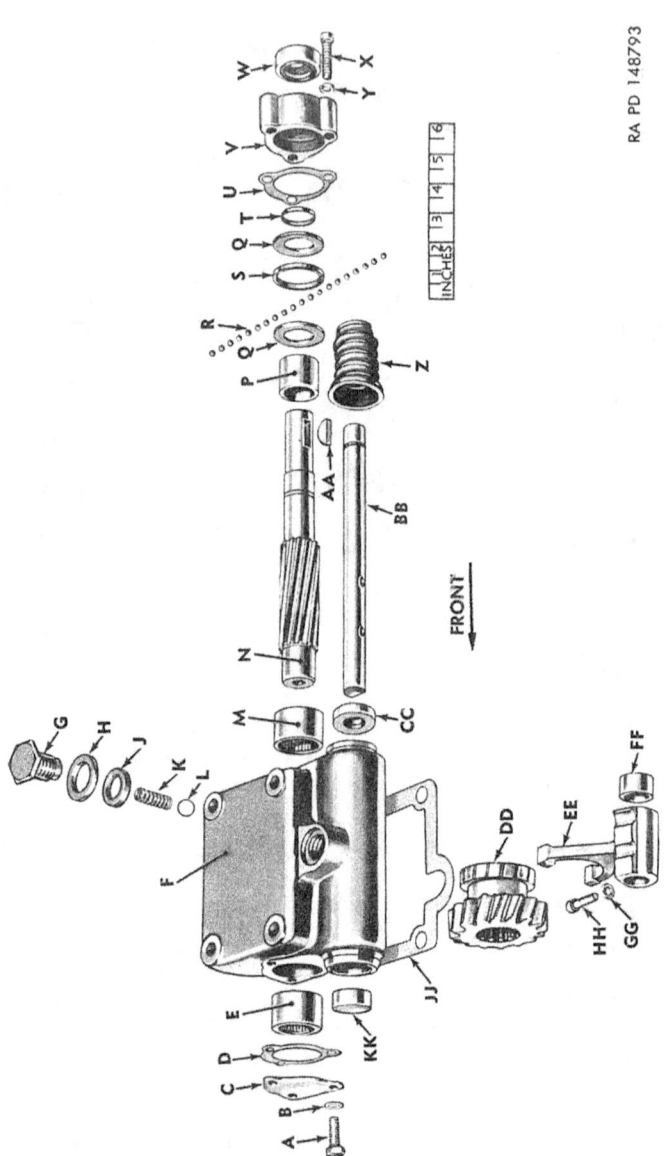

A—SCREW, CAP, 5/16–18 X 1/2—180073
B—WASHER, LOCK, 5/16-IN.—120214
C—COVER, FRONT BEARING, DRIVE SHAFT—CHU-1458
D—GASKET, COVER, FRONT BEARING—CHU-407
E—BEARING, FRONT, DRIVE SHAFT, ASSY—709460
F—HOUSING—CHU-1450
G—RETAINER, SPRING, BALL—CHU-1468
H—WASHER, BALL SPRING RETAINER—7413234
J—SEAL, RETAINER, BALL SPRING—7413233
K—SPRING, BALL, SHIFTER POPPET—CHU-152
L—BALL, SHIFTER POPPET—104921
M—BEARING, REAR, DRIVE SHAFT, ASSY—TR-BR1616
N—SHAFT, DRIVE—CHU-1460
P—RACE, INNER (PART OF REAR BEARING ASSY)—CHU-1454
Q—WASHER, THRUST—CHU-1455
R—BALL, BEARING, DRIVE SHAFT—104916
S—RING, RETAINER, BEARING BALL—CHU-1457
T—RING, LOCKING—CHU-463A
U—GASKET, CAP, DRIVE SHAFT REAR BEARING—CHU-407
V—CAP, BEARING, REAR, DRIVE SHAFT—CHU-1451
W—SEAL, OIL, DRIVE SHAFT, ASSY—CHU-1464
X—SCREW, CROSS-RECESS, 5/16–18 X 1 1/4—154000
Y—WASHER, LOCK, SPECIAL, 5/16-IN.—CHU-1087
Z—BOOT, SHIFTER SHAFT—CHU-1462
AA—KEY, WOODRUFF—112139
BB—SHAFT, SHIFTER—CHU-1463
CC—SEAL, OIL, SHIFTER SHAFT, ASSY—CHU-1465
DD—GEAR, SLIDING—7412888
EE—FORK, SHIFTER—CHU-1456
FF—SPACER, SHIFTER FORK—CHU-1470
GG—WASHER, LOCK, 1/4-IN.—120380
HH—SCREW, CAP, 1/4–20 X 5/8—180018
JJ—GASKET, ACCESSORY DRIVE-TO-POWER TAKE-OFF—7412884
KK—PLUG, HOUSING—CHU-1467

*Figure 104. Components of accessory drive unit for dump truck.*

(E) will remain in housing. The inner race (P) of the drive shaft rear bearing assembly (M) will remain on drive shaft. Remove sliding gear (DD) from housing.

*d.* Remove locking ring (T) from drive shaft. Remove two thrust washers (Q), bearing ball retainer ring (S), and 21 drive shaft bearing balls (R).

*e.* Remove shifter shaft boot (Z).

*f.* Remove ball spring retainer (G), ball spring retainer washer (H), ball spring retainer seal (J), shifter poppet ball spring (K), and shifter poppet ball (L).

*g.* Remove ¼-20 x ⅝ cap screw (HH) and ¼-inch lock washer (GG) which attach shifter fork (EE) to shifter shaft (BB).

*h.* With soft hammer applied at rear end of shifter shaft (BB), drive shaft toward front of housing until housing plug (KK) is forced from housing. Shaft can then be removed from housing. Remove shifter fork (EE) from housing.

*i.* Drive shaft front (E) and rear (M) bearing assemblies may remain in housing until inspected (par. 150). Drive shaft oil seal assembly (W) and shifter shaft oil seal assembly (CC) may remain in place until after inspection (par. 150).

## 150. Cleaning and Inspection of Dump Truck Accessory Drive

*Note.* Key letters noted in parentheses are in figure 104 unless otherwise indicated.

*a. Cleaning.*—Clean all parts thoroughly with dry-cleaning solvent or volatile mineral spirits. Thoroughly scrub drive shaft rear bearing assembly (M) and drive shaft front bearing assembly (E) so that needle bearings can be inspected. Do not clean shifter shaft boot (Z) with solvent.

*b. Inspection.*

    (1) *Housing.*—Thoroughly inspect housing (F) for cracks and damaged threads. Examine mounting surface for scores or roughness. Small nicks may be honed out. Check clearance of shifter shaft (BB) in passages in housing. Replace shifter shaft or housing if clearance is excessive (par. 345).

    (2) *Drive shaft front and rear bearings.*—Examine needle bearings in drive shaft front bearing assembly (E), and drive shaft rear bearing assembly (M). If needle bearings are checked, bent, or otherwise damaged, replace with new parts (par. 151). The inner race (P), which is a part of drive shaft rear bearing assembly (M), is pressed on drive shaft (N). Examine outer surface of race for roughness or scores. If race is damaged, the rear bearing assembly (with inner race) must be replaced (par. 151).

(3) *Oil seals.*—Examine shifter shaft oil seal assembly (CC) for damage or looseness in housing. Examine drive shaft oil seal assembly (W) in drive shaft rear bearing cap (V) for similar damage. Replace if necessary as described in paragraph 151.

(4) *Shifter shaft.*—Inspect shifter shaft (BB) for damage. Check outside diameter and clearance as described in (1) above. Refer to paragraph 345 for dimensions. Inspect shifter shaft boot (Z) for stretched or damaged condition. The boot must fit tightly in shifter shaft groove and groove on housing (refer to figure 98). Replace boot if damaged.

(5) *Sliding gear.*—Check inner splines of sliding gear (DD) for roughness or damage. Check teeth for chipped, rough, or damaged condition. Small nicks or burs on teeth may be honed out. Check condition of shifter fork groove on sliding gear. Check clearance of shifter fork pads to groove ((6) below). If groove of gear is excessively worn (par. 345), replace gear.

(6) *Shifter fork.*—Inspect shifter fork (EE) for damaged or sprung fork legs and worn fork pads. Check clearance of fork pads in sliding gear groove ((5) above). If clearance is excessive (par. 345), check width of pads for excessive wear (par. 345).

(7) *Drive shaft.*—Check spiral splines on drive shaft (N) for damaged, chipped, or worn condition. Replace shaft if splines are damaged. Check outer surface of shaft at front end. The drive shaft front bearing assembly (E) operates directly on shaft. Check surface for excessive wear (par. 345). Check keyway for damage. Check outer surface of rear end of shaft where drive shaft bearing balls (R) contact. Surface must not be grooved, rough, or worn (par. 345).

(8) *Drive shaft thrust washers and bearing balls.*—Inspect thrust washers (Q) for excessive wear (par. 345). Inspect drive shaft bearing balls (R) for roughness or wear. There are 21 bearing balls. Replace balls if any damage is evident. Inspect bearing ball retainer ring (S) for damage or excessively worn inside surface.

(9) *Drive shaft rear bearing cap.*—Check drive shaft rear bearing cap (V) for damage. Replace drive shaft oil seal assembly (W) if necessary as described in paragraph 151. Drive shaft rear bearing cap gasket (U) must be replaced with a new part at assembly.

(10) *Drive shaft front bearing cover.*—Drive shaft front bearing cover (C) must not be distorted. Front bearing cover gasket (D) must be replaced at assembly.

(11) *Shifter poppet ball and retainer.*—Inspect ball spring retainer (G) for stripped or damaged threads. Ball spring retainer seal (J) must be replaced with new part at assembly. Check compression of shifter poppet ball spring (K). Replace if not up to standard (par. 345). Examine shifter poppet ball (L) for roughness, excessive wear, or checks. Replace if damaged.

(12) *Housing plug.*—Housing plug (KK) may be sprung or damaged when shifter shaft is removed. Replace plug with new part at assembly.

## 151. Repair of Dump Truck Accessory Drive

*Note.* Key letters noted in parentheses are in figure 104 unless otherwise indicated.

a. *Drive Shaft Front Bearing Assembly Replacement.*
  (1) With a suitable tool, remove drive shaft front bearing assembly (E) from housing. Use care not to damage housing bore.
  (2) Drive new bearing into case, using plastic hammer in manner illustrated in figure 100. Drive bearing assembly into housing one-sixteenth of an inch from outer surface of housing.

b. *Drive Shaft Rear Bearing Assembly Replacement.*
  (1) Inner race (P) is pressed onto drive shaft (N). Remove old inner race from shaft, using care not to damage shaft.
  (2) Press new inner race (P) onto shaft until it bottoms on shoulder of shaft. Refer to figure 103 for location of inner race.
  (3) With suitable tool, remove drive shaft rear bearing assembly (M) from housing.
  (4) Press new bearing into housing until it is flush with outer surface of housing. Make certain that tool used to install bearing assembly contacts entire edge surface of bearing to prevent damage to bearing assembly.

c. *Shifter Shaft Oil Seal Assembly Replacement.*
  (1) With a suitable tool, drive out old shifter shaft oil seal assembly (CC) from housing.
  (2) Coat outer surface of seal assembly with plastic type gasket cement. Do not coat inner surface.
  (3) Apply small quantity of universal gear lubricant (GO) on lip of seal.

(4) Install seal in housing with lip of seal toward outer end until seal bottoms in housing. Refer to item X, figure 98 for correct position of oil seal assembly. After installation, wipe surplus gasket cement from exposed surface of seal assembly.

*d. Drive Shaft Oil Seal Assembly Replacement.*

(1) Drive shaft oil seal assembly (W) is pressed into drive shaft rear bearing cap (V). With suitable tool, remove old seal assembly.

(2) Coat outer surface of seal assembly with plastic type gasket cement. Do not coat inner surface.

(3) Apply small quantity of universal gear lubricant (GO) on lips of seal.

(4) Press seal assembly into drive shaft rear bearing cap (V) with spring-loaded leather lip toward inside. Press seal flush with outer surface of cap. Refer to item N, figure 98 for correct position of oil seal assembly. After installation, wipe surplus gasket cement from exposed surface of seal assembly.

### 152. Assembly of Dump Truck Accessory Drive

*Note.* Key letters noted in parentheses are in figure 104 unless otherwise indicated. The sectional view (fig. 98) shows the various components in their correct positions.

*a.* With shifter fork (EE) in position in housing, insert shifter shaft (B) through rear hole in housing and through shifter fork spacer (EE) and shifter fork; then insert shifter shaft into rear hole in housing. Shaft must be installed in housing with threaded holes in shaft in position to attach shifter fork.

*b.* Install $\frac{1}{4}$-20 x $\frac{5}{8}$ cap screw and $\frac{1}{4}$-inch lock washer which attach fork to shaft. Note items V and W, figure 98, for correct installation of cap screw and lock washer. Tighten cap screw firmly. Refer also to figure 101 which shows fork installed.

*c.* Insert shifter poppet ball (L) into housing to engage notch on shifter fork (EE). Refer to figure 102 for correct installation of poppet ball and retaining parts. Insert shifter poppet ball spring (K), new ball spring retainer seal (J), ball spring retainer washer (H), and ball spring retainer (G). Tighten ball spring retainer firmly.

*d.* Install new housing plug (KK) into housing, bottoming plug on housing shoulder.

*e.* With drive shaft (N) positioned as shown in figure 103, install a thrust washer (Q); then install bearing ball retainer ring (S).

*f.* Lubricate drive shaft bearing balls (R); then place 21 bearing balls into retainer ring.

*g.* Place second thrust washer (Q) over bearing balls. Install new locking ring (T) onto shaft groove of shaft.

*h.* Thoroughly lubricate drive shaft front (**E**) and rear (**M**) bearing assemblies. With sliding gear (**DD**) in place on shifter fork (groove toward rear of housing), insert assembled drive shaft in housing through drive shaft rear bearing assembly (**M**), then through sliding gear and drive shaft front bearing assembly (**E**). Make certain that thrust washer (**Q**) at front bottoms against drive shaft rear bearing assembly.

*i.* With new drive shaft rear bearing cap gasket (**U**) in place, install drive shaft rear bearing cap (**V**) with drive shaft oil seal assembly (**W**) in place. Use three $5/16$–18 x $1\frac{1}{4}$ fillister head cross-recess screws and $5/16$-inch special lock washers. Tighten screws firmly.

*j.* With new drive shaft front cover gasket (**D**) in place, install drive shaft front bearing cover (**C**) with three $5/16$–18 x $\frac{1}{2}$ cap screws and $5/16$-inch lock washers. Tighten cap screws firmly.

*k.* Install shifter shaft boot (**Z**) over drive shaft and on housing. Clamp both ends of boot with wire if necessary.

# CHAPTER 7

# FRONT AXLE

## Section I. DESCRIPTION AND DATA

### 153. Description and Operation

*a. General.*—Front axle assembly (fig. 105) is hypoid, single-reduction type consisting of a housing, differential and carrier assembly, axle shaft and universal joint assemblies, and steering knuckle support assemblies. Power is transmitted from transfer to drive pinion through a tubular propeller shaft. Power is transmitted from drive pinion to drive gear and differential assembly, then to the wheels through axle shaft and universal joint assemblies. Action of universal joints permits delivery of power to the wheels when they are turned from straightahead position. Front axle is automatically engaged and disengaged by action of a jaw-type clutch located in transfer unit. Normally, front axle is disengaged except when tractive effort is required.

*b. Axle Housing.*—The axle housing is of the conventional one-piece banjo type with carrier assembly and cover openings near center of housing. The spherical shaped housing outer ends, torque rod brackets, spring seats, and steering knuckle stops are welded to the axle housing. Oil seals are used at outer ends of housing to prevent lubricant losses, also thrust washers are installed at outer ends of housing to absorb end thrust of universal joint assembly. External surface of housing outer ends are machined and polished to provide smooth surface for housing outer end oil and dust seals.

*c. Axle Shaft and Universal Joint Assemblies.*—The axle shafts are full-floating type with constant-velocity universal joints at steering knuckles. Each assembly consists of inner and outer shafts with integral yokes which form a universal joint around five steel balls. Outer shafts are the same for right and left sides and are splined at outer ends to engage drive flange. Inner shafts are of different lengths and are splined at inner ends to engage side gears at differential. Universal joint outer balls are select fit and the center ball is stand-

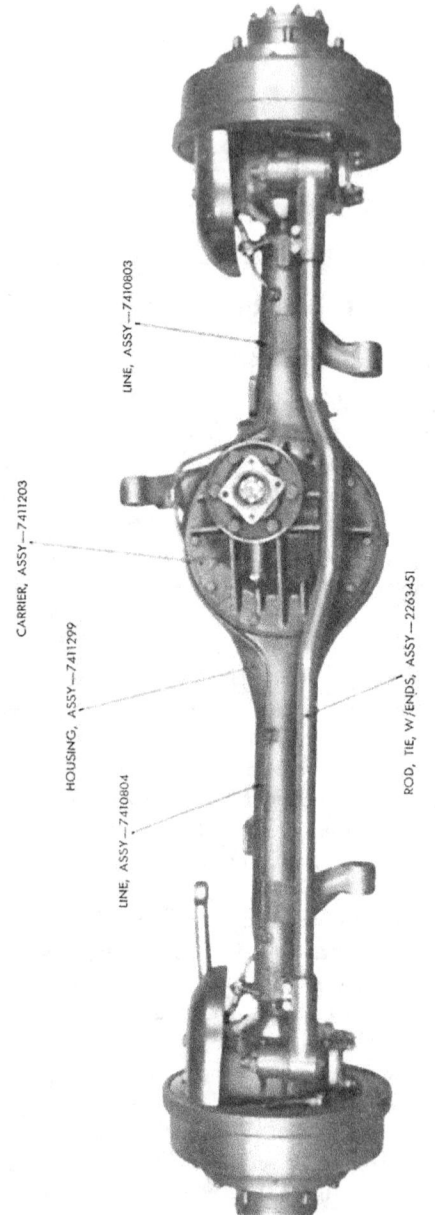

*Figure 105. General view of front axle assembly.*

ard size. Universal joint assemblies are completely enclosed within the steering knuckle supports.

*d. Steering Knuckle Supports.*—Steering knuckle supports are supported at outer ends of housing by tapered roller bearings. Supports are held in position on the bearings by steering knuckle trunnions, which permit steering knuckle supports to turn as front wheels are turned to right or left. Bearing adjustment is accomplished by use of shim pack between steering knuckle support and trunnions. Steering arms, to which tie rod ends attach, are integral with steering knuckle supports. Housing outer end oil and dust seals are attached to inner side of steering knuckle supports and are held in place by suitable retainers.

*e. Steering Knuckles.*—Steering knuckles are attached to steering knuckle supports by bolts and lock washers, which also serve to attach brake backing plate and brake anchor blocks. Steering knuckles act as spindles for mounting wheel hubs and bearings. Bushing type bearings are pressed inside steering knuckle which supports outer axle shaft; thrust washer, staked in place at inner side of steering knuckle flange, absorbs end thrust of universal joint assembly. A brake oil shield is installed at outer side of steering knuckle flange which prevents any escaping lubricant reaching brake linings.

*f. Tie Rod Assembly.*—The tie rod is a solid rod, threaded at each end and double offset to clear the differential carrier assembly. Rod has finer threads (16 per in.) on the left end than on the right end (12 per in.) to permit a finer degree of toe-in adjustment. The tie rod is attached to integral arm on steering knuckle supports by tapered stud installed in tie rod end. Tapered stud is held to support arm by nut and in tie rod end by snap ring. Tie rod is threaded into tie rod ends and securely held by clamp bolts, also a lock at left end. In addition to controlling toe-in, tie rod also transmits the turning force from the left steering knuckle support to the right steering knuckle support.

*g. Differential and Carrier Assembly.*—The differential and carrier assembly used in the front axle is the same as used in rear axles, except for the method of installation. Complete description and service information is given in chapter 8 and will not be repeated in this chapter.

## 154. Data

| | |
|---|---|
| Manufacturer | GM Corporation |
| Type | hypoid, single-reduction |
| Ratio | 6.17 to 1 |
| Universal joints | Bendix-Weiss |

## Section II. FRONT AXLE ALINEMENT

### 155. Front Axle Alinement

*a. General.*—Front axle alinement factors, such as camber, caster, turning angle, and toe-in, have a major effect on steering from the standpoint of control, ease of steering, and safety. Front axle misalinement is a major cause of premature and uneven tire wear.

*b. Caster.*—Front axle caster is the inclination of the center line through the upper and lower steering knuckle support trunnion bearings toward the rear of the vehicle ((L), fig. 106). Caster is established by design; therefore no adjustment can be made. The axle is given this caster angle to provide a "castering" action at the front wheels when the vehicle is in motion. When the front axle has proper caster, the wheels will tend to point straightahead as long as the vehicle is in motion. Caster angle is affected by a twisted axle housing, loose spring "U" bolts, or sagging springs. Insufficient caster will permit front wheels to wander out of straightahead position. Excessive caster will cause hard steering when turning. Caster angle must be checked with the axle installed on the vehicle, using wheel alinement indicator 41–I–130 for this purpose (fig. 107). Refer to paragraph 156 for caster angle.

*c. Camber.*—Camber is the sidewise inclination of the front wheels. Positive camber is the outward inclination of the wheels as viewed from the front of the vehicle; that is, the wheels are farther apart at the top than at the bottom (H) minus (G), fig. 106). Camber is established by design; therefore no adjustment can be made. A bent axle housing, bent steering knuckle, loose steering knuckle support trunnion bearings, or loose wheel bearings will affect camber. Unequal camber will cause vehicle to pull toward side having most camber. Camber may be measured with a square and rule in manner illustrated in figure 106; however, a more accurate determination can be made by use of wheel alinement indicator 41–I–130 for this purpose (fig. 107). Camber dimensions given in paragraph 156 are for straight-ahead position only and must be checked with axle installed on vehicle.

*d. Toe-In.*—Toe-in is the amount which the front wheels are closer together at the front than at the rear ((A) minus (B), fig. 106). An adjustable tie rod, connecting the two steering knuckle supports, is used to adjust toe-in. Camber causes both wheels to tend to turn outward from the vehicle; however, by adjusting tie rod to give wheels proper toe-in, the tendency to turn outward is counteracted and the wheels roll straightahead with no scuffing action on tires. Toe-in is affected by loose wheel bearings, bent axle housing, bent steering knuckle, loose steering knuckle support trunnion bearings, or a bent or improperly adjusted tie rod. Improper toe-in causes excessive tire wear or "scuffing." Unequal toe-in may cause the vehicle to pull

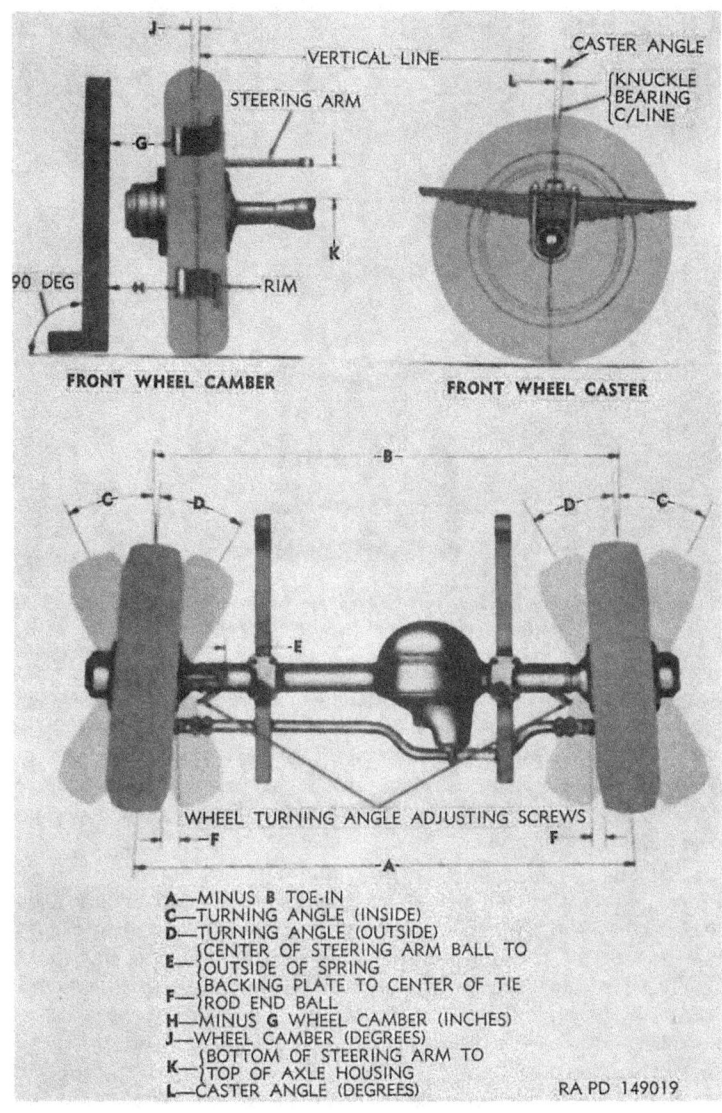

*Figure 106. Front wheel and axle alinement chart.*

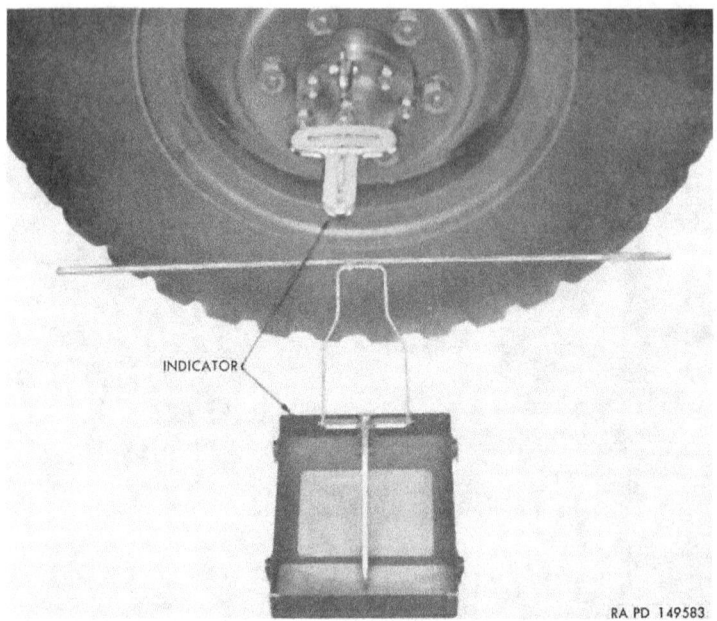

*Figure 107. Checking camber, caster, king pin inclination, and turning angle with wheel alinement indicator 41–I–130.*

toward the side having the least toe-in. When wheels are turned from straightahead to either right or left, toe-in changes, until at extreme right or left positions they are farther apart at the front than at the rear. This condition is termed toe-out. Always measure toe-in with wheels in straightahead position, either by actually measuring A minus B, figure 106 or refer to TM 9–819A. Toe-in dimensions are given in paragraph 156.

*e. Turning Angle.*—The turning angle is the maximum angle through which the front wheels may be turned to right or left from the straightahead position. This angle is greater for the inside wheel than the outside wheel on a turn. The turning angle for the inside wheel is shown as C, figure 106, and the turning angle for the outside wheel is shown as D, figure 106. Turning angle should be checked with the axle installed on the vehicle, using wheel alinement indicator 41–I–130 for this purpose (fig. 107). Stop plugs, threaded and welded in housing (fig. 108), are provided to limit the angle through which the inside wheel can turn. Refer to paragraph 156 for inside and outside turning angle, and to *f* below for adjustment procedure.

*f. Turning Angle Adjustment.*
   (1) *Remove stop plug (screw).*—Cut braze (weld) attaching steering knuckle support stop plug (screw) (fig. 108) to axle

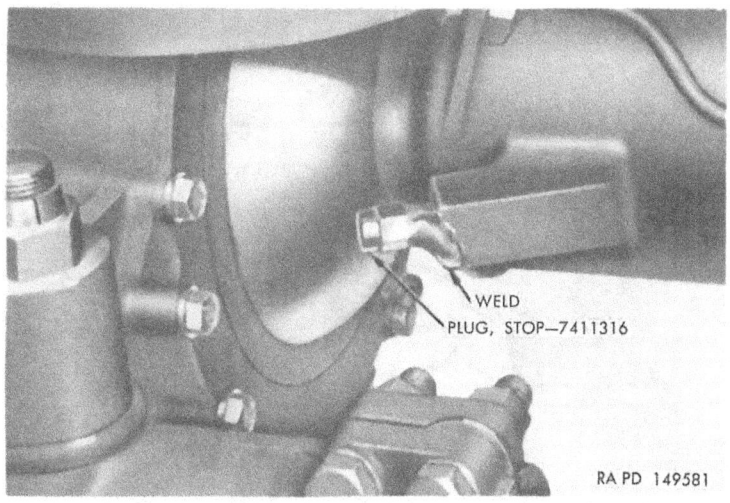

*Figure 108. Steering knuckle support stop plug installed in housing.*

housing. If plug head is damaged, plug should be replaced.

(2) *Install new stop plug.*—Install new steering knuckle support stop plug (screw) (fig. 108) by threading it into axle housing.

(3) *Adjust stop plug.*—Adjust stop plug by threading into or out of housing until angle C, figure 106, is correct for left wheel as indicated in paragraph 156. Use wheel alinement indicator 41–I–130 to obtain correct angle (fig. 107).

(4) *Check angle "D."*—With wheels in position giving correct angle for "C" on left wheel, check angle D, figure 106, for the right wheel. If this angle is different than given in paragraph 156, toe-out on turns will not be correct. Inspect for bent, loose, or twisted tie rod and correct as necessary.

(5) *Braze stop plug.*—When correct settings have been obtained, braze (weld) stop plug (fig. 108) to housing.

(6) At right wheel, repeat operations previously described in (1) through (5) above.

### 156. Alinement Data

| | | |
|---|---|---|
| A minus B | toe-in (at hub C/C) | $5/32$ to $7/32$ in. |
| C | turning angle—inside | 28 deg +1 deg, −0 deg |
| D | turning angle—outside | 26 deg |
| E | center line of steering arm ball to outside of spring | $3\frac{1}{4}$ in. |
| F | backing plate to center of tie rod end pin | $2\frac{7}{8}$ in. |
| H minus G | wheel camber | $27/64$ in. to 0 in. |
| J | wheel camber | $3/4$ deg to 0 deg |
| K | bottom of steering arm to top of axle housing | $3\frac{3}{8}$ in. |
| L | caster angle | 1 deg 45 min |

177

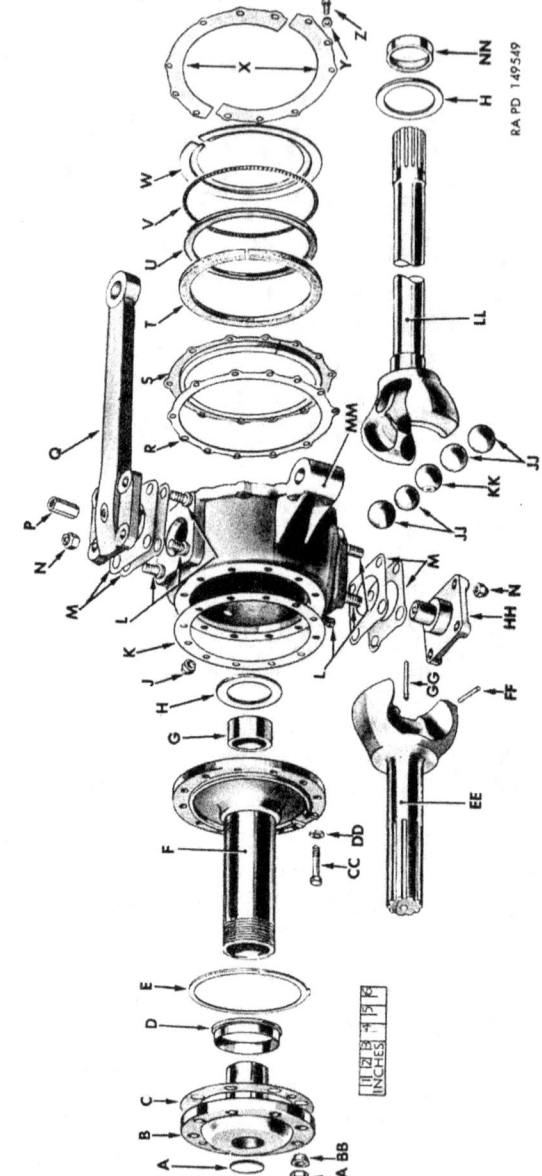

Figure 109. Components of steering knuckle, support, oil seals, and universal joint (left side shown).

178

A—PLUG, EXPANSION, 1³⁄₁₆-IN.—541411
B—FLANGE, DRIVE, HUB—YT-2283030
C—GASKET—7411265
D—SLEEVE, OIL SEAL—7411433
E—SHIELD, OIL, BRAKE—7411431
F—KNUCKLE, STEERING—YT-2275666
G—BEARING, BUSHING TYPE—7411312
H—WASHER, THRUST—7411310
J—PLUG, PIPE, ⅜-IN.—143980
K—GASKET—7411313
L—STUD, ½-13–20 X 1¹³⁄₁₆—7411451
M—SHIM, TRUNNION BEARING
    0.002-IN—7377411
    0.005-IN—7377241
    0.010-IN—7377412
    0.020-IN—CV-3678167
N—NUT, ½-20—442801
P—{SPACER (INNER)—7538373
    SPACER (OUTER REAR)—7411315
Q—ARM, STEERING—7411311
R—GASKET—CV-3659641
S—RETAINER, OIL SEAL, OUTER—7411301
T—SEAL, OIL—CV-3659639
U—SEAL, DUST—7411303
V—SPRING, DUST SEAL—7411305
W—RETAINER, DUST SEAL—7411302
X—RETAINER, OIL SEAL, INNER—7411300
Y—WASHER, LOCK, ⁵⁄₁₆-IN.—120638
Z—SCREW, CAP, ⁵⁄₁₆-18 X ⅝—180075
AA—NUT, ¼-20—442801
BB—WEDGE, DOWEL, TAPERED—7412264
CC—{SCREW, CAP ⅜-16 X 1⁵⁄₁₆—7412113
    SCREW, CAP ⅜-16 X 1¹¹⁄₁₆—7412112
DD—WASHER, LOCK, ⅜-IN.—120632
EE—SHAFT, OUTER—YT-2283033
FF—PIN, TAPER-GROOVE, ³⁄₁₆ X 1⅝—187798
GG—PIN, CENTER BALL—7377358
HH—TRUNNION, STEERING KNUCKLE—7411454
JJ—BALL, UNIVERSAL JOINT, OUTER
    1.372 DIAM—CV-3660105
    1.373 DIAM—7377050
    1.374 DIAM—7377051
    1.375 DIAM—CV-3660108
    1.376 DIAM—7377049
    1.377 DIAM—7377046
    1.378 DIAM—CV-3660111
KK—BALL, UNIVERSAL JOINT, CENTER—YT-2056647
LL—{SHAFT, INNER, LEFT—YT-2277364
    SHAFT, INNER, RIGHT—YT-2277363
MM—{SUPPORT, STEERING KNUCKLE, LEFT—7411453
    SUPPORT, STEERING KNUCKLE, RIGHT—7411452
NN—SEAL, OIL, SHAFT, ASSY—7411307

## Section III. DISASSEMBLY OF FRONT AXLE INTO SUBASSEMBLIES

### 157. General

*a.* The following procedures are based on the assumption that the axle assembly has been removed from the vehicle in accordance with instructions outlined in TM 9-819A. Many of the following operations can be performed with the axle assembly installed on the vehicle; however, for maximum accessibility and efficiency, the axle should be removed from the vehicle and placed on a suitable work stand whenever available.

*b.* Before cleaning or dissembling the axle, make careful visual inspection for evidence of lubricant leakage which migh not otherwise be visible after the assembly or parts have been cleaned. Make a note of all such points so that the cause may be determined either during disassembly or at time of inspection after dissambly. Thoroughly clean the assembly, using steam or other suitable method, to remove all accumulated dirt or other foreign material which might injure parts if it were not removed.

### 158. Preliminary Disassembly Operations

*a. Drain Lubricant.*—Place suitable receptacle under axle; then remove plug and gasket from axle housing to permit lubricant to drain.

*b. Remove Wheel Hub, Drums, and Bearings.*—Remove wheel hubs, brake drums, and bearings as directed in paragraph 256*b*.

*c. Remove Brake Assembly.*—Remove brake backing plate and shoe assembly, also brake hose and shield as directed in paragraph 229.

### 159. Removal of Axle Shaft and Universal Joint Assembly

*a. Remove Steering Knuckle.*—Use brass or soft hammer (fig. 110) to tap steering knuckle ((F), fig. 109) on all sides to loosen knuckle from steering knuckle support ((MM), fig. 109). When loose, pull knuckle straight outward to complete removal. Remove and discard gasket ((K), fig. 109) from knuckle or support.

*b. Remove Axle Shaft and Universal Joint Assembly.*—Grasp outer shaft (fig. 111) and pull shaft and universal joint assembly straight out of axle housing.

*Note.* In some instances it may be necessary to use a small block of wood as a pry between universal joint and steering knuckle support to loosen splined inner shaft from side gear splines.

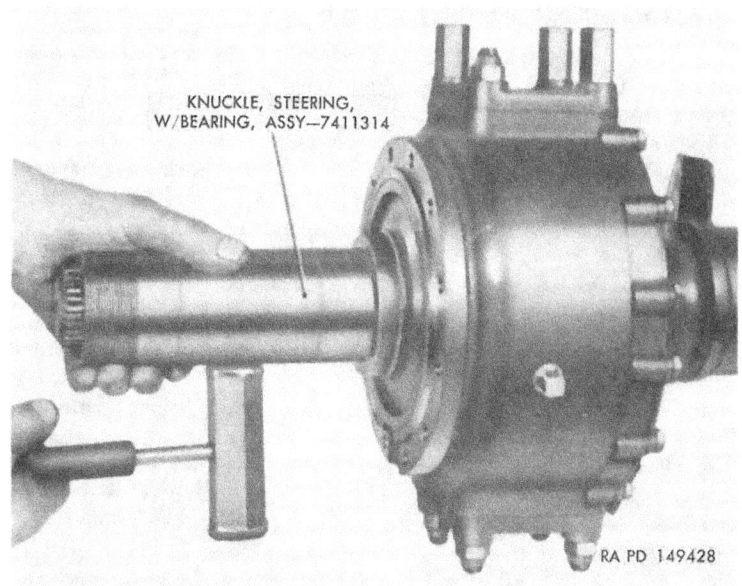

*Figure 110. Removing steering knuckle.*

*Figure 111. Removing or installing universal joint assembly.*

### 160. Removal of Tie Rod Assembly

*a.* Remove nuts ((A), fig. 124) attaching tie rod studs to right and left steering knuckle supports. Remove tie rod stud seal ((B), fig. 124) from each stud.

*b.* Strike arm on steering knuckle support with heavy hammer as downward pressure is applied to tie rod end with a pinch bar. Several sharp blows at arm may be required to loosen tapered tie rod stud from arm. Tie rod assembly can now be completely removed.

### 161. Removal of Steering Knuckle Support

*Note.* Steering knuckle support and associated parts are illustrated in figure 109 and key letters noted in parentheses refer to this illustration unless otherwise indicated.

*a. General.*—The following instructions should be performed at both ends of axle housing.

*b. Remove Seals and Retainers.*—Remove twelve $5/16$–18 x $5/8$ cap screws (Z) and $5/16$-inch lock washers (Y) attaching oil seal inner retainers (X) to steering knuckle support (MM); then remove two retainers. Remove dust seal retainer (W), oil seal (T), and oil seal outer retainer (S), all of which are split and can be completely removed at this time. Dust seal (U) and dust seal spring (V) cannot be removed until steering knuckle support (MM) is removed.

*c. Remove Steering Knuckle Trunnions.*—Remove four $1/2$–20 nuts (N) attaching steering knuckle trunnions (HH) to bottom of steering knuckle support (MM). Remove $1/2$–20 nut (N) and three spacers (P) attaching steering knuckle trunnion ((L), fig. 121) to top of steering knuckle support at right side, or steering arm (Q) to support at left side. The steering knuckle trunnion is pressed into the steering arm (Q) at left side. Remove each trunnion and shim pack. Attach shims to their respective trunnions and tag so that they can be installed in their original location.

*d. Remove Steering Knuckle Support and Bearings.*—Lift steering knuckle support (MM) off housing outer end and catch lower trunnion bearing cone ((T), fig. 121) as support is lifted off. Remove upper trunnion bearing cone ((C), fig. 121) from support. Attach identification tag to upper and lower bearing cones so that they can be installed in their original position.

### 162. Removal of Housing Cover, and Differential and Carrier Assembly

*a.* Remove ten $7/16$–20 nuts ((E), fig. 121) attaching cover ((H), fig. 121) to housing. Tap cover lightly with soft hammer to loosen; then remove from $7/16$–20 x 2 studs ((P), fig. 121) in housing. Remove and discard gasket ((J), fig. 121).

b. Remove ten 7/16–20 nuts ((AA), fig. 132) attaching carrier ((Z), fig. 132) to housing. Tap carrier lightly with soft hammer to loosen carrier from housing; then withdraw differential and carrier assembly from housing. Remove and discard gasket ((Y), fig. 132).

### 163. Rebuild of Front Axle Differential and Carrier Assembly

Since this procedure is the same as for rear axle differential and carrier assembly, this information will not be repeated in this section. Refer to paragraphs 197 through 202 for rebuild of differential and carrier assembly.

### Section IV. REBUILD OF AXLE SHAFT AND UNIVERSAL JOINT

### 164. General

The following procedures cover the disassembly, cleaning, inspection, and reassembly of axle shaft and universal joint assembly. Refer to paragraph 159 for removal procedures.

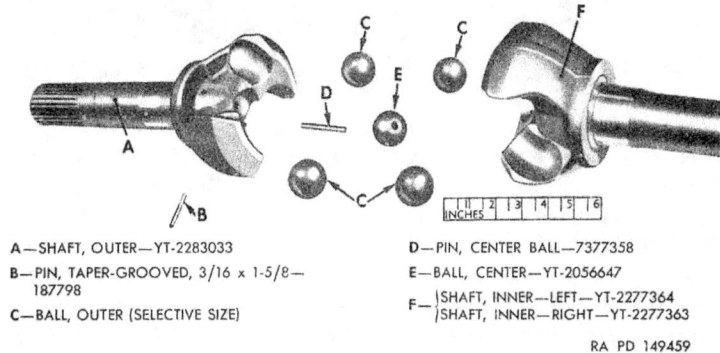

A—SHAFT, OUTER—YT-2283033
B—PIN, TAPER-GROOVED, 3/16 x 1-5/8—187798
C—BALL, OUTER (SELECTIVE SIZE)
D—PIN, CENTER BALL—7377358
E—BALL, CENTER—YT-2056647
F—{SHAFT, INNER—LEFT—YT-2277364
{SHAFT, INNER—RIGHT—YT-2277363

RA PD 149459

*Figure 112. Axle shaft and universal joint components.*

### 165. Disassembly

*Note.* Key letters noted in parentheses are in figure 112 unless otherwise indicated.

a. *Cleaning.*—Immerse assembly in dry-cleaning solvent or volatile mineral spirits to remove oil, grease, or other deposits.

b. *Check Universal Joint for Play or Backlash.*—Before disassembling, it should be determined if excessive play or backlash exists in the universal joint. Place the assembly in a vise in a vertical position, with the outer (short) shaft (A) up, and the vise jaws gripping the inner shaft (F) just below the machined surface on the shaft.

**183**

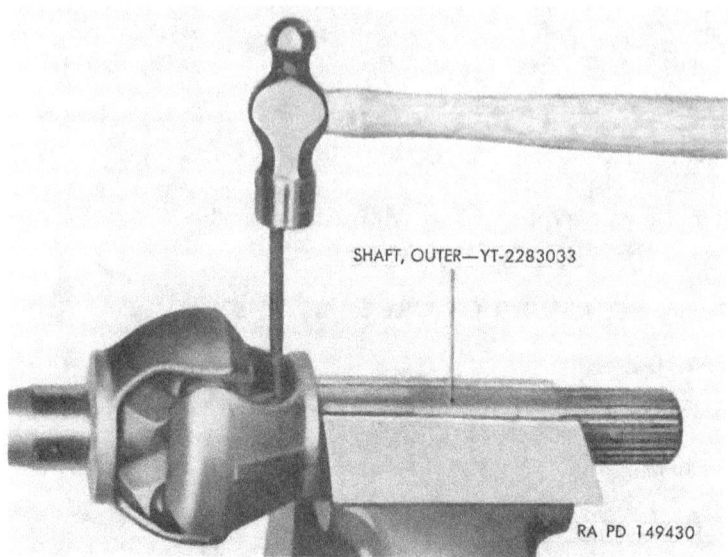

*Figure 113. Universal joint center ball lock pin removal or installation.*

Soft metal or wood protectors should be used in jaws of vise. Firmly push down on the outer shaft so that it rests on the center ball, and at the same time attempt to twist the joint in both directions. If any play or backlash is evident, oversize outer balls should be installed at assembly (par. 169).

*c. Remove Center Ball Pin Lock Pin.*—Position axle shaft and universal joint in a vise (fig. 113) or lay on a bench. Use a suitable punch and hammer to drive $3/16$ x $1 5/8$ taper-grooved pin (B) out of outer shaft (fig. 113).

*d. Dislodge Center Ball Pin.*—Remove assembly from vise and hold in a vertical position with outer shaft (A) down. Bounce the outer shaft on a block of wood (fig. 114) to dislodge center ball pin (D), allowing the pin to drop downward in drilled passage in outer shaft.

*e. Remove Balls.*—With the assembly in a vertical position, inner shaft (F) up, clamp the outer shaft (A) in a vise, using soft metal jaw plates. Swing the outer shaft to one side and at the same time raise it slightly to pull the two shafts apart and loosen the center ball. Turn the center ball with thumb and finger, so that groove in center ball lines up with one of the outer balls (fig. 115). Outer ball can be removed by pushing it past the center ball groove, using thumb (fig. 115). Bend outer shaft sharply in opposite direction to release three

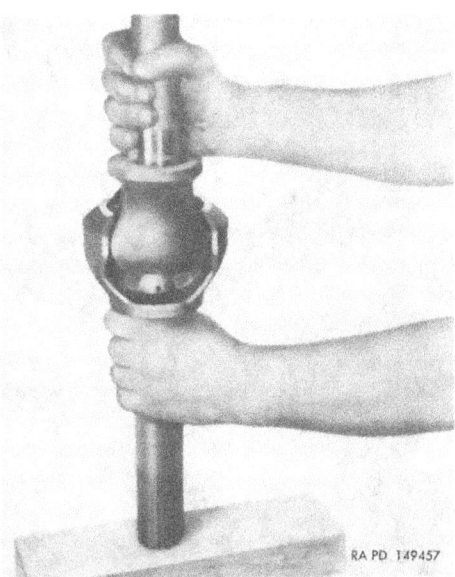

*Figure 114. Dislodging universal joint center ball pin.*

remaining outer balls. Separate the outer shaft from the inner shaft; then remove center ball. Remove outer shaft from vise; then remove center ball pin from hole in outer shaft.

## 166. Cleaning

Immerse balls, shafts, and pins in dry-cleaning solvent or volatile mineral spirits to loosen all grease or other deposits. Remove balls from cleaning solution and dry thoroughly, being sure hole in center ball is clean. Remove shafts and clean any remaining deposits in splines and ball races with cleaning solution and bristle brush. Be sure drilled passages in outer shaft are clean.

## 167. Inspection

*Note.* Key letters noted in parentheses are in figure 112 unless otherwise indicated.

*a. Shafts.*—Inspect shaft splines for wear, twist, or other damage. Examine ball races in yokes for excessive wear, roughness, or other damage. Carefully examine joints for cracks. Inspect shafts for twisted or bent condition. Whenever either the inner or outer shaft is damaged, a complete new shaft and universal joint assembly must be installed.

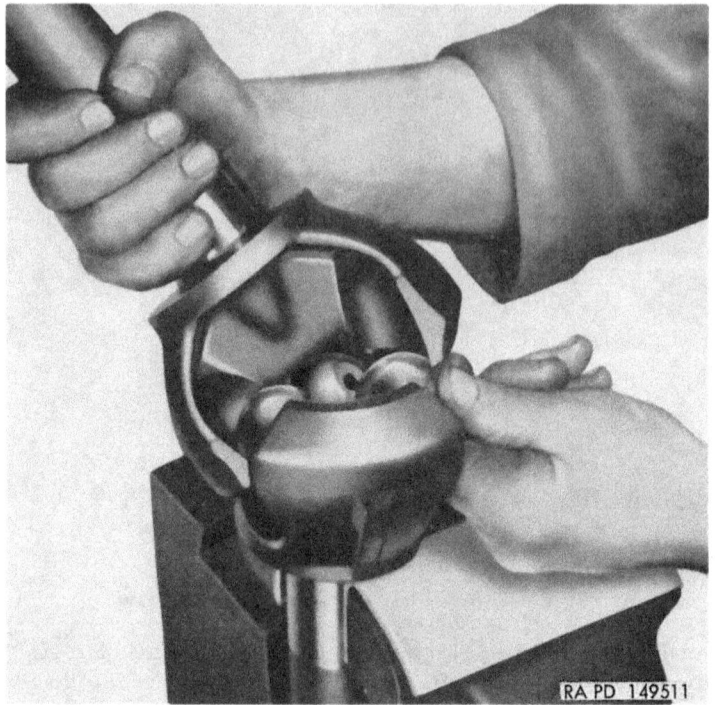

*Figure 115. Universal joint outer balls removal or installation.*

*b. Balls.*—Carefully examine balls for chips, cracks, or rough spots. Use a micrometer and check balls for out-of-round condition. Replace damaged balls with new balls of same diameter unless check made in paragraph 165*b* indicated that oversize balls are necessary.

*c. Pins.*—Examine center ball pin (D) for damage. Check freeness of pin in drilled hole in outer shaft, also in center ball. Center ball pin is locked with a $3/16$ x $1 5/8$ taper-grooved pin (B) which should be replaced if not tight in drilled hole in outer shaft.

## 168. Repair

Whenever inspection of shafts or universal joints indicated that these parts are worn or damaged, they must be replaced with new parts; therefore, no repair is recommended.

## 169. Assembly

*Note.* Key letters noted in parentheses are in figure 112 unless otherwise indicated.

*a. Select Correct Size Balls.*—Whenever check (par. 165*b*) indicates that play or blacklash exists, this condition can be corrected by installation of larger outer balls. Outer balls (C) are available in seven sizes: 0.001-, 0.002-, and 0.003-inch undersize, standard, and 0.001-, 0.002-, and 0.003-inch oversize. Measure diameter of the original balls with a micrometer to determine the size of each ball. Select one or two balls 0.001 inch larger than the smallest ball originally used in the assembly. It is desirable to keep the balls within 0.001 inch of the same size and the variation should not exceed 0.002 inch. As the universal joint is being assembled, the two largest balls should be installed diagonally across from each other.

*b. Position Inner Shaft.*—Place inner shaft (F) in vise, using soft jaw plates, with universal joint end up. Be sure vise does not grasp on a ground surface.

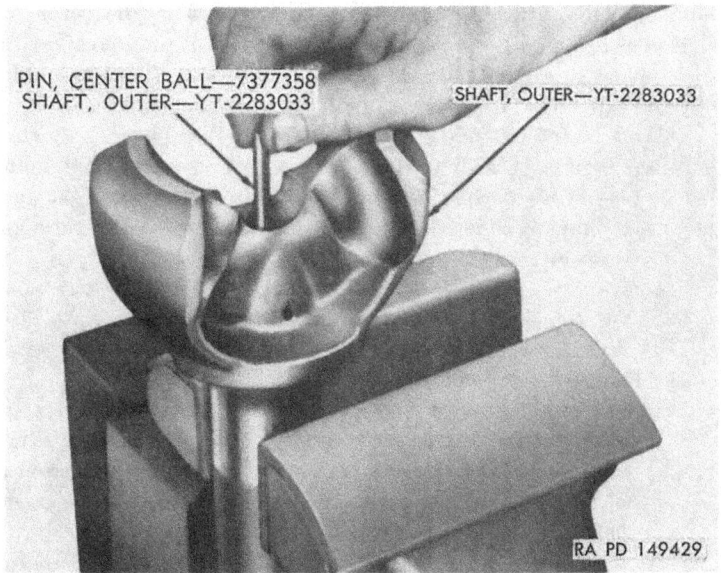

*Figure 116. Universal joint center ball pin installation.*

*c. Position Center Ball.*—Place center ball (E) on the seat at the center of the inner shift.

*d. Install Center Ball Pin.*—Insert center ball pin (D) into hole in outer shaft (fig. 116). Check pin several times to be sure it slides freely and does not stick.

*e. Position Outer Shaft.*—Place outer shaft (A) over inner shaft (F) with outer shaft resting on center ball (E). Be sure center ball pin (E) does not drop out of outer shaft during this operation.

*f. Install Three Outer Balls.*—Bend outer shaft at joint to an extreme angle as necessary to slip three of the outer balls (C) into the races. Be sure the two largest balls are diagonally across from each other.

*g. Install Fourth Outer Ball.*—Bend outer shaft at joint in opposite direction to obtain necessary clearance to install fourth outer ball. Rotate center ball to line up groove in ball with race for the remaining ball (fig. 115). Slip the fourth ball past the center ball and into race; then bend outer shaft to a straight position.

*h. Position Center Ball Pin.*—Raise outer shaft only sufficient to free the center ball; then rotate ball until center ball pin drops into drilled hole in ball.

*i. Install Center Ball Pin Lock Pin.*—Install a new 3/16 x 1 5/8 taper-grooved pin (B) into drilled hole in outer shaft and drive into position (fig. 113). Remove assembly from vise and lay on bench or anvil so as to support one end of pin. Strike end of pin sharply with prick punch to expand end and lock it in position. Turn assembly over and expand opposite end of pin in the same manner.

*j. Check Universal Joint Play or Backlash.*—When oversized outer balls have been installed in used races, it is only necessary to determine that no play or backlash exists when shaft is in vertical position, and that a maximum of 35 pounds pull is required to move shaft through its normal operating range.

    (1) *Position Assembly in Vise.*—Install assembly in vise in a vertical position with the outer shaft at the top and the vise jaws gripping the inner shaft just below the universal joint. Use soft jaw plates to protect shaft.

    (2) *Determine Play or Backlash.*—Firmly push down on outer shaft so that it rests on the center ball, at the same time, attempt to twist the joint in both directions. The presence of play or backlash indicates the need of still large outer balls.

    (3) *Determine Pull Required to Move Shift Through Its Normal Operating Range.*—With assembly still mounted in vise, attach a spring scale within one inch of end of outer shaft. With spring scale, pull shaft through its normal operating range and note reading on spring scale. A pull of more than 30 pounds indicates that outer balls of too large an oversize have been installed. Ideal conditions when oversize outer balls are installed in used universal joints are as follows: Vertical or straight position—no play or backlash; 10 to 15 degree turn—30 pound maximum drag; 15 to 32 degree or full turn—free with slight lash permissible.

## Section V. REBUILD OF STEERING KNUCKLE, SUPPORT, TRUNNIONS, SEALS, AND BEARINGS

### 170. General

All parts covered in this section have been disassembled during their removal from axle assembly; therefore, no disassembly instructions are included in this section.

### 171. Cleaning

Immerse all parts in dry-cleaning solvent or volatile mineral spirits to loosen and remove all accumulations of grease, dirt, or other foreign deposits. Remove each part separately and use bristle brush to remove all accumulated deposits. Whenever available, compressed air may also be used to remove deposits and to dry parts. Particular attention should be given to trunnion bearing cones. Slush the bearings up and down in cleaning fluid. Use a bristle brush to clean thoroughly, repeating immersion and brushing until all dirt is removed. Dry bearings with compressed air, directing air in such a manner so as not to spin bearing.

### 172. Inspection

*Note.* Key letters noted in parentheses are in figure 109 unless otherwise indicated.

*a. General.*—Before any attempt is made to inspect parts, they must have been thoroughly cleaned as directed in paragraph 171. Refer to chapter 21 for dimensional data or other necessary inspection data.

*b. Steering Knuckle.*—Inspect threads for damage and clean up with thread restoring tool, otherwise install new steering knuckle (F) assembly. Inspect bushing type bearing (G) inside steering knuckle for wear, roughness, or other damage. Refer to paragraph 346 for dimensional data, also to paragraph 173*a* whenever inspection indicates that new part should be installed. Inspect thrust washer (H) for wear, roughness, or other damage. Refer to paragraph 346 for thrust washer thickness and paragraph 173*b* for instructions necessary to install new part. Inspect brake oil shield (E) for bent or damaged condition, also for proper installation (fig. 117). Inspect oil seal sleeve (D) for roughness or grooves and replace as directed in paragraph 173*c* if necessary.

*c. Steering Knuckle Support.*—Inspect steering knuckle (MM) for cracked, broken, or distorted condition. Inspect for stripped or damaged threads in tapped holes. Inspect hole in integral boss to which the tie rod stud attaches for evidence of wear due to loose stud. Install new support if any of the above conditions exist. Inspect $\frac{1}{2}$–13–20 x $1\frac{15}{16}$ studs (L) for looseness in support, also for damaged or stripped

189

threads and tighten or replace as necessary. Inspect shim and gasket surfaces for smoothness and clean up with fine file if necessary.

*d. Steering Knuckle Trunnion.*—Inspect steering knuckle trunnions (HH) for distortion or other damage. Inspect shim surface for smoothness. Replace with new part if damaged or clean up shim surface with fine file.

*e. Steering Arm.*—Inspect steering arm (Q) for bent or other damaged condition. Inspect drag link tapered hole in end of arm for wear due to loose stud. Inner or drag link end of arm has an upward offset of twenty-seven thirty seconds of an inch, when shim surface is flat on a face plate. Replace with new part if bent or otherwise defective. Clean up shim surface with fine file if necessary. If trunnion which is pressed into steering arm requires replacement, it can be removed and installed using an arbor press.

*f. Trunnion Bearings.*—Inspect trunnion bearing cones ((C), fig. 121) for chipped, cracked, or worn condition. Replace with new parts if damaged or worn.

*g. Seals and Retainers.*—Felt type oil seal (T), rubber dust seal (U), and dust seal spring (V) should be discarded and new parts installed. Inspect outer oil seal retainer (S), dust seal retainer (W), and inner oil seal retainer (X) for bent or damaged condition that

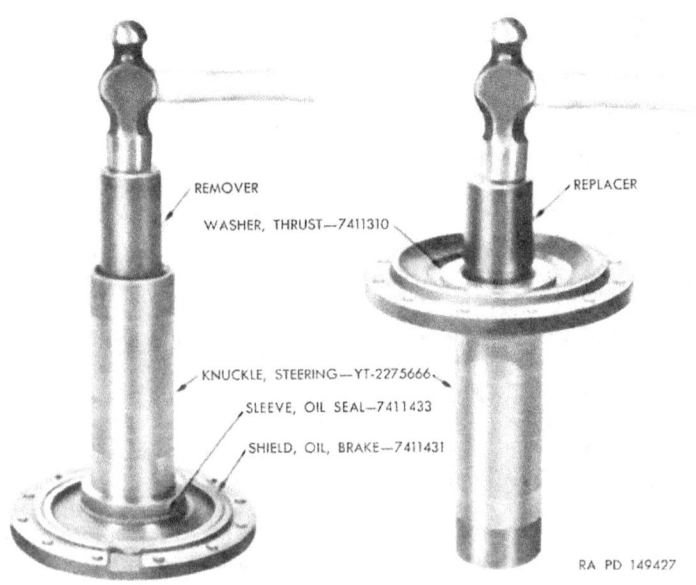

*Figure 117. Removing or installing steering knuckle bushing type bearing, using remover 41-R-2369-725 or replacer 41-R-2388-250.*

would render these parts unfit for further use. Straighten or replace, whichever is necessary.

### 173. Repair

*a. Steering Knuckle Bearing.*
(1) *Removal.*—Place steering knuckle in arbor press or on bench and use steering knuckle bearing remover 41–R–2369–725 to press or drive bushing type bearing (G) from steering knuckle (fig. 117).
(2) *Installation.*—Place steering knuckle in arbor press or on bench and use steering knuckle bearing replacer 41–R–2388–250 to press or drive bushing type bearing (G) into steering knuckle (fig. 117). Shoulder inside steering knuckle properly locates bearing in its correct position (fig. 118). Burnish or grind bearing to 1.786 to 1.788-inch diameter.

*b. Steering Knuckle Thrust Washer.*
(1) *Removal.*—Use sharp chisel to remove metal stakes retaining thrust washer (fig. 119) to steering knuckle; then remove

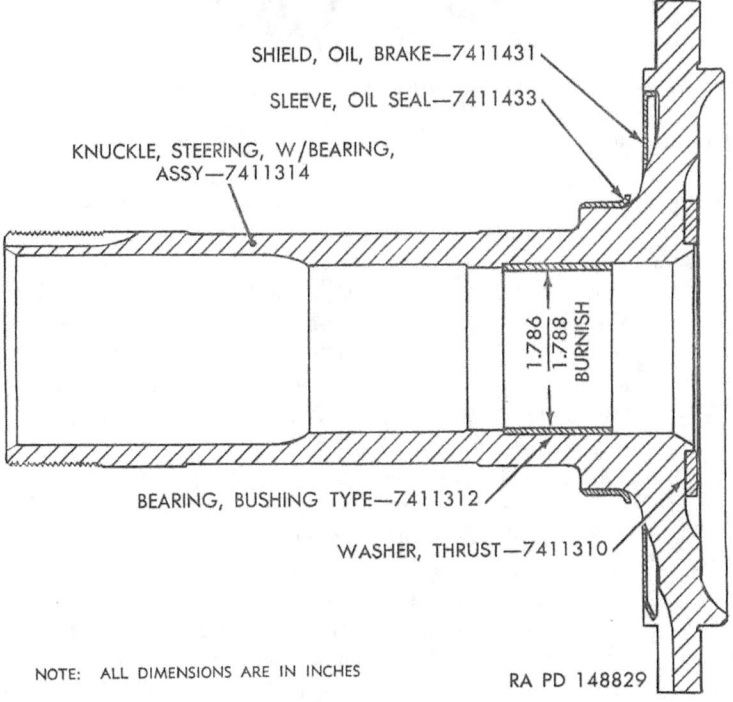

*Figure 118. Sectional view of steering knuckle assembly.*

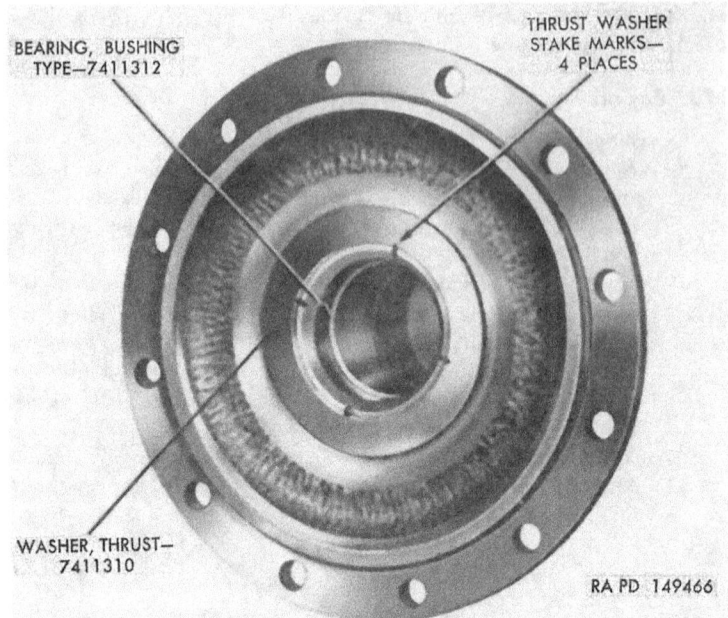

*Figure 119. Steering knuckle bushing type bearing and thrust washer installed.*

thrust washer. Use fine file to remove sharp stake points from steering knuckle.

(2) *Installation.*—With steering knuckle standing on outer end on bench, position thrust washer (H) at inside flange of steering knuckle, with chamfered outer edge toward knuckle. Use block of hard wood or plastic hammer to drive thrust washer into place until fully seated. Use sharp chisel and stake steering knuckle (fig. 119) at four places to hold thrust washer. Use fine stone to remove any metal from face of thrust washer as a result of staking operation.

*c. Steering Knuckle Oil Seal Sleeve.*

(1) *Removal.*—Use ball peen hammer and tap entire circumference of outer surface of oil seal sleeves (D). Peening of this surface will cause metal in sleeve to stretch until sleeve can be removed from steering knuckle.

(2) *Installation.*—With steering knuckle standing on flange end, position oil seal sleeve (D) on steering knuckle. Use replacer 41–R–2395–518 in manner illustrated in figure 120 to drive sleeve onto knuckle until edge of sleeve is flush with shoulder on steering knuckle (fig. 118).

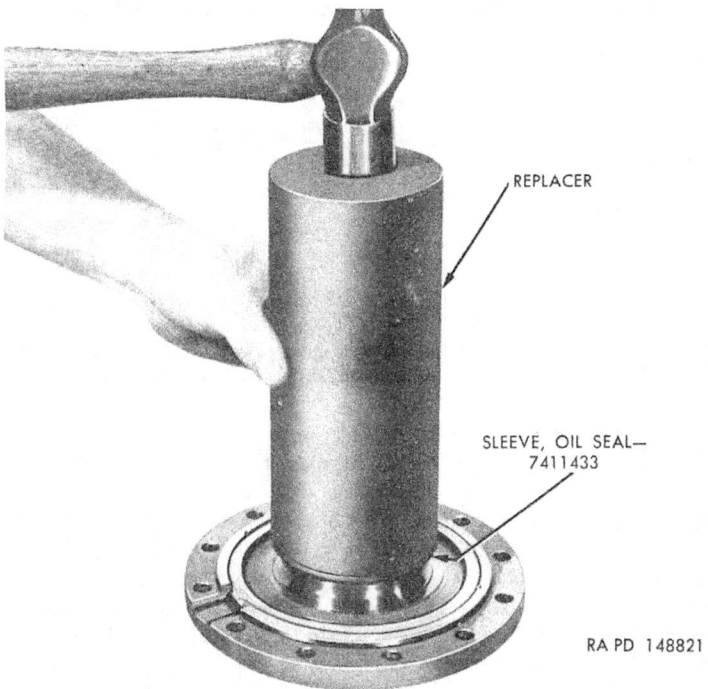

*Figure 120. Installing oil seal sleeve on steering knuckle using replacer 41-R-2395-518.*

d. *Steering Knuckle Brake Oil Shield.*
 (1) *Removal.*—Insert screw driver on other similar tool under brake oil shield (E) at slot in steering knuckle; then pry shield from steering knuckle (fig. 117).
 (2) *Installation.*—Position brake oil shield (E) on steering knuckle with lip of shield in steering knuckle slot (fig. 117). Use soft hammer and tap shield until fully seated in knuckle. If necessary, stake knuckle in several places to hold shield in knuckle.

e. *Steering Knuckle Trunnion Studs.*
 (1) *Removal.*—Remove damaged $\frac{1}{2}$–13–20 × $1^{15}/_{16}$ studs (L) from steering knuckle support, using a stud remover.
 (2) *Installation.*—Install new $\frac{1}{2}$–13–20 × $1^{15}/_{16}$ studs (L), using stud replacing tool. Height of all studs should be $1\frac{3}{8}$ inches, except two at upper outside of left steering knuckle (where dowel rings are used) (fig. 127), which should be driven to a height of $1^{3}/_{16}$ inches.

## Section VI. REBUILD OF FRONT AXLE HOUSING

### 174. General

The following procedures cover disassembly, cleaning, inspection, and repair of axle housing after axle assembly has been disassembled into subassemblies as directed in paragraphs 157 through 163.

### 175. Disassembly

*Note.* Key letters noted in parentheses are in figure 121 unless otherwise indicated.

*a. Remove Trunnion Bearing Cups.*

*Note.* Bearing cups need not be removed unless inspection (par. 177*b*) indicates that these parts should be replaced with new parts, then proceed as follows:

Use brass rod and hammer to drive trunnion bearing cup (B) and trunnion bearing oil retainer (A) upward to remove from housing. Discard trunnion bearing oil retainer, since this part has probably been damaged during removal. Use brass rod and hammer to drive trunnion bearing cup (S) downward to remove from housing.

*b. Remove Brake Line Assemblies.*—Remove three cap screws, lock washers, and clips attaching brake lines to axle housing. Loosen brake line fitting nut attaching each line to junction block located on top of axle housing, then remove right and left brake line assemblies (fig. 105).

*c. Remove Thrust Washer.*—Remove thrust washer (V) from housing outer end by driving a sharp chisel between thrust washer and housing at stake marks. Use care not to damage thrust washer seat in housing.

*d. Remove Shaft Oil Seal.*—Position remover 41–R–2371–850 through shaft oil seal (U); then strike remover several sharp blows with hammer to remove seal (fig. 122).

### 176. Cleaning

*Note.* Key letters noted in parentheses are in figure 121 unless otherwise indicated.

*a. General.*—Immerse parts in dry-cleaning solvent or volatile mineral spirits to loosen and remove all accumulated grease, dirt, or other deposits. Use of bristle brush and repeated use of cleaning solvent will remove all deposits.

*b. Housing.*—Clean housing (D) inside and out, using long-handled brush or swab to remove dirt and grease from inside. Be sure that all particles of gaskets are removed, also remove sealing compound if present. Clean polished surfaces at housing outer ends. Clean oil seal and thrust washer seating surfaces at outer ends of housing.

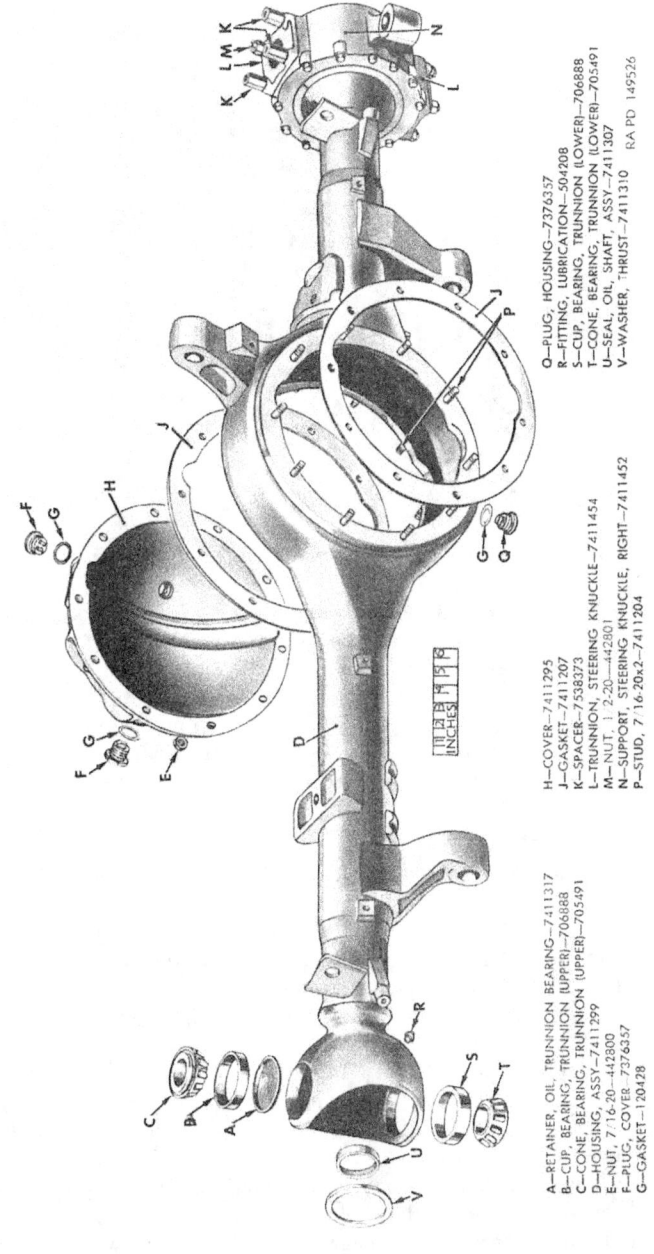

*Figure 121. Front axle housing and associated parts.*

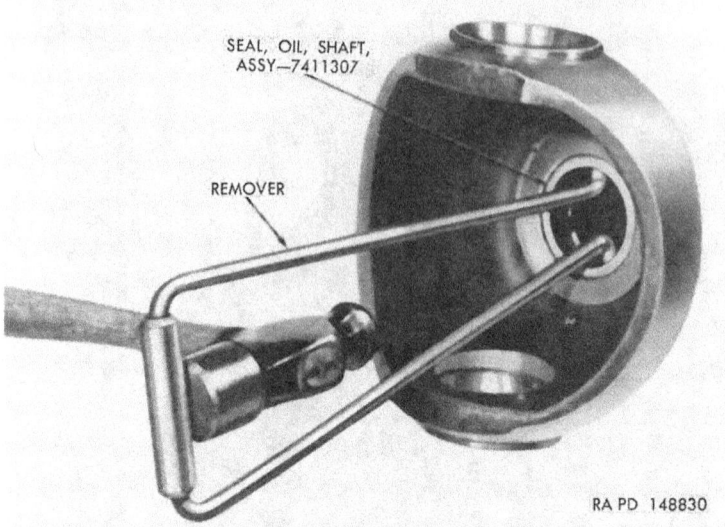

*Figure 122. Removing shaft oil seal assembly using remover 41-R-2371-850.*

   *c. Housing Cover.*—Clean inside and outside of housing cover (H) to remove all dirt and grease. Be sure that all particles of gasket and sealing compound are removed.

   *d. Magnetic Plugs.*—Clean housing and cover plugs (F) and (Q) to remove all particles of metal adhering to plug magnet.

   *e. Brake Line Assemblies.*—Clean inside of brake lines of any deposits, using compressed air to remove all obstructions.

**177. Inspection**

   *Note.* Key letters noted in parentheses are in figure 121 unless otherwise indicated.

   *a. General.*—Before inspecting, all parts must have been cleaned as directed in paragraph 176. Refer to paragraph 346 for dimensional data or other necessary inspection data.

   *b. Trunnion Bearing Cups.*—Carefully examine each trunnion bearing cup (B) and (S) for evidence of wear, pits, cracks, or other defects. Replace with new part if defective.

   *c. Brake Lines.*—Inspect brake line tubing for bends, cracks, or other defects. Inspect tubing nuts for stripped threads. Replace line assembly if defective.

   *d. Housing.*—Housing assembly (D) should be carefully inspected visually or use locally available checking equipment to determine if bent, twisted, or otherwise damaged. Repair or replace, whichever is the more practical, dependent upon the nature of the defect and the availability of adequate straightening equipment. Inspect spherical surfaces at each end of housing for evidence of scratches or other marks that might impair efficiency of oil and dust seals. Slight imperfections can sometimes be cleaned up with crocus cloth or fine stone; if not, replace housing. Inspect machined gasket surfaces to be sure they are smooth. Inspect housing cover and carrier studs (P) to be sure that they are tight and that threads are not stripped. Replace damaged studs. Inspect all parts welded to housing to be sure they are not damaged or bent, also that welding is not broken.

*e. Housing Cover.*—Carefully inspect cover (H) for evidence of distortion or cracks. Gasket surface should be flat to provide good cover-to-housing seal. Inspect cover plug threads for damage or stripping. Install new cover if damaged.

*f. Thrust Washer.*—Inspect thrust washer (V) for roughness at thrust surface, also check thickness to determine if worn. Refer to paragraph 346 for thrust washer thickness.

*g. Shaft Oil Seal.*—Shaft oil seal, assembly (U) installed in outer end of housing should always be replaced whenever axle is disassembled.

### 178. Repair

*a. Stop Plug Replacement.*—Stop plugs, used to control turning radius, are welded in place to prevent unauthorized adjustment.
    (1) *Removal.*—Use torch to loosen tack weld holding threaded plug in its correct position; then remove plug by threading out of bracket welded to housing.
    (2) *Installation.*—Thread new stop plug into bracket. Check turning angle as described in paragraph 155*e*. When proper turning angle (fig. 106) given in paragraph 156 has been obtained as described in paragraph 155*f*, plugs should be tack welded to prevent loosening and tampering.

*b. Stud Replacement.*
    (1) *Removal.*—Remove damaged studs (P) from axle housing, using a stud remover.
    (2) *Installation.*—Install new studs (P), using stud replacing tool. Height of all studs should be $1\frac{3}{8}$ inch.

*c. Straightening Housing.*—Axle housing assembly (D) can sometimes be straightened provided suitable straightening equipment and trained operating personnel are available. Each straightening job presents its own problem, therefore no specific instructions can be given.

### 179. Assembly

*a. Shaft Oil Seal Installation.*—Apply thin coating of plastic type gasket cement to outer surface of seal contacting housing. Position

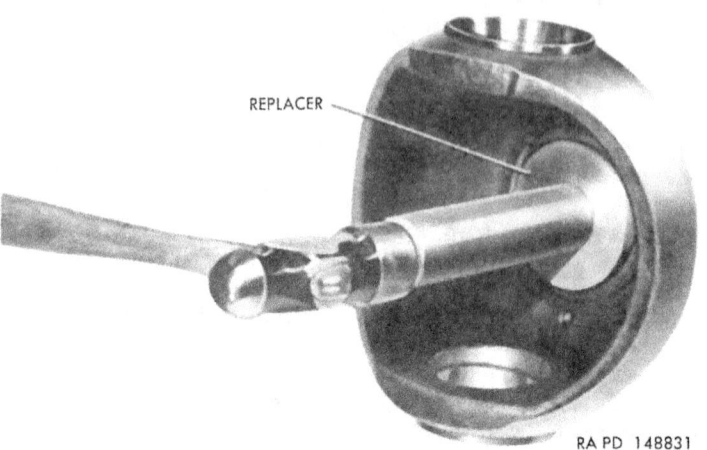

*Figure 123. Installing shaft oil seal assembly using replacer 41-R-2392-640.*

shaft oil seal assembly (U) into end of housing with lip of seal toward housing. Use oil seal replacer 41-R-2392-640 to drive and properly position seal into housing (fig. 123).

*b. Thrust Washer Installation.*—Position thrust washer (V) into end of housing with chamfered outer edge toward housing. Use block of hard wood or plastic hammer to drive thrust washer into place until fully seated. Use sharp chisel and stake thrust washer to housing at four points. Use fine stone to remove any protruding metal as a result of staking.

*c. Trunnion Bearing Cup Installation.*—At top of housing only, install trunnion bearing oil retainer (A), being sure that it rests in counterbore below bearing cup seat. Position trunnion bearing cups (B and S) squarely in housing, then use block of hard wood or plastic hammer to drive cups into housing until fully seated.

*d. Brake Line Installation.*—Position brake line assemblies on housing, thread line fitting nuts into junction block on top of axle housing, and tighten firmly. Secure lines to housing with three clips, three ¼-inch lock washers, and three ¼-28 x ½ cap screws. Tighten cap screws firmly.

## Section VII. REBUILD OF TIE ROD

### 180. General

The following procedures cover disassembly, cleaning, inspection, repair, and assembly of tie rod after removal from axle assembly as directed in paragraphs 157 through 163.

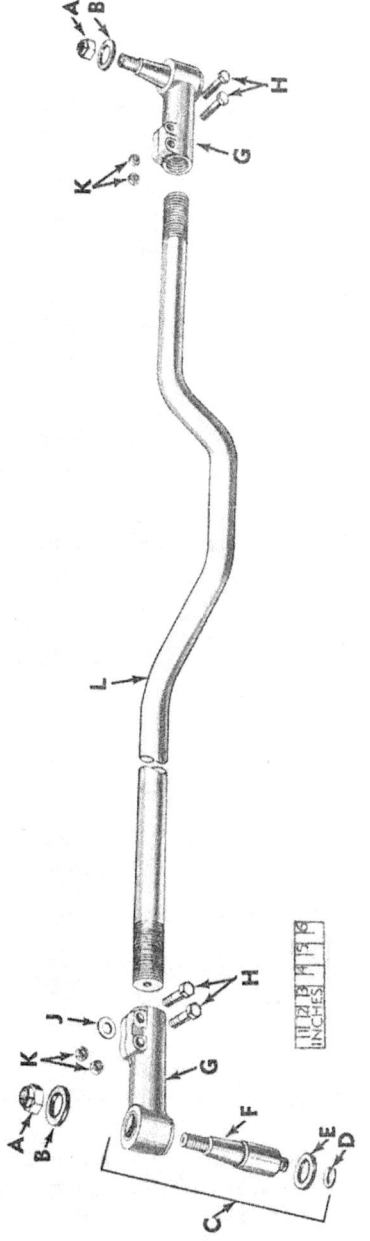

A—NUT, 7/8-14—442805
B—SEAL, TIE ROD STUD—7411418
C—{END, TIE ROD, LEFT, ASSY—7411416
END, TIE ROD, RIGHT, ASSY—7411412
D—RING, SNAP, STUD—YT-2287559
E—WASHER, STUD—YT-2287559
F—STUD, TIE ROD, ASSY—YT-2275616
G—{END, TIE ROD, LEFT—YT-2271940
END, TIE ROD, RIGHT—YT-2271939
H—SCREW, CAP, 1/2-20x2-1/2—181709
J—LOCK, TIE ROD—7411419
K—NUT, 1/2-20—442801
L—ROD, TIE—7411417

RA PD 149469

*Figure 124. Components of front axle tie rod assembly.*

199

## 181. Disassembly

*Note.* Key letters noted in parentheses are in figure 124 unless otherwise indicated.

*a. Tie Rod End Removal.*—Remove ½–20 nut (K) from ½–20 x 2 cap screws (H), then remove cap screws, also tie rod lock (J) at tie rod left end. Remove tie rod ends (G) from tie rod (L) by unscrewing ends from tie rod.

*b. Tie Rod Stud Removal.*—Use snap ring pliers to remove stud snap ring (D), then remove stud washer (E). Place tie rod end assembly on bed of arbor press and press tie rod stud assembly (F) from tie rod end (G).

*Note.* Pressure should be applied at threaded end of stud.

## 182. Cleaning

Immerse all parts in dry-cleaning sovent or volatile mineral spirits to remove all grease, dirt, or other foreign matter.

## 183. Inspection

*Note.* Key letters noted in parentheses are in figure 124 unless otherwise indicated.

*a. Tie Rod.*—Examine tie rod (L) for stripped or damaged threads. If threads are stripped, tie rod should be replaced, or if damaged, they can sometimes be repaired, depending upon the nature of the damage. Inspect for bent condition, which can sometimes be detected by measuring length of rod. Rod should measure between $55\frac{7}{8}$ and $56\frac{3}{8}$ inches, otherwise it is probably bent at offset.

*b. Tie Rod Ends.*—Inspect tie rod ends (G) for cracked or broken condition. Inspect for damaged or stripped threads. Replace if any of the foregoing conditions exist.

*c. Tie Rod Stud.*—Inspect tie rod stud (F) for stripped or damaged threads. Inspect for looseness of stud in bushing. Replace if threads are damaged or if stud is loose in bushing.

*d. Tie Rod Stud Seal.*—Tie rod stud seal (B) is made of rubber and should always be replaced with new part.

## 184. Assembly

*Note.* Key letters noted in parentheses are in figure 124 unless otherwise indicated.

*a. Tie Rod Stud Installation.* Position tie rod end on bed of arbor press with clamp bolt holes down; then start tie rod stud (F) through tie rod end bore. Install piece of pipe or ring over stud in such a manner as to press against the stud bushing.

*Caution:* During this operation do not press on the stud as this may damage the bonding material between the stud and bushing. Press stud and bushing into tie rod end until bushing is flush with tie rod end. Install stud washer (E) and secure with stud snap ring (D), using snap ring pliers. Install tie rod stud seal (B) over tapered end of tie rod stud.

   *b. Tie Rod End Installation.*

   *Note.* Right tie rod end assembly has coarse threads (12 per in) while left tie rod end assembly has fine threads (16 per in).

Thread each tie rod end assembly (C) onto tie rod (L) an equal distance. Each tie rod end will engage tie rod a distance of three inches when in approximately correct location.

   *c. Tie Rod Length Adjustment.*—Carefully measure distance between center line of tie rod end studs and thread tie rod ends on or off tie rod until dimension is $62^{17}/_{32}$ inches. Since right and left tie rod ends have different thread sizes, a finer degree of adjustment can be obtained. Install $1/2$–20 x $2 1/2$ cap screws (H) and $1/2$–20 nuts (K) used to clamp tie rod end to tie rod; also install tie rod lock (J) at tie rod left end.

## Section VIII. ASSEMBLY OF FRONT AXLE FROM SUBASSEMBLIES

### 185. General

   *a.* Assembly procedures, are arranged in logical sequence. All subassemblies should be rebuilt, repaired, and adjusted before beginning operations described herein.

   *b.* Whenever necessary, procedures are given for checking fits and determining clearances or adjustments. Specifications of new parts, clearances, and repair or rebuild standards are listed in chapter 21.

   *c.* Cleanliness is important in handling axle components. All parts should be arranged on clean surface and protected from dirt until assembled. Tools and equipment must be clean to prevent contaminating parts during assembly procedures.

   *d.* Gaskets, seals, snap rings, and lock washers must be replaced as indicated in following procedures.

### 186. Installation of Steering Knuckle Support and Trunnion Bearings

   *Note.* Sectional view of steering knuckle, support, and universal joint assembly installed is shown in figure 125.

   *a. Dust Seal and Spring Installation.* Install dust seal spring and dust seal by stretching these parts over housing outer end as illustrated in figure 126.

   *Note.* Be sure seal is installed so that bevel at inner surface of seal will fit contour of housing outer end.

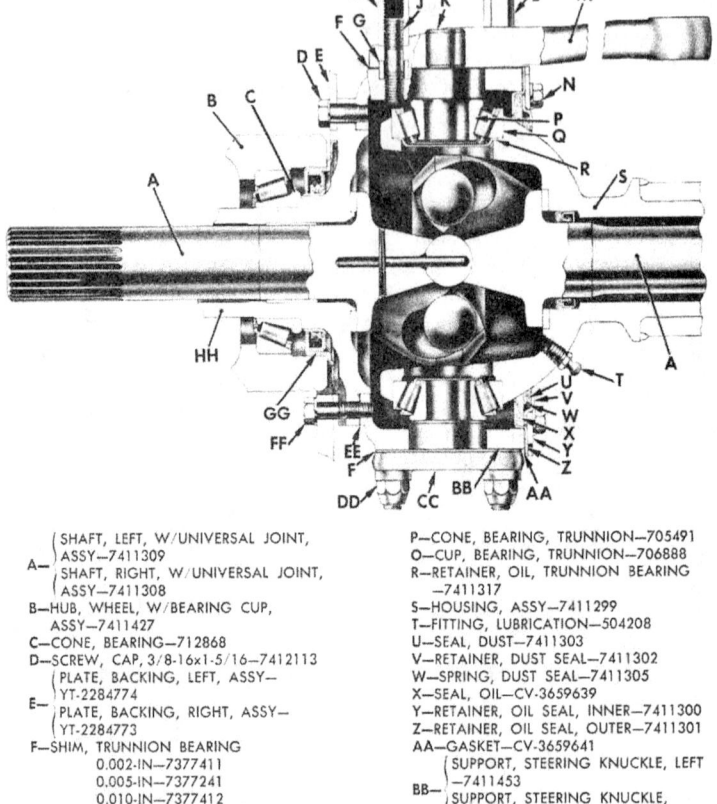

```
                  {SHAFT, LEFT, W/UNIVERSAL JOINT,        P—CONE, BEARING, TRUNNION—705491
                  { ASSY—7411309                          O—CUP, BEARING, TRUNNION—706888
             A—  {                                        R—RETAINER, OIL, TRUNNION BEARING
                  { SHAFT, RIGHT, W/UNIVERSAL JOINT,         —7411317
                  { ASSY—7411308                          S—HOUSING, ASSY—7411299
             B—HUB, WHEEL, W/BEARING CUP,                 T—FITTING, LUBRICATION—504208
                ASSY—7411427                              U—SEAL, DUST—7411303
             C—CONE, BEARING—712868                       V—RETAINER, DUST SEAL—7411302
             D—SCREW, CAP, 3/8-16x1-5/16—7412113          W—SPRING, DUST SEAL—7411305
                  { PLATE, BACKING, LEFT, ASSY—           X—SEAL, OIL—CV-3659639
             E—  { YT-2284774                             Y—RETAINER, OIL SEAL, INNER—7411300
                  { PLATE, BACKING, RIGHT, ASSY—          Z—RETAINER, OIL SEAL, OUTER—7411301
                  { YT-2284773                            AA—GASKET—CV-3659641
             F—SHIM, TRUNNION BEARING                         { SUPPORT, STEERING KNUCKLE, LEFT
                    0.002-IN—7377411                      BB—{   —7411453
                    0.005-IN—7377241                          { SUPPORT, STEERING KNUCKLE,
                    0.010-IN—7377412                          { RIGHT—7411452
                    0.020-IN—CV-3678167                   CC—TRUNNION, STEERING KNUCKLE—
             G—RING, DOWEL—7411319                           7411454
             H—SPACER, 1/2-20 x                           DD—NUT, 1/2-20—442801
                  1-7/16—7411315                          EE—GASKET—7411313
             J—STUD, 1/2-13 - 20x1-15/16—7411451          FF—SCREW, CAP, 3/8-16x1-11/16—
             K—TRUNNION, STEERING KNUCKLE—                   7412112
                7411455                                   GG—SEAL, OIL, ASSY—7411429
             L—SPACER, 1/2-20x1-1/4—7538373               HH—KNUCKLE, STEERING, ASSY—7411314
             M—ARM, STEERING—7411311
             N—SCREW, CAP, 5/16-18x5/8—180075
                                                                           RA PD 148835
```

*Figure 125. Sectional view of steering knuckle, support, and universal joint assembly.*

b. *Trunnion Bearing Lubrication.*—Make sure that trunnion bearing cones ((C), fig. 121) are clean. Pack bearings with automotive and artillery grease (GAA) by hand pack method or with bearing lubricator until grease is forced between all rollers. Coat trunnion bearing cups with same lubricant.

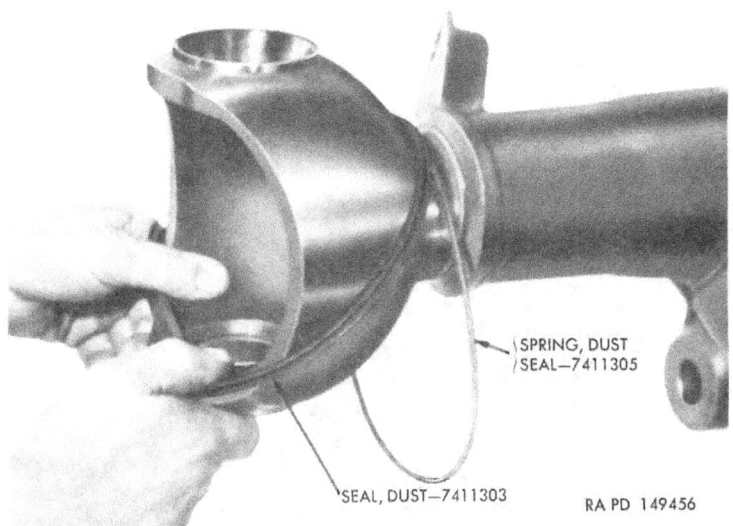

*Figure 126. Housing outer dust seal and spring installation.*

*c. Steering Knuckle Support Installation.*—Install trunnion bearing cone ((C), fig. 121) in trunnion bearing cup ((B), fig. 121) at top of housing, being sure that cone is in its original location (unless new bearings are installed) as identified by tag at time of disassembly.

*Note.* Be sure that steering knuckle support with dowel rings (fig. 127) is installed at left (steering arm) end of housing.

Position steering knuckle support ((MM), fig. 109) over upper trunnion bearing cone, being sure that tie rod integral arm is toward rear. Position trunnion bearing cone ((T), fig. 121), steering knuckle trunnion ((HH), fig. 109), and shims ((M), fig. 109), removed at time of disassembly, at bottom of steering knuckle support and secure temporarily with four ½–20 nuts (fig. 128). At right side install steering knuckle trunnion ((L), fig. 121) or steering arm ((Q), fig. 109) at left side, using shims (fig. 128) removed at time of disassembly. Secure right upper trunnion with one ½–20 nut ((M), fig. 121) at outer front stud and three 1¼-inch long spacers ((K), fig. 121) at other three studs. Secure steering arm ((Q), fig. 109) with ½–20 nut ((N), fig. 109) at front outer stud, one 1⁷⁄₁₆-inch long spacer ((P), fig. 109) at rear outer stud, and two 1¼-inch long spacers at two inner studs.

*d. Trunnion Bearing Adjustment.*—Tighten spacers or nuts ((P) and (N), fig. 109) attaching steering knuckle trunnions or steering arm to steering knuckle support to torque of 70 to 80 pound-feet. Use torque wrench as illustrated in figure 129 and measure pound-

*Figure 127. Location of dowel rings in left steering knuckle support.*

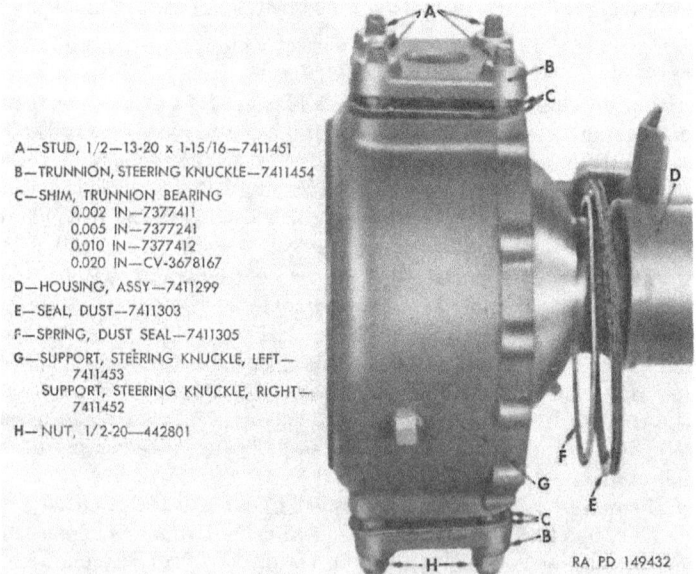

A—STUD, 1/2—13-20 x 1-15/16—7411451
B—TRUNNION, STEERING KNUCKLE—7411454
C—SHIM, TRUNNION BEARING
   0.002 IN—7377411
   0.005 IN—7377241
   0.010 IN—7377412
   0.020 IN—CV-3678167
D—HOUSING, ASSY—7411299
E—SEAL, DUST—7411303
F—SPRING, DUST SEAL—7411305
G—SUPPORT, STEERING KNUCKLE, LEFT—
   7411453
   SUPPORT, STEERING KNUCKLE, RIGHT—
   7411452
H—NUT, 1/2-20—442801

*Figure 128. Steering knuckle support trunnions and bearing adjusting shims.*

feet torque required to turn steering knuckle support. Torque wrench reading should be 8 to 12 pound-feet with steering knuckle support in motion. Add or remove trunnion bearing shims ((M), fig. 109) as necessary to obtain correct torque reading.

*Caution:* Be sure that the same total shim thickness is used at upper and lower trunnions. Shim pack should be measured with a micrometer. Adjust right and left trunnion bearings separately.

*Figure 129. Checking steering knuckle trunnion support bearing adjustment.*

   *e. Housing Outer Seal Installation.*
     (1) *Gasket installation.*—Apply light coating of plastic type gasket cement to both sides of gasket ((R), fig. 109); then position gasket on steering knuckle support.
     (2) *Oil seal retainer installation.*—Outer oil seal retainer ((S), fig. 109) is positioned over housing by spreading at split. Position retainer against steering knuckle support with split in retainer at top.
     (3) *Dust seal retainer installation.*—Inner dust seal retainer ((W), fig. 109) is positioned over housing by spreading at split.
     (4) *Lubricate parts.*—Apply film of automotive and artillery grease (GAA) to spherical surface of housing, oil seal, and dust seal to provide initial lubrication when unit is first placed in service, also as a preservative after installation.
     (5) *Oil seal installation.*—Position oil seal ((T), fig. 109) into outer oil seal retainer ((S), fig. 109) with split in seal at right angles to split in retainer.

(6) *Dust seal and spring installation.*—Fit dust seal spring ((V), fig. 109) into groove in edge of dust seal ((U), fig. 109); then position seal and spring assembly against oil seal (felt), as dust seal retainer ((W), fig. 109) is positioned to retain seals. While holding dust seal retainer against seals, install two inner oil seal retainers ((X), fig. 109), with splits on a horizontal line, and secure with twelve $5/16$–18 x $5/8$ cap screws and $5/16$-inch lock washers. Cap screws should be drawn up only enough to hold parts in place until seals and retainers can be properly positioned; then make final tightening of cap screws.

### 187. Differential Carrier Installation

*a. Gasket Installation.*—Apply light coating of plastic type gasket cement to both sides of gasket ((J), fig. 121); then position gasket over studs and against housing.

*b. Carrier Assembly Installation.*—Position differential and carrier assembly on housing, being sure that drive pinion is above the center line of axle housing (fig. 105). Install ten $7/16$–20 nuts on studs and tighten to torque of 45 to 55 pound-feet.

### 188. Housing Cover Installation

*a. Gasket Installation.*—Apply light coating of plastic type gasket cement to both sides of gasket ((J), fig. 121); then position gasket over studs and against housing.

*b. Cover Installation.*

Note. Covers used on front and rear axles are the same, however, they are installed differently; therefore, follow instructions carefully.

When installing cover on front axle, make certain that oil filler hole marked "FRONT OIL LEVEL" is upright (not upside down). Position cover over studs and install ten $7/16$–20 nuts on studs and tighten to torque of 45 to 55 pound-feet.

### 189. Tie Rod Installation

*a. Tie Rod Adjustment.*—Measure distance between center line of tie rod end studs (par. 184c) to be sure dimension is $62{17/32}$ inches.

*b. Tie Rod Installation.*—Be sure seal is placed over tapered stud, then insert tie rod end studs into tapered holes of arms on steering knuckle support. Install nut on each tie rod end stud and tighten to torque of 175 to 200 pound-feet (fig. 130).

*c. Check Installation.*—Swing left steering knuckle support against steering knuckle stop plug. Measure clearance between tie rod and rib on differential carrier. Clearance must be at least one-eighth of an inch.

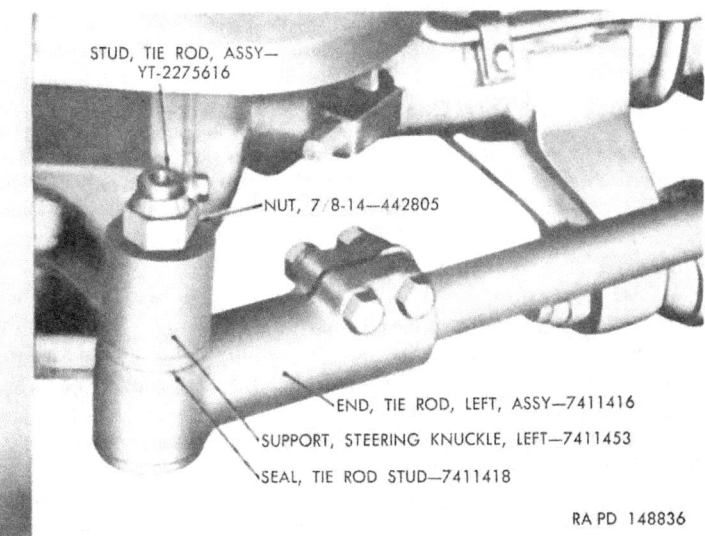

*Figure 130. Steering tie rod—Rod installed.*

## 190. Axle Shaft and Universal Joint Installation

*a. Lubricate.*—Using universal gear lubricant (GO), apply to oil seal and thrust washer in end of axle housing, also apply generously to universal joint balls.

*b. Assembly Installation.*

*Note.* Right and left assemblies have different length inner shafts; long shaft is used at left side while shorter shaft is used at right side.

Insert inner shaft of assembly carefully through oil seal at outer end of housing so as not to damage seal, guiding splined end of inner shaft into differential side gear splines until thrust face of inner shaft is against thrust washer at outer end of housing (fig. 111).

## 191. Steering Knuckle Installation

*a. Lubricate.*—Using universal gear lubricant (GO), apply generously to thrust washer at flange end, also to bushing type bearing inside knuckle.

*b. Gasket Installation.*—Apply light coating of plastic type gasket cement to both sides of gasket ((K), fig. 109); then position gasket against outer surface of steering knuckle support, being careful to aline gasket with cap screw holes.

*c. Knuckle Installation.*—Position steering knuckle ((F), fig. 109) over outer shaft ((EE), fig. 109); then rotate as necessary to position lip in brake oil shield ((E), fig. 109) downward to drain any lubricant

207

leakage. When oil shield lip is down and cap screw holes are in alinement, push steering knuckle against steering knuckle support.

## 192. Final Assembly Operations

*a. Brake Assembly Installation.*—Install brake backing plate and shoe assembly, also brake hose and shield as directed in paragraph 234.

*b. Wheel Hub, Drum, and Bearings Installation.*—Install wheel hub, drum, and bearings as directed in paragraph 256e, also adjust bearings as directed in paragraph 256f.

*c. Leakage Test.*—Install a suitable air pressure gage in axle vent line hole in torque rod bracket on top of axle housing. Using a suitable adapter, attach air line to one of filler plug holes in axle housing cover. Fill axle with air to a pressure of 15 p.s.i. Air must not escape faster than 5 pounds in 45 seconds.

**Caution:** Do not apply more than 15 pounds air pressure. Remove air line and gage.

*d. Lubrication.*—Lubricate universal joint and steering knuckle, also axle differential, with type and quantity of lubricant specified in TM 9–819A or on official lubrication order.

# CHAPTER 8

# REAR AXLE

## Section I. DESCRIPTION AND DATA

### 193. Description and Operation

*a. General.*—Both rear axles (fig. 131) are hypoid, single-reduction, full-floating type equipped with banjo type housing. Differential and carrier is mounted as an assembly in housing. Forward rear axle and rear rear axle are mounted in tandem with upper and lower torque rods connecting each axle to vehicle frame. The two rear axles are similar in design and construction, the major difference between the two is that the opening for the differential carrier in the forward rear axle is off-center. Forward rear axle housing also supports the propeller shaft pillow block assembly which transmits drive to rear rear axle.

*b. Operation.*—Power is transmitted from the transfer by two propeller shafts, one to each of the rear axles. Forward rear axle is driven direct from transfer by a single shaft, while drive to rear rear axle is through two propeller shafts and a pillow block mounted on a bracket attached to forward rear axle housing. Driving force is transmitted from axles to vehicle frame by six torque rods. Three torque rods are attached to each axle and take all the driving and braking load at rear axles.

### 194. Data

```
Manufacturer_____ GM Corporation
Type_____ hypoid, single-reduction
Ratio_____ 6.17 to 1
```

## Section II. DISASSEMBLY OF REAR AXLE INTO SUBASSEMBLIES

### 195. General

*a.* The following procedures are based on the assumption that the axle assembly has been removed from the vehicle in accordance with instructions contained in TM 9-819A. Some of the following opera-

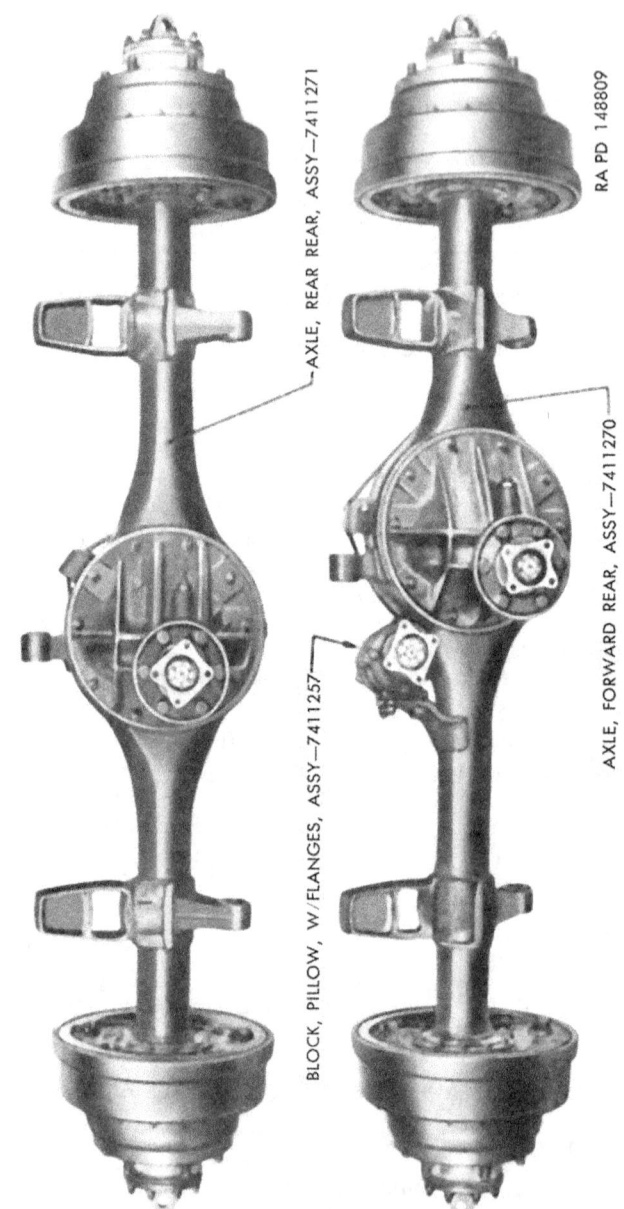

*Figure 131. General view of rear axles.*

tions can be performed with the axle assembly installed on the vehicle, however, for maximum accessibility and efficiency, the axle should be removed from the vehicle and placed on a suitable work stand.

*b.* Before cleaning or disassembling the axle, make a careful visual inspection for evidence of lubricant leakage which might otherwise not be visible after the assembly or parts have been cleaned. Make a note of any such points, so that the cause may be determined either during disassembly or at time of inspection after disassembly. Thoroughly clean the assembly, using steam or other suitable method, to remove all accumulated dirt or other foreign material.

### 196. Disassembly Operations

*a. Drain Lubricant.*—Place suitable receptacle under axle, then remove plug and gasket from axle housing to permit lubricant to drain.

*b. Removal of Wheel Hubs, Drums, and Bearings.*—Remove wheel hubs, brake drums, and bearings as directed in paragraph 256.

*c. Removal of Brake Assembly.*—Remove brake backing plate and shoe assembly as directed in paragraph 229.

*d. Removal of Housing Cover.*—Remove ten $7/16$–20 nuts ((G), fig. 148) attaching cover ((H), fig. 148) to housing. Tap cover lightly with soft hammer to loosen, then remove from $7/16$–20 x 2 studs ((K), fig. 148) in housing. Remove and discard gasket ((J), fig. 148).

*e. Removal of Differential and Carrier Assembly.*—Remove ten $7/16$–20 nuts ((AA), fig. 132) attaching carrier assembly ((Z), fig. 132) to housing. Tap carrier with soft hammer to loosen carrier assembly from housing, then withdraw differential and carrier assembly from housing. Remove and discard gasket ((Y), fig. 132) used between carrier and housing.

*f. Removal of Brake Line Assemblies.*—Remove two cap screws, lock washers, and clips attaching each line to axle housing. Loosen brake line fitting nut attaching each line to elbow in junction block located on top of axle housing, then remove right and left brake line assemblies.

### Section III. REBUILD OF DIFFERENTIAL AND CARRIER ASSEMBLY

### 197. General

The following procedures cover disassembly, cleaning, inspection, and repair of differential and carrier assembly, after unit has been removed from axle housing as directed in paragraph 196.

### 198. Disassembly

*Note.* Key letters noted in parentheses are in figure 132 unless otherwise indicated.

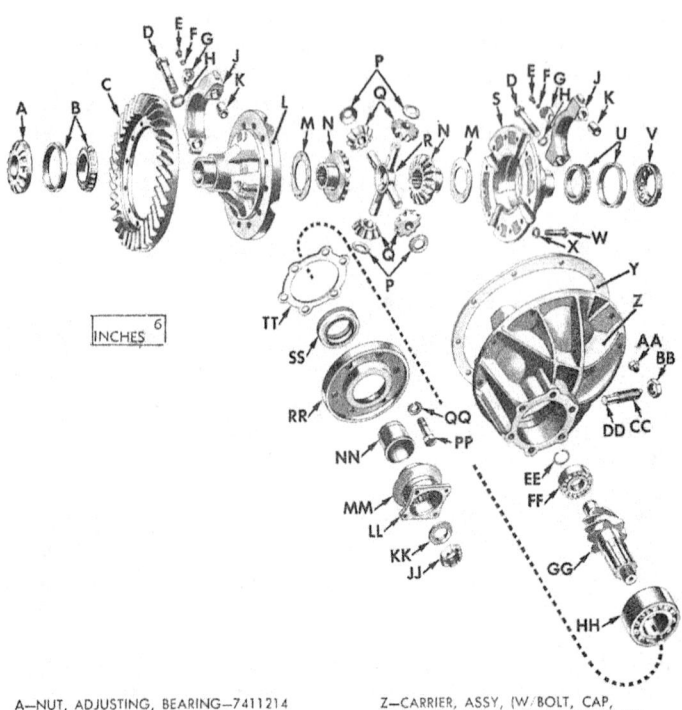

A—NUT, ADJUSTING, BEARING—7411214
B—BEARING, ASSY—707695
C—GEAR, DRIVE—CV-3652522
D—SCREW, CAP, 11/16-11x3-5/8—CV-3652254
E—SCREW, CAP, 5/16-18x5/8—180075
F—WASHER, LOCK, 5/16-IN—120638
G—LOCK, BEARING ADJUSTING NUT—7411212
H—WASHER, LOCK, 11/16-IN—131141
J—CAP, BEARING—CV-3652275
K—DOWEL, CAP—7411225
L—CASE—CV-3661872
M—WASHER, THRUST (SIDE GEAR)—7411222
N—GEAR, SIDE—7411209
P—WASHER, THRUST (SPIDER PINION)—7411221
Q—PINION, SPIDER—7411217
R—SPIDER—7411223
S—COVER, CASE—CV-3661873
T—CAP, BEARING—CV-3652275
U—BEARING, ASSY—707695
V—NUT, ADJUSTING, BEARING—7411214
W—SCREW, CAP, 1/2-20x1-1/8—7411211
X—WASHER, LOCK, 1/2-IN—131101
Y—GASKET—7411207

Z—CARRIER, ASSY, (W/BOLT, CAP, DOWEL, AND WASHER)—CV-605925
AA—NUT, 7/16-20—442800
BB—NUT, 7/8-14—124954
CC—SCREW (THRUST PAD), 7/8-14x3-3/4—6196340
DD—PAD, THRUST—7411216
EE—LOCK, PINION INNER BEARING—7411213
FF—BEARING (DRIVE PINION INNER)—707652
GG—PINION, DRIVE—CV-3652549
HH—BEARING (DRIVE PINION OUTER)—710143
JJ—NUT (DRIVE PINION) 1-1/8-18—7411215
KK—WASHER, PLAIN, 1-5/32 IDx2-1/16 OD—7411224
LL—FLANGE, W/DEFLECTOR, AND SLEEVE, ASSY—7411206
MM—DEFLECTOR, FLANGE—7377182
NN—SLEEVE, FLANGE—7411220
PP—SCREW, CAP, 5/8-11x1-3/4—7411219
QQ—WASHER, LOCK, 5/8-IN—131140
RR—RETAINER, PINION OIL SEAL—7413230
SS—SEAL, OIL, PINION—7413229
TT—GASKET—7411208

RA PD 149548

*Figure 132. Components of differential and carrier assembly.*

*a. Mount Carrier Assembly.* Mount assembly in a suitable work stand to facilitate disassembly operations.

*b. Removal of Differential Assembly.*

(1) *Removal of thrust pad and screw.*—Loosen $7/8$–14 nut (BB) which locks $7/8$–14 x $3\frac{3}{4}$ screw (CC) in differential carrier; then remove $7/8$–14 x $3\frac{3}{4}$ screw (CC) and thrust pad (DD) from carrier.

(2) *Removal of bearing adjusting nut.*—Remove $5/16$–18 x $5/8$ cap screws (E), $5/16$-inch lock washers (F), and bearing adjusting nut locks (G). Loosen $11/16$–11 x $3\frac{5}{8}$ cap screws (D) several turns to free adjusting nuts. Use spanner wrench 41–W–3247–500 in manner illustrated in figure 144 to remove adjusting nuts.

(3) *Removal of bearing caps.*

*Note.* Use prick punch to mark each bearing cap (J) and carrier assembly (Z) so that caps can be installed in their original location.

Remove four $11/16$–11 x $3\frac{5}{8}$ cap screws (D) and $11/16$-inch lock washers (H) attaching bearing caps (J) to carrier. Tap bearing caps with soft hammer sufficiently to loosen and remove.

(4) *Removal of differential assembly.*—Carefully withdraw differential assembly from carrier and lay aside for further disassembly. Remove differential side bearing cups and tag for location identification at time of assembly.

*c. Disassembly of Differential Assembly.*

(1) *Removal of drive gear.*—Determine that a match mark (fig. 133) is clearly visible on case and case cover. If mark is not clearly legible, make a new mark. Stand assembly on end to prevent damage to differential side bearings. Loosen and remove twelve $1/2$–20 x $1\frac{1}{8}$ cap screws (W) and $1/2$-inch lock washers (X) attaching drive gear (C) to case (L); then remove drive gear.

(2) *Separate Differential Case and Cover.*—Tap lightly with soft hammer to separate case cover (S) from case (L); then remove side gears (N), thrust washers (M) and (P), spider pinions (Q), and spider (R).

(3) *Removal of differential side bearings.*

*Note.* If inspection (par. 200d) indicates that bearings are satisfactory for further service, they need not be removed.

With case or cover resting on bench, install adapter 41–A–18–293 and remover 41–R–2367–950 in manner illustrated in figure 134, being sure that remover is adjusted to inner race to prevent damage to bearing cage and rollers.

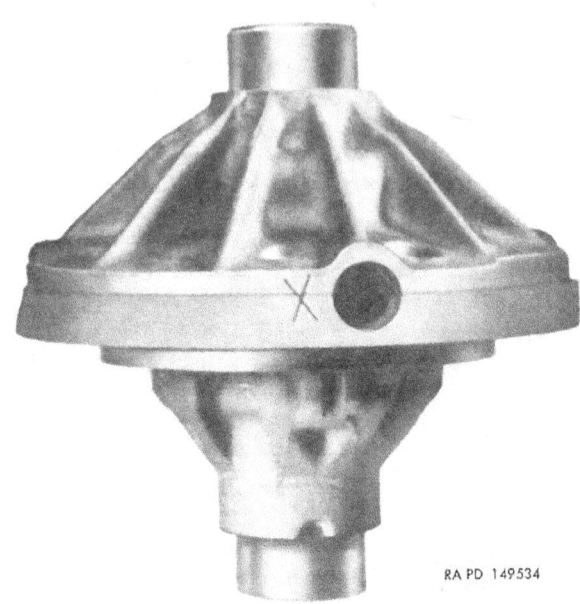

*Figure 133. Differential case and cover match marks.*

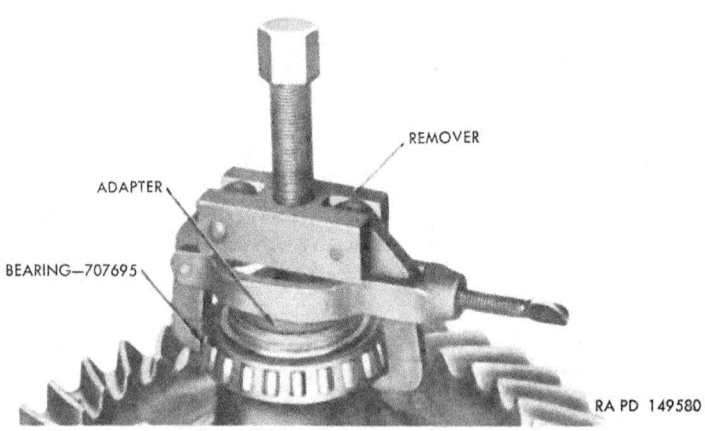

*Figure 134. Use of differential side bearing remover 41-R-2367-950 and adapter 41-A-18-293.*

*d. Removal of Drive Pinion Assembly.*
   (1) *Removal of propeller shaft flange.*—Use tool 41-T-3215-910 to prevent flange with deflector and sleeve assembly (LL) from turning while nut (JJ) is removed, also remove plain washer (KK). Install remover 41-R-2367-950 in manner illustrated in figure 135; then thread screw into puller until flange is removed.

*Figure 135. Removal of propeller shaft flange with remover 41-R-2367-950.*

   (2) *Removal of pinion oil seal retainer.*—Remove six ⅝-11 x 1¾ cap screws (PP) and ⅝-inch lock washers (QQ) attaching pinion oil seal retainer (RR) to carrier; then tap retainer with soft hammer to remove. Remove and discard gasket (TT).
   (3) *Removal of pinion assembly from carrier.*—Place brass rod against inner end of drive pinion (GG), then drive drive pinion and bearings assembly from carrier assembly (Z).

215

*e. Disassembly of Drive Pinion.*

*Note.* Following procedures need not be accomplished until inspection of bearings (par. 200c) indicate the need for replacement.

(1) *Removal of pinion inner bearing.*—Use suitable snap ring pliers in manner illustrated in figure 136 to remove pinion inner bearing lock (EE). Install attachment 41–A–345–328 (part of puller 41–P–2905–60) onto drive pinion; then place in arbor press to press pinion out of bearing (fig. 137).

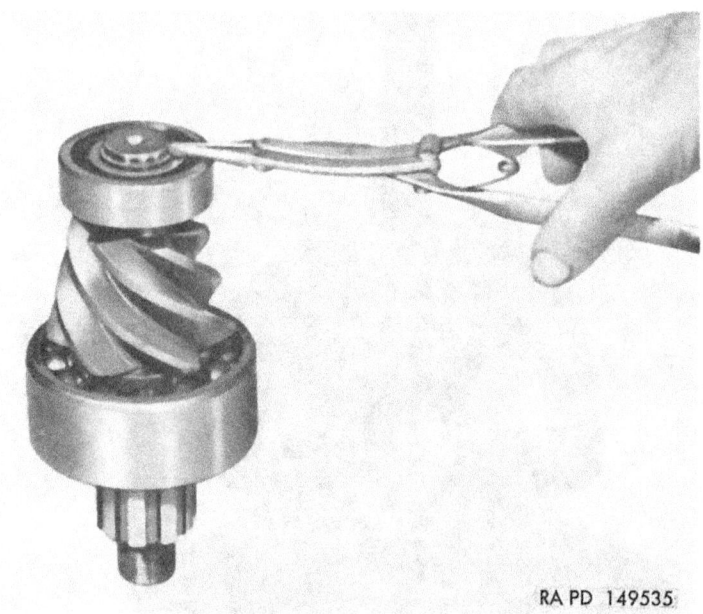

*Figure 136. Removal or installation of pinion inner bearing lock with snap ring pliers.*

(2) *Removal of pinion outer bearing.*—Install attachment 41–A–345–328 (part of puller 41–P–2905–60) onto drive pinion, then place in arbor press to press drive pinion (GG) out of bearing (fig. 138).

## 199. Cleaning

*a. General.*—Immerse all parts in dry-cleaning solvent or volatile mineral spirits to loosen and remove all grease or other deposits. Remove each part and examine carefully for cleanliness; use bristle brush if necessary to remove deposits. Remove sealing compound and pieces of gaskets from gasket surfaces.

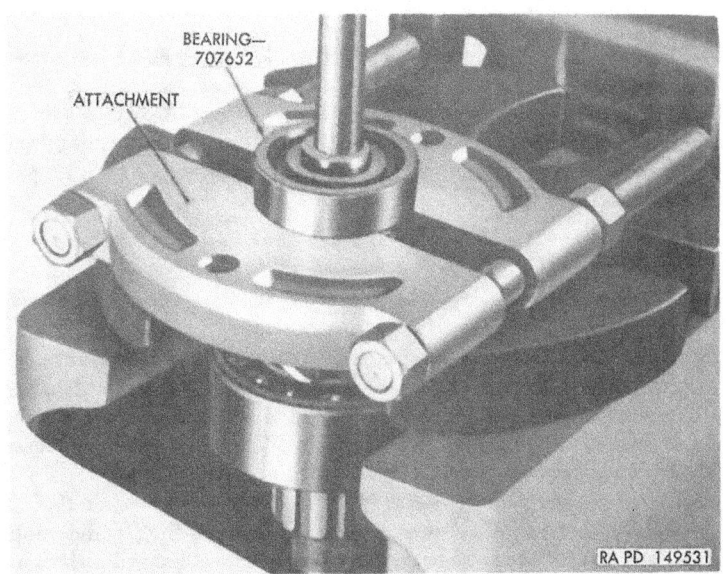

*Figure 137. Removal of drive pinion inner bearing with attachment 41-A-345-328 (part of puller set 41-P-2905-60).*

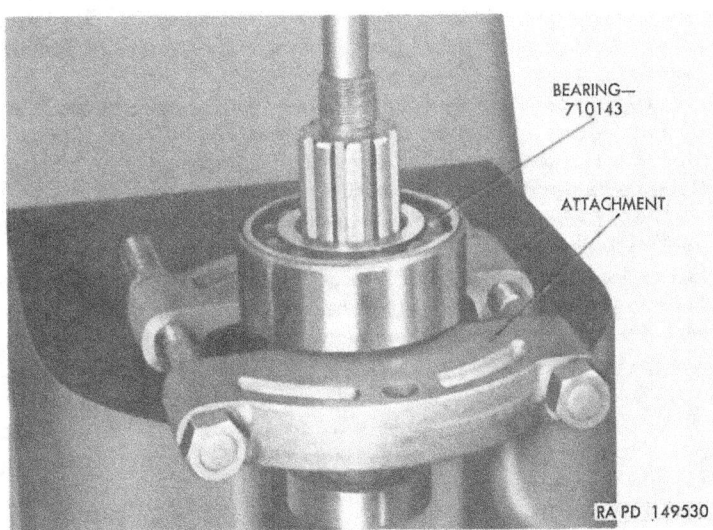

*Figure 138. Removal of drive pinion outer bearing with attachment 41-A-345-328 (part of puller set 41-P-2905-60).*

*b. Bearings.*—Particular attention should be given to differential side bearings, also drive pinion bearings. Slush bearings in cleaning fluid and use bristle brush to remove all traces of dirt; repeat immersion and brushing until bearings are clean. Dry bearings with compressed air, directing air through bearing in such a manner so as not to spin bearings.

### 200. Inspection

*Note.* Key letters noted in parentheses are in figure 132 unless otherwise indicated.

*a. General.*—Before inspection, all parts must have been thoroughly cleaned as previously directed in paragraph 199. Whenever available, the magnaflux method should be applied to all steel parts, except ball or roller bearings. Refer to paragraph 347 for new part dimensions or other necessary inspection data.

*b. Differential Carrier.*—Carefully inspect carrier assembly (Z) for cracks, breaks, or distortion, or other damage that would render the carrier unfit for further service. Inspect for damaged or stripped threads, and repair or replace carrier, whichever is necessary. Inspect gasket surfaces for smoothness and clean up with fine file or stone if necessary.

*c. Drive Pinion Bearings.*—Rotate each bearing slowly and carefully by hand and note if any roughness is evident, also visually inspect balls and rollers for chipped, cracked, or pitted condition. Replace if defective. Apply universal gear lubricant (GO) to each bearing; then wrap in clean cloth or paper.

*d. Differential Side Bearings.*—Inspect bearing assemblies (B and U) for evidence of chipped, cracked, or worn condition of rollers and cup. Replace if defective, otherwise apply universal gear lubricant (GO) and wrap in clean cloth or paper.

*e. Side Gears.*—Inspect side gears (N) for chipped or worn gear teeth. Check diameter of pilot on gear at point of contact with case (L) and case cover (S) to determine if worn. Install gears on axle shaft to determine if worn at splines. Check thrust washer face of gear for scratched, scuffed, or worn condition. Replace worn or damaged gears.

*f. Spider Pinion.*—Inspect spider pinion (Q) for chipped or worn teeth. Install pinion on spider (R) to determine if worn. Check thrust washer face of gear for scratched, scuffed, or worn condition. Replace worn or damaged gears.

*g. Spider.*—Inspect spider (R) for cracks or wear by installing spider pinion (Q) on each spider arm. Replace if worn or damaged.

*h. Differential Case and Case Cover.*—Inspect case (L) and case cover (S) for cracked or broken condition. Inspect thrust washer surfaces for worn or scored condition. Replace both parts if either is damaged.

*i. Drive Gear and Drive Pinion.*—Inspect drive gear (C) and drive pinion (GG) for chipped, broken, or scuffed teeth. Inspect cap screw threads in drive gear. Inspect splines and threads on drive pinion. Replace both parts if either is damaged.

*j. Pinion Oil Seal.*—Inspect pinion oil seal for wear or cuts on lip of seal. If wear or damage is evident, replace seal as directed in paragraph 201*a*.

*k. Propeller Shaft Flange.*—Inspect flange with deflector and sleeve assembly (LL) for damage or wear at splines. Replace if worn or damaged. Inspect flange sleeve (NN) for scratched, worn, or grooved condition and replace as directed in paragraph 201*b*. Inspect for damaged flange deflector (MM) and replace as directed in paragraph 201*c*.

*l. Thrust Pad.*—Inspect thrust pad (DD) for wear and replace if necessary.

## 201. REPAIR

*Note.* Key letters noted in parentheses are in figure 132 unless otherwise indicated.

a. *Pinion Oil Seal*
   (1) *Removal.*—Use suitable drift or rod and hammer to drive pinion oil seal (SS) from retainer. Clean seal contact surface in retainer to remove any accumulated sealing compound or other matter.
   (2) *Installation.*—Place retainer on flat surface with outer side down. Coat outer surface of seal with a light film of plastic type gasket cement. Position seal in retainer with lip of seal up. Use pinion shaft oil seal replacer 41-R-2393-175 in manner illustrated in figure 139 to drive and properly locate seal in retainer.

b. *Propeller Shaft Flange Sleeve.*
   (1) *Removal.*—Use ball peen hammer and tap entire circumference of outer surface of flange sleeve (NN). Peening of this surface will cause metal in sleeve to stretch until sleeve can be removed from flange.
   (2) *Installation.*—Position flange sleeve (NN) over flange; then use oil seal sleeve replacer 41-R-2395-515 in manner illustrated in figure 140 to install and properly locate flange sleeve.

c. *Deflector Replacement.*
   (1) *Removal.*—Press or drive flange deflector (MM) from flange.
   (2) *Installation.*—Press flange deflector (MM) onto flange to dimension given in figure 141. Stake deflector into groove in

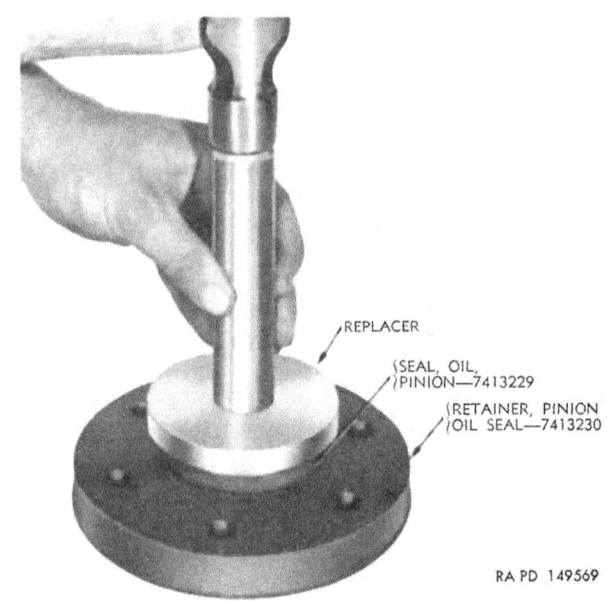

*Figure 139. Drive pinion oil seal installation using replacer 41-R-2393-175.*

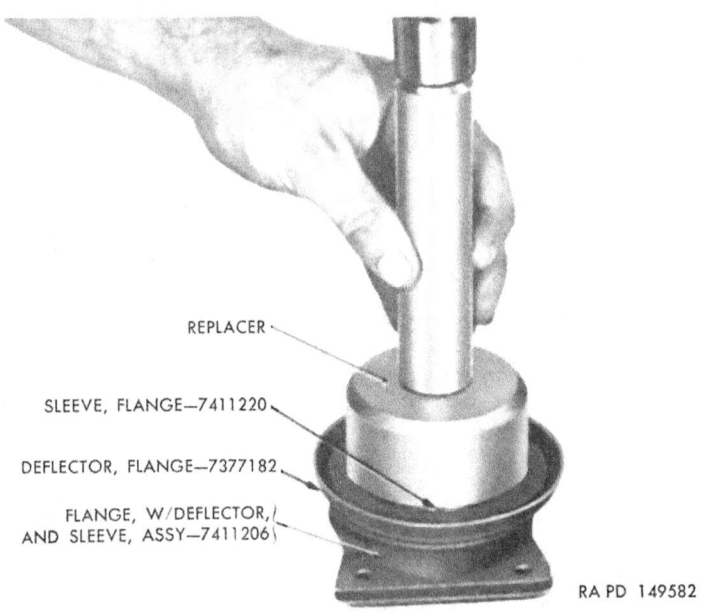

*Figure 140. Flange sleeve installation with replacer 41-R-2395-515.*

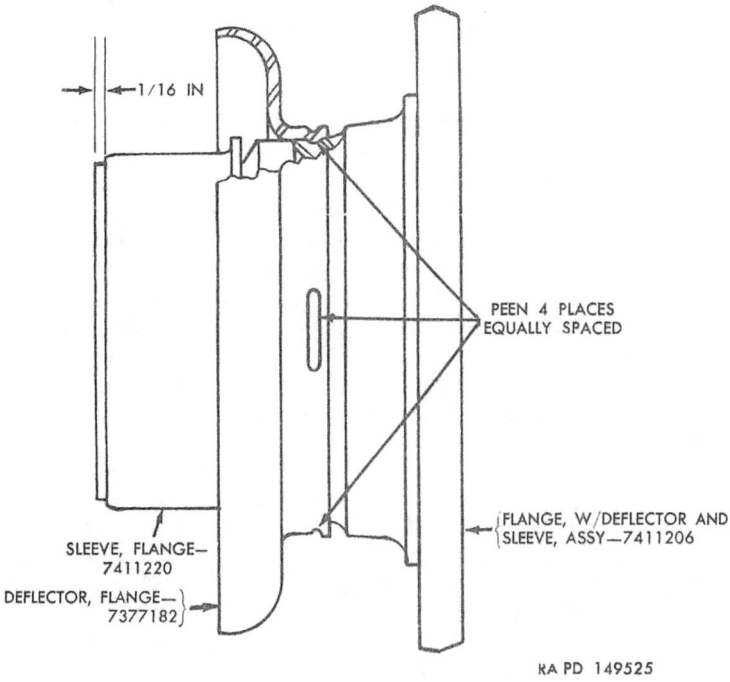

*Figure 141. Flange sleeve and deflector installation.*

flange at four equally spaced points. Each stake should be approximately one-half of an inch in length (fig. 141).

## 202. Assembly

*Note.* Key letters noted in parentheses are in figure 132 unless otherwise indicated.

*a. General.*—During assembly operations each part must be thoroughly lubricated with universal gear lubricant (GO).

*b. Mount Carrier.*—Mount carrier assembly (Z) in suitable work stand to facilitate assembly operations.

*c. Assembly of Drive Pinion.*—If inspection indicated that drive pinion (GG) is to be replaced, it will also be necessary to install a new drive gear (C), as these parts are serviced in matched sets.

    (1) *Installation of pinion inner bearing.*—Position drive pinion (GG) in arbor press and locate bearing (FF) on pinion; then press bearing onto pinion until inner race is seated against shoulder on pinion.

        *Note.* Do not apply pressure to bearing outer race as this would damage bearing rollers.

Use snap ring pliers (fig. 136) to spread pinion inner bearing lock (EE); then install lock in groove, being sure that lock is fully seated.

(2) *Installation of pinion outer bearing.*—Position drive pinion (GG) in arbor press and locate bearing (HH) on pinion.

*Note.* Inner race extends beyond outer race on one side only and this side must be toward pinion teeth.

Press bearing onto pinion until seated against heel of teeth, being careful that pressure is applied at inner race only as bearing may be damaged by pressing on outer race.

d. *Installation of Drive Pinion Assembly.*—Be sure that bearing bore in carrier assembly (Z) is thoroughly cleaned before installing pinion assembly. Insert pinion assembly into carrier, being careful to aline bearing with carrier bore. Use suitable driver or soft hammer at outer race to drive assembly into carrier until bearing is fully seated against shoulder in carrier.

(1) *Installation of oil seal retainer.*—Apply light film of plastic type gasket cement to both sides of gasket (TT); then position gasket to carrier. Install pinion oil seal retainer (RR); then secure with six 5/8–11 x 1 3/4 cap screws (PP) and 5/8-inch lock washers (QQ). Tighten cap screws to a torque of 160 to 180 pound-feet.

(2) *Installation of propeller shaft flange.*—Be sure that oil seal and flange sleeve are lubricated to facilitate assembly. Install flange with deflector and sleeve assembly (LL) over pinion splines; 15/32 ID x 2 1/16 OD plain washer (KK) and 1 1/8–18 drive pinion nut (JJ). Use flange holding tool 41–T–3215–910 to prevent flange turning while tightening nut to a torque of 160 to 280 pound-feet. Aline slot in nut with hole in pinion; then install and bend cotter pin.

e. *Assembly of Differential.*

(1) *Installation of spider pinion and side gears.*—Clean inside of case and case cover thoroughly and lubricate all parts. Install new thrust washer (M) in case cover with side having oil grooves toward case cover; then install side gear (N) in case cover. Install spider pinion (Q) and new thrust washer (P) over each arm of spider (R); then install spider and pinion assembly over side gear previously installed. Install remaining side gear (N) and new thrust washer (M) over spider. Position case (L) over previously assembled parts, being sure that match marks are in alinement (fig. 133).

(2) *Installation of drive gear.*—Cut heads off two old 1/2–20 x 1 1/8 cap screws; then taper the shank of screws slightly and cut slot in end of screw with hack saw. Install two screws in drive gear to serve as guide pins (fig. 142) and facilitate

alinement of drive gear with case. Position drive gear (C) over case (L), using guide pins to properly aline gear, case, and case cover. Install ten 1/2–20 x 1 1/8 cap screws (W) and 1/2-inch lock washers (X); then remove two guide pins and install two remaining 1/2–20x1 1/8 cap screws and 1/2-inch lock washers. Tighten 12 cap screws evenly and alternately to a torque of 85 to 95 pound-feet.

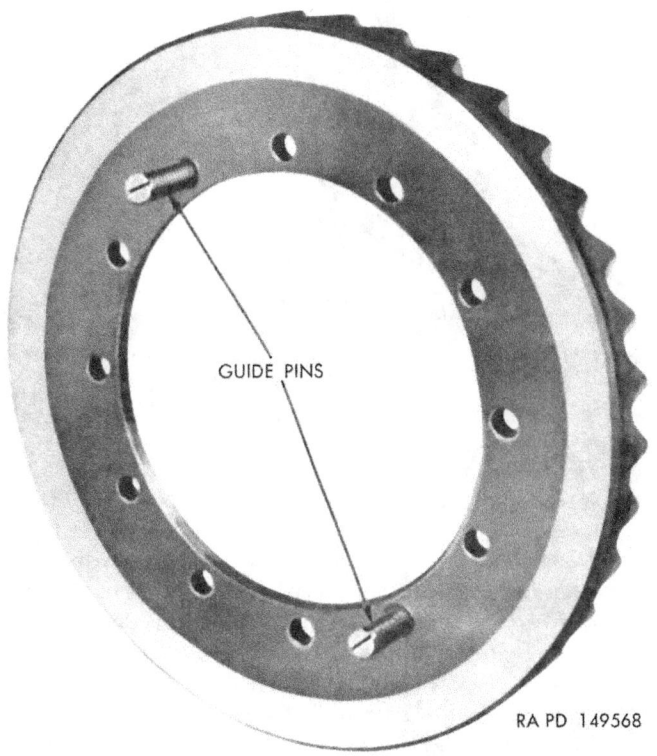

*Figure 142. Use of guide pins in drive gear.*

(3) *Installation of differential side bearings.*—Install cone of bearing assemblies (B) and (U) in case and case cover, using replacer 41–R–2381–220 in manner illustrated in figure 143. Bearing must seat solidly against shoulder of case and case cover.

*f. Installation of Differential Assembly.*

(1) *Position carrier.*—Rotate carrier in holding fixture so that bearing caps are on top. Observe that cap dowels (K) are

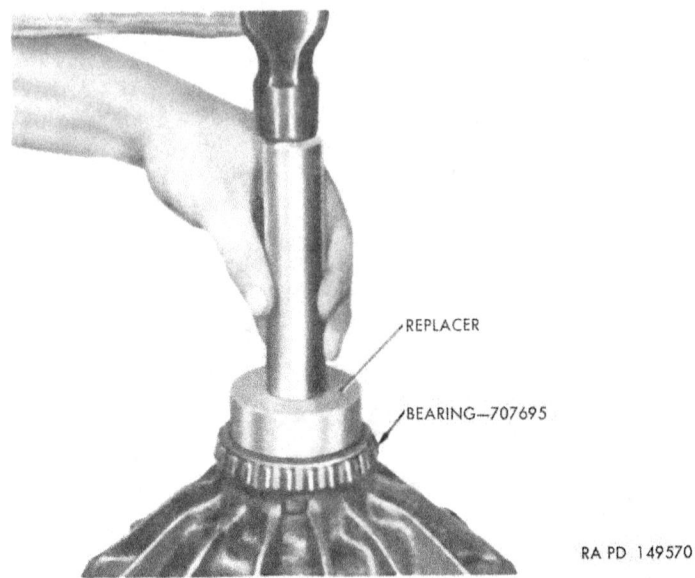

*Figure 143. Installing differential side bearing with replacer 41-R-2381-220.*

    in cap or carrier and cap is properly identified by punch marks made at time of disassembly.

  (2) *Position differential assembly.*—Place cup on each differential side bearing cone; then carefully position differential assembly in carrier.

  (3) *Installation of bearing caps.*—Select proper bearing cap (J) according to identification marks made at time of disassembly and install in proper position, being sure that cap dowels (K) are in place. Install $^{11}/_{16}$–11 x $3\frac{5}{8}$ cap screws (D) and $^{11}/_{16}$-inch lock washers (H). Tighten cap screws only until lock washers just start to flatten, being certain that bearing cups are not cocked.

  (4) *Installation of bearing adjusting nuts.*—Use extreme care when starting bearing adjusting nuts (A) and (V) to be sure that they are not cross-threaded. When certain that nuts are properly started, tighten $^{11}/_{16}$–11 x $3\frac{5}{8}$ cap screws (D) until bearing adjusting nuts (A) and (V) can just be turned. Use adjustable spanner wrench 41-W-3247-500 in manner illustrated in figure 144 and thread each nut into carrier alternately and equally until tight. During tightening, rotate differential to be sure that bearings are seating.

*Figure 144. Adjusting drive gear and pinion with spanner wrench 41-W-3247-500.*

g. *Adjustment of Drive Gear and Drive Pinion Backlash.*

(1) *Removal of backlash.*—Back off bearing adjusting nut (A) on tooth side of drive gear, then tighten bearing adjusting nut (V) until all backlash between drive gear and drive pinion is removed.

*Note.* Make each change in small steps and revolve differential each step.

(2) *Seating bearings.*—Back off bearing adjusting nut (V) on plain side of bevel gear approximately two notches, stopping nut in position to permit bearing adjusting nut lock (G) to engage nut. Tighten bearing adjusting nut (A), on tooth side of drive gear, to seat bearings. Back off same nut until free of bearing; then tighten enough to eliminate all end play in bearings.

(3) *Preload bearings.*—When bearing adjusting nut (A), on tooth side of drive gear, has been tightened to remove all play in bearings ((2) above), tighten an additional one or two more notches to preload bearings, stopping at a position where bearing adjusting nut lock can be installed.

(4) *Checking backlash.*—Install a dial indicator in manner illustrated in figure 145. Oscillate drive gear slightly and note dial indicator reading. Backlash should be 0.005 to 0.008 inch.

(5) *Adjusting backlash.*—If backlash is less than 0.005 inch, loosen bearing adjusting nut (V) on plain side of bevel gear

*Figure 145. Checking drive gear and drive pinion backlash with dial indicator 41-I-100.*

one notch; then tighten opposite bearing adjusting nut (A) one notch to maintain preload adjustment. Revolve differential assembly a number of times to assure bearing alinement, then take another backlash reading at dial indicator. If backlash is more than 0.008 inch, loosen bearing adjusting nut (A) on tooth side of drive gear one notch, then tighten opposite bearing adjusting nut (V) one notch to maintain preload adjustment. Revolve differential assembly a number of times to assure bearing alinement; then take another backlash reading at dial indicator.

(6) *Final bearing cap tightening.*—Tighten two $11/16$–11 x $3\frac{5}{8}$ cap screws (D) at each bearing cap to a torque of 130 to 160 pound-feet.

(7) *Checking drive gear runout.*—Install a dial indicator 41–I–100 in manner illustrated in figure 146. Rotate differential assembly slowly and note total runout. Drive gear runout limit is 0.004 inch.

(8) *Installation of adjusting nut locks.*—Install one bearing adjusting nut lock (G) at each bearing cap, being sure that ears on lock engage bearing adjusting nut. Secure lock with $5/16$–18 x $5/8$ cap screw (E) and $5/16$-inch lock washer (F) and tighten cap screw.

(9) *Installation of thrust pad and screw.*—Install new thrust pad (DD) on thrust pad screw (CC) and thread screw into carrier until thrust pad just contacts drive gear. Rotate differential slowly and note any variations in contact of thrust pad against drive gear. Adjust thrust pad at point of maxi-

*Figure 146. Checking drive gear runout with dial indicator 41-1-100.*

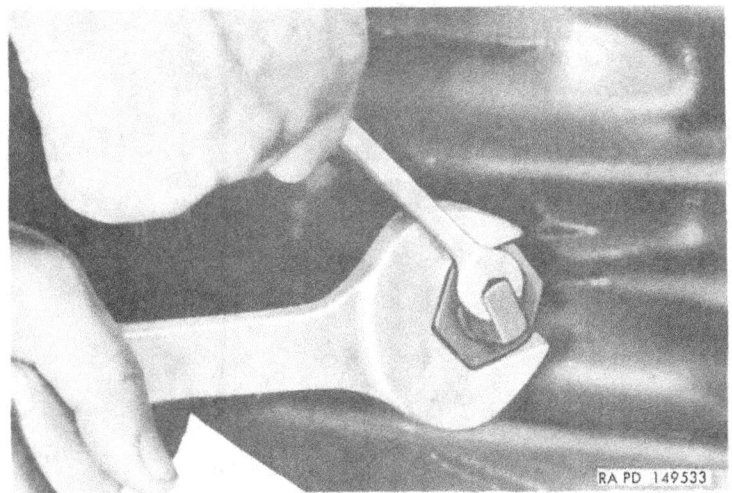

*Figure 147. Thrust pad screw adjustment.*

mum contact. Back-off thrust pad screw one-twelfth of a turn and tighten nut (fig. 147).

### Section IV. CLEANING AND INSPECTION OF REAR AXLE HOUSING AND AXLE SHAFTS

#### 203. Cleaning

*Note.* Key letters noted in parentheses are in figure 148 unless otherwise indicated.

*a. General.*—Immerse all parts in dry-cleaning solvent or volatile mineral spirits to loosen and remove all accumulated grease, dirt, or

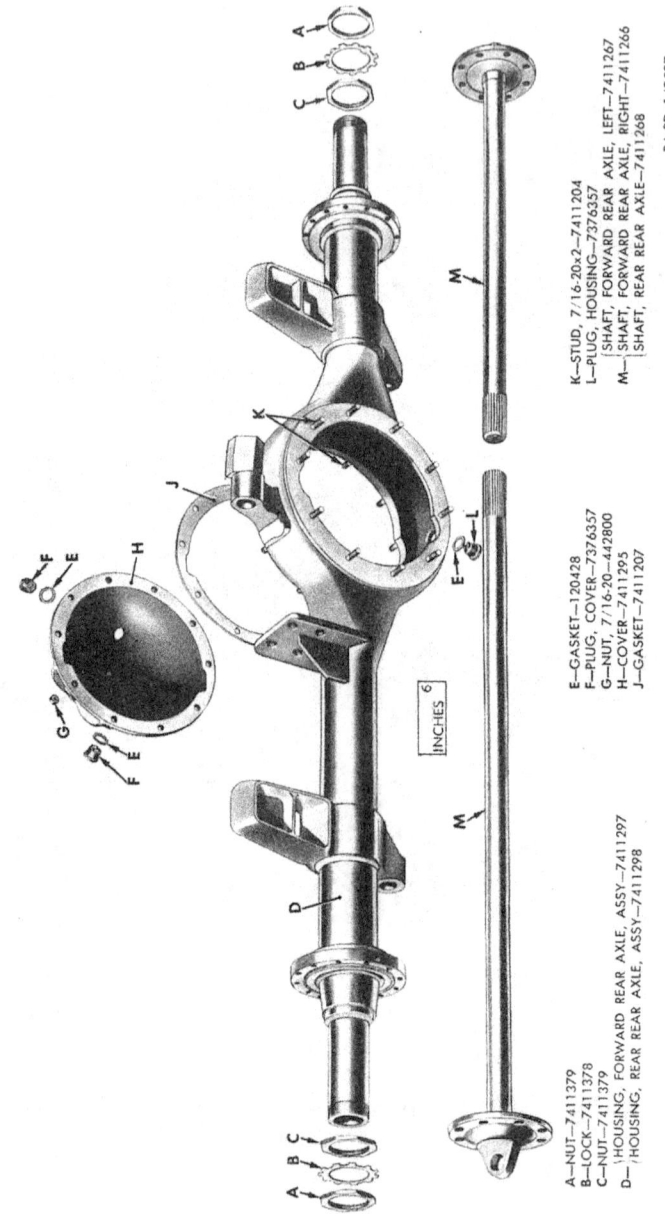

*Figure 148. Rear axle housing and associated parts.*

other deposits. Use of bristle brush and repeated use of cleaning solvent should remove all deposits.

*b. Housing.*—Clean rear axle housing inside and out using long-handled brush or swab to remove idrt and grease from inside of housing. Be sure that all particles of gasket and sealing compound are removed and that bolting surfaces are clean.

*c. Housing Cover.*—Clean inside and outside of cover (H) to remove all dirt and grease. Be sure that all particles of gasket and sealing compound are removed and that bolting surface is clean.

*d. Magnetic Plugs.*—Clean housing plug (L) and cover plugs (F) to remove all particles of metal adhering to plug magnet.

*e. Brake Line Assemblies.*—Clean inside of brake lines of any deposits, using compressed air to remove all obstructions.

*f. Axle Shafts.*—Clean axle shafts of all accumulated grease and dirt. Be sure splines are cleaned thoroughly. Remove all gasket particles from inside of flange.

## 204. Inspection

*Note.* Key letters noted in parentheses are in figure 148 unless otherwise indicated.

*a. General.*—Before inspecting, all parts must have been cleaned as directed in paragraph 203.

*b. Housing.*—Housing assembly should be carefully inspected visually or use locally available checking equipment to determine if bent, twisted, or otherwise damaged. Replace or repair, whichever is the more practical, depending upon the nature of the defect and availability of adequate straightening equipment. Inspect machined gasket surfaces to be sure they are smooth and that gasket will form a tight seal. Inspect adjusting nut threads at each end of housing, and if necessary, clean up with thread restoring tool. Inspect cover and carrier studs (K) to be sure they are tight and threads are not stripped. Replace damaged studs. Inspect all parts welded to housing to be sure they are not damaged, cracked, or bent, also that welding is not broken.

*c. Housing Cover.*—Carefully inspect cover (H) for evidence of distortion or cracks. Gasket surface should be smooth and free of imperfections that would prevent good gasket seal. Inspect cover plug threads for damage or stripping. Install new cover if damaged.

*d. Adjusting Nuts and Locks.*—Inspect nuts (A) and (C) for damaged or stripped threads and replace if necessary. Inspect locks (B) for broken locking tabs and replace if necessary.

*e. Axle Shafts.*—Inspect axle shafts for twisted, worn, or damaged splines. Inspect tapered dowel wedge holes in flange of axle shaft for evidence of excessive wear by fitting new dowel wedge in each hole. Dowel should protrude, otherwise wear is indicated at tapered hole. Shaft flange and splines must be square with each other within 0.003-inch total dial indicator reading.

*f. Brake Lines.*—Inspect brake line tubing for bends, cracks, or other defects. Inspect tubing nuts for stripped threads. Replace line assembly if defective.

## Section V. ASSEMBLY OF REAR AXLE FROM SUBASSEMBLIES

### 205. General

*a.* Assembly procedures are arranged in logical sequence. All subassemblies should be rebuilt, repaired, and adjusted before beginning operations described in text following.

*b.* Cleanliness is important in handling axle components. All parts should be protected from dirt during assembly operations. Tools and equipment must be clean to prevent contaminating parts during assembly procedures.

*c.* Gaskets, seals, and lock washers must be replaced with new parts wherever required.

### 206. Installation of Differential Carrier Assembly

*a. Installation of gasket.*—Apply light coating of plastic type gasket cement to both sides of gasket ((J), fig. 148); then position gasket over studs and against housing.

*b. Installation of carrier assembly.*—Position differential carrier assembly over studs and tightly against housing. Install ten $7/16$–20 nuts ((AA), fig. 132) on studs and tighten to 45 to 55 pound-feet.

### 207. Installation of Housing Cover

*a. Installation of gasket.*—Apply light coating of plastic type gasket cement to both sides of gasket ((J), fig. 148); then position gasket over studs and against housing.

*b. Installation of Cover.*

*Note.* Covers used on front and rear axles are the same; however, they are installed differently; therefore, follow instructions carefully.

When installing cover ((H), fig. 148) on rear axle, make certain that oil filler hole marked "REAR OIL LEVEL" is upright (not upside down). Position cover over studs and install ten $7/16$–20 nuts ((G), fig. 148) on studs and tighten to 45 to 55 pounds.

### 208. Installation of Brake Lines

Position right and left brake lines on housing and attach line fitting to elbow in junction block. Secure each line to housing with two clips, $1/4$–28 x $1/2$ cap screws, and $1/4$-inch lock washers. Tighten clip cap screws and line fittings.

## 209. Final Assembly Operations

*a. Brake Assembly Installation.*—Install brake backing plate and shoe assembly as directed in paragraph 234.

*b. Wheel Hub, Drum, and Bearings Installation.*—Install wheel, hub, drum, and bearings as directed in paragraph 256*e*, also adjust bearings as directed in paragraph 256*f*.

*c. Lubrication.*—Lubricate axle differential with type and quantity of lubricant specified in TM 9-819A or on official lubrication order.

# CHAPTER 9

# PROPELLER SHAFTS, UNIVERSAL JOINTS, AND PILLOW BLOCK

## Section I. DESCRIPTION

### 210. Arrangement of Propeller Shafts

Propeller shaft and universal joint assemblies interconnect the transmission and transfer, and the transfer and front and rear axles in arrangement illustrated in figure 149. The propeller shafts transmit power from the transmission to the transfer, and from the transfer to front and rear axles. Replacement procedures for all propeller shaft assemblies and pillow block assembly are contained in TM 9-819A.

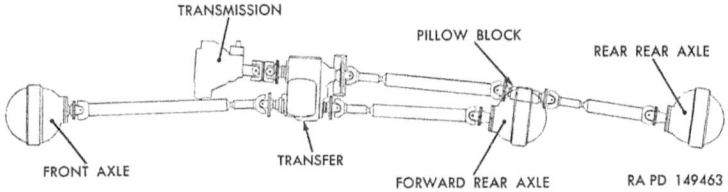

Figure 149. *Arrangement of propeller shafts and universal joints.*

### 211. Axle Propeller Shafts and Universal Joints

The four axle propeller shaft assemblies (fig. 150) are tubular type equipped with fixed yoke universal joint assembly at one end, and slip yoke universal joint at the opposite end. Each joint is equipped with a flange which bolts to a companion flange at the axle differential carriers, transfer output shafts, or pillow block assembly as illustrated in figure 149. Universal joint assemblies at fixed and slip joint ends are identical, and are needle roller bearing type. Each roller bearing assembly is retained in place on journals in yokes by a snap ring. Rebuild procedures for axle propeller shafts are contained in section II of this chapter.

### 212. Transmission-to-Transfer Propeller Shaft
(fig. 152)

Power from transmission to transfer is through two universal joint assemblies which are bolted together, thus eliminating shaft usually used. Slip joint between transmission and transfer is through splined sleeve yoke (Q) which engages transmission output shaft. The universal joints are needle roller bearing type, with bearings retained in place with journal bearing caps (C) secured to yokes with $\frac{5}{16}$–24 x $\frac{1}{2}$ cap screws (A). Rebuild procedures for this propeller shaft assembly are contained in paragraphs 218 through 221.

### 213. Pillow Block Assembly

The pillow block assembly (fig. 155), mounted on top of housing of forward rear axle (fig. 149), connects and supports the two propeller shafts required to transmit power from transfer to rear rear axle. Rebuild procedures for pillow block assembly are contained in paragraphs 222 through 225.

### Section II. REBUILD OF AXLE PROPELLER SHAFT AND UNIVERSAL JOINTS

### 214. Description

The four axle propeller shaft assemblies ((C), (D), (E), and (F), fig. 150) are identical in construction except lengths of tubular shaft and stub shaft assemblies. Each shaft has a slip joint at one end to permit telescopic action of shaft during operation. One half of the

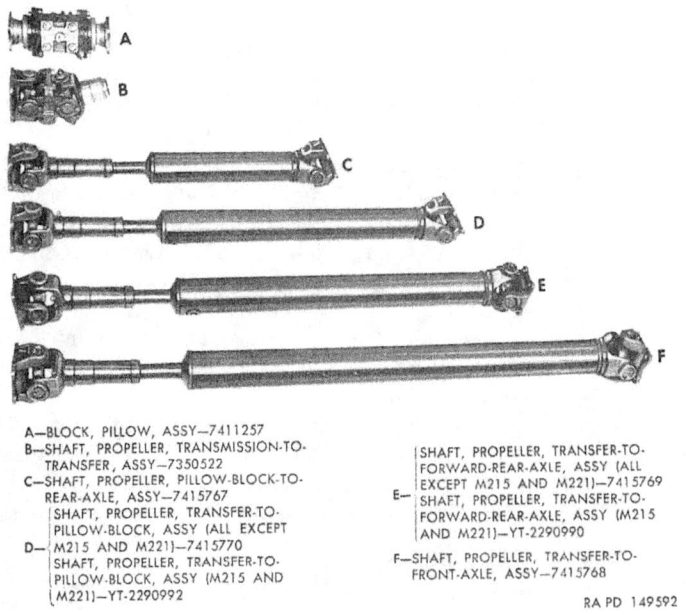

A—BLOCK, PILLOW, ASSY—7411257
B—SHAFT, PROPELLER, TRANSMISSION-TO-TRANSFER, ASSY—7350522
C—SHAFT, PROPELLER, PILLOW-BLOCK-TO-REAR-AXLE, ASSY—7415767
D—{ SHAFT, PROPELLER, TRANSFER-TO-PILLOW-BLOCK, ASSY (ALL EXCEPT M215 AND M221)—7415770
{ SHAFT, PROPELLER, TRANSFER-TO-PILLOW-BLOCK, ASSY (M215 AND M221)—YT-2290992
E—{ SHAFT, PROPELLER, TRANSFER-TO-FORWARD-REAR-AXLE, ASSY (ALL EXCEPT M215 AND M221)—7415769
{ SHAFT, PROPELLER, TRANSFER-TO-FORWARD-REAR-AXLE, ASSY (M215 AND M221)—YT-2290990
F—SHAFT, PROPELLER, TRANSFER-TO-FRONT-AXLE, ASSY—7415768

RA PD 149592

*Figure 150. Propeller shaft assemblies and pillow block assembly removed.*

slip joint is a splined solid stub shaft welded to tubular shaft ((L), fig. 151). The slip joint sleeve yoke ((G), fig. 151) fits over the stub shaft splines, thus forming the slip joint.

## 215. Disassembly

*Note.* Key letters noted in parentheses are in figure 151. Disassembly procedures for all axle propeller shaft assemblies are identical.

*a.* Loosen dust cap (K) which is threaded to slip joint sleeve yoke assembly (G). Slide sleeve yoke from stub shaft.

*b.* Pinch ends of bearing snap rings (B) together; then remove four snap rings from each universal joint.

*c.* Strike the yokes sharply with lead hammer to force the roller bearing assembly (C) from yokes far enough to permit removal.

*Note.* The roller bearings are loose in the bearing retainer and may fall out when bearing assembly is removed.

*d.* Remove opposite roller bearing from yoke. Push universal joint journal assembly (M) to one side as far as possible. Tilt yoke and remove journal from yoke. Repeat procedures on opposite yoke. Remove bearing gaskets (D) and bearing gasket retainers (E) from journal.

*e.* Opposite universal joint assembly can be disassembled as explained in *b* through *d* above.

## 216. Cleaning and Inspection

*Note.* Key letters noted in parentheses are in figure 151.

*a. Cleaning.*—Clean all parts thoroughly with dry-cleaning solvent or volatile mineral spirits.

*b. Inspection.*
   (1) *Shaft assemblies.*—The shaft assemblies with stub shaft and yoke (L) are of different lengths, depending upon location. Examine shaft assembly for bent condition. Each shaft is balanced and no attempt should be made to repair the shafts. If shaft is bent, or if yoke or splines are damaged, replace with new part.
   (2) *Universal joint flange yokes.*—Inspect universal joint flange yokes (A) for damage. Replace with new parts if damaged. Do not attempt to repair yokes.
   (3) *Slip joint sleeve yoke assembly.*—Remove relief valve (P) and lubrication fitting (F) and thoroughly clean out interior of slip joint sleeve yoke assembly (G). Examine yoke for damage. Replace if yoke is distorted or damaged in any manner. The dust cap cork washer (H) should be replaced with new part at assembly.
   (4) *Universal joint journal assemblies.*—Inspect bearing surfaces on universal joint journal assemblies (M) for roughness

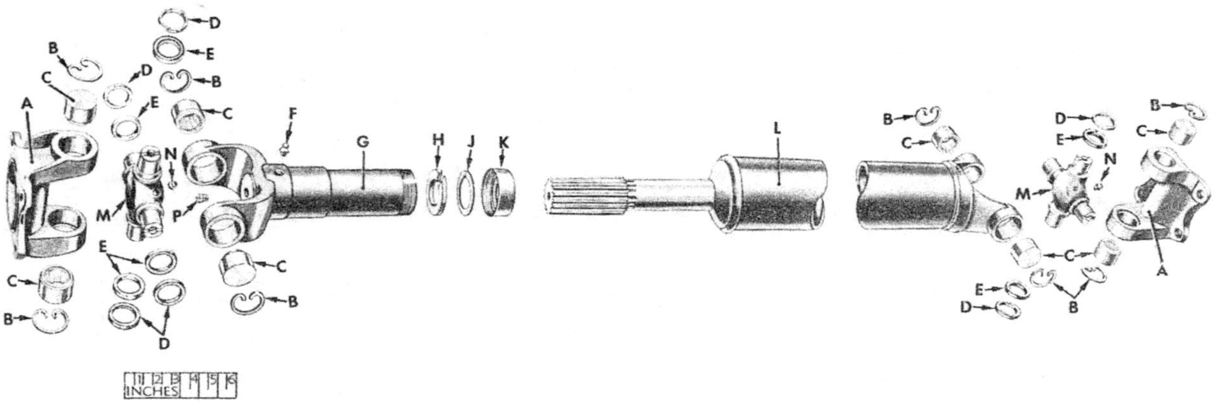

A—YOKE, FLANGE, UNIVERSAL JOINT—7350550
B—RING, SNAP, BEARING—A281978
C—BEARING, ROLLER, ASSY—7521040
D—GASKET, BEARING—7350542
E—RETAINER, BEARING GASKET—7730545
F—FITTING, LUBRICATION—504208
G—YOKE, SLEEVE, SLIP JOINT, ASSY—7415765
H—WASHER, CORK, DUST CAP—A160310
J—WASHER, STEEL, DUST CAP—SP-3-1/2—15-53
K—CAP, DUST—5160305
L { SHAFT, PILLOW-BLOCK-TO-REAR-REAR-AXLE, ASSY—YT-2287725
SHAFT, TRANSFER-TO-FORWARD-REAR-AXLE, ASSY (ALL EXCEPT M215 AND M221)—YT-2287727
SHAFT, TRANSFER-TO-FORWARD-REAR-AXLE, ASSY (M215 AND M221)—YT-2290991
SHAFT, TRANSFER-TO-FRONT-AXLE, ASSY—YT-2287731
SHAFT, TRANSFER-TO-PILLOW-BLOCK, ASSY (ALL EXCEPT M215 AND M221)—YT-2287729
SHAFT, TRANSFER-TO-PILLOW-BLOCK, ASSY (M215 AND M221)—YT-2290993 }
M—JOURNAL, UNIVERSAL JOINT, ASSY—SP-3-5-268X
N—FITTING, LUBRICATION, JOURNAL—110347
P—VALVE, RELIEF—587310

RA PD 149556

*Figure 151. Axle propeller shaft and universal joint components.*

or scoring. Make certain that lubricant passages in journals are clean and free. Upon installation, use new bearing gasket retainers (E) and bearing gaskets (D) on journals. Make certain that journal lubrication fittings (N) are clean and firmly installed.

(5) *Roller bearing assemblies.*—Examine rollers in roller bearing assemblies (C) for bent or scuffed condition. There are 34 rollers mounted in the retainer. The rollers are not held in place with a retainer. If the rollers are damaged, replace with new parts. If outer retainer is damaged or scuffed, replace entire bearing assembly. If the bearing snap rings (B) have been damaged or stretched during disassembly, replace with new parts at assembly.

### 217. Assembly

*Note.* Key letters noted in parentheses are in figure 151. All parts should be thoroughly lubricated during assembly with automotive and artillery grease (GAA).

*a.* Install a new bearing gasket retainer (E) and a new bearing gasket (D) on each end of the universal joint journal assembly (M). Position journal in place in universal joint flange yoke (A).

*b.* Install 34 rollers in bearing retainer. Apply small quantity of grease to hold rollers in place. Install a roller bearing assembly (C) into yoke over each end of journal.

*c.* Install relief valve (P) into end of slip joint sleeve yoke (G). Position journal in place in yoke of slip joint sleeve yoke. Install a roller bearing assembly into yoke over each end of journal.

*d.* Install bearing snap ring (B) at each bearing, making certain snap ring engages groove in yokes.

*e.* Universal joint at fixed end of shaft may be assembled in the same manner described in *a* through *d* above.

*f.* One splined tooth in slip joint sleeve yoke assembly (G) is blank. This blank spline must match with a similar blank spline on stub shaft of propeller shaft to assure proper universal joint alinement.

*g.* Position dust cap (K) over stub shaft; then install dust cap steel washer (J) and a new dust cap cork washer (H). Insert slip joint sleeve yoke assembly (G) over stub shaft splines, matching blank spline of yoke with blank spline of shaft.

*h.* Thread dust cap (K) onto sleeve yoke. Do not use a wrench. Tighten with fingers.

*i.* Install lubrication fitting (F) in slip joint sleeve yoke, and journal lubrication fittings (N) in universal joint journals (M). Thoroughly lubricate universal joints with automotive and artillery grease (GAA).

## Section III. REBUILD OF TRANSMISSION-TO-TRANSFER PROPELLER SHAFT ASSEMBLY

### 218. Description

The propeller shaft assembly between transmission and transfer (B, fig 150) consists of two universal joint assemblies bolted together to form a propeller shaft assembly. The sleeve yoke assembly (Q, fig. 152) fits on splined output shaft of transmission to form a slip joint. The flange yoke (P, fig. 152) bolts to a companion flange on transfer input shaft. Needle type roller bearing assemblies are used at each universal joint.

### 219. Disassembly

*Note.* Key letters noted in parentheses are in figure 152.

   a. Remove the four 1/2–20 safety nuts (K) from 1/2–20 x 1 3/8 special bolts (N). Separate flange yokes (L) and (M).

   b. Remove lubrication fitting (J) from universal joint journal assembly (H) at the sleeve yoke end of the propeller shaft.

   c. Straighten lugs on lock straps (B) at the four journal bearing caps (C). Remove two 5/16–24 x 1/2 cap screws (A) from each journal bearing cap. Remove lock straps (B) and journal bearing caps (C).

   d. Strike either flange yoke or sleeve yoke sharply with lead hammer to force a journal bearing assembly (D) out of yoke far enough to permit removal. Remove opposite bearing assembly from yoke.

   e. Push universal joint journal assembly (H) to one side as far as possible. Tilt yoke and remove yoke from journal. Repeat procedures on opposite yoke. Remove journal bearing gaskets (E) and gasket retainers (F) from journal.

   f. Remove relief valve (G) from universal joint journal (H).

   g. Opposite universal joint assembly can be disassembled as explained in *b* through *f* above.

### 220. Cleaning, Inspection, and Repair

*Note.* Key letters noted in parentheses are in figure 152 unless otherwise indicated.

   a. *Cleaning.* Thoroughly clean all parts with dry-cleaning solvent or volatile mineral spirits.

   b. *Inspection and Repair.*
      (1) *Flange yokes.*—Inspect flange yokes (L), (M), and (P) for distortion, damage, or stripped threads. Replace yokes if these conditions exist.
      (2) *Sleeve yoke assembly.*—Inspect splines of sleeve yoke assembly (Q) for distortion, chipped condition, or excessively worn condition. If these conditions exist, replace sleeve yoke assembly. Examine sleeve yoke sleeve (R) for scoring,

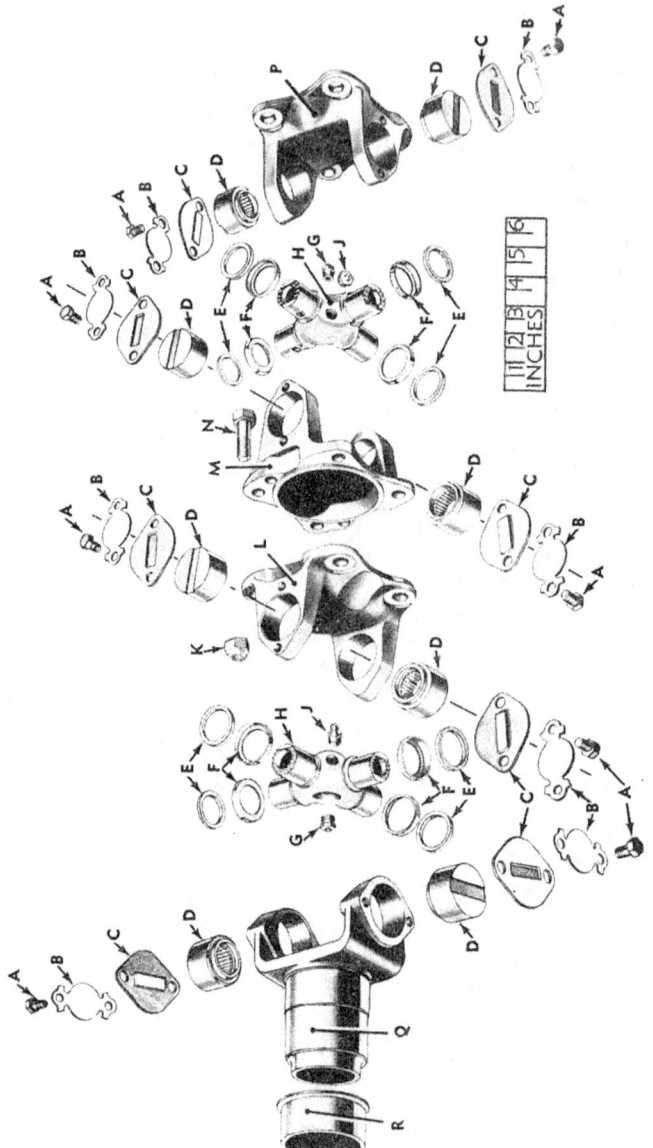

Figure 152. Transmission-to-transfer propeller shaft components.

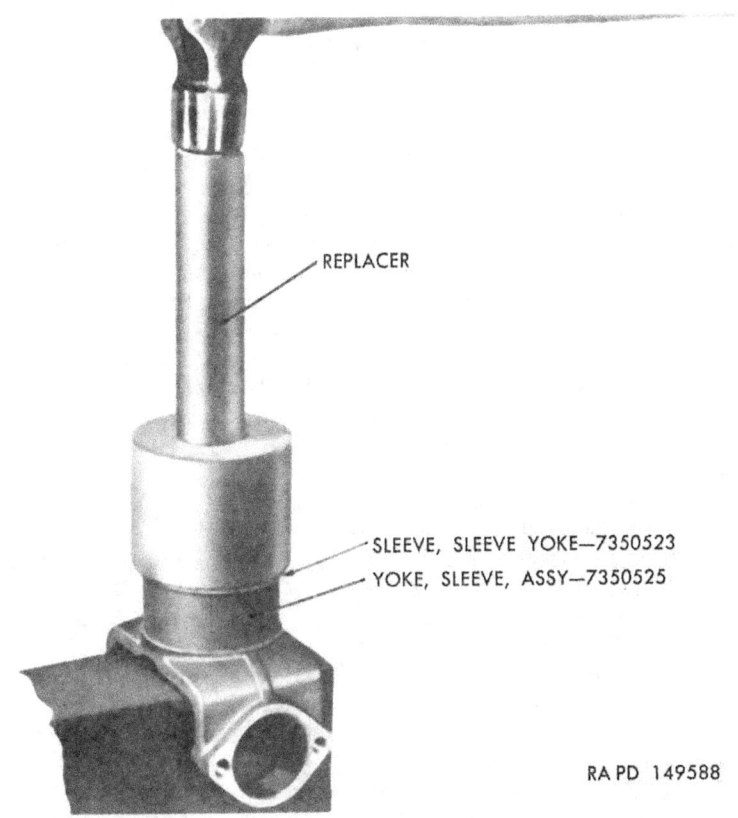

*Figure 153. Installing yoke oil seal sleeve with replacer 41-R-2395-535.*

*Figure 152.—Continued.*

A—SCREW, CAP, 5/16-24 x 1/2—5284326
B—STRAP, LOCK—5281981
C—CAP, JOURNAL BEARING—5283074
D—BEARING, JOURNAL, ASSY—A281980
E—GASKET, JOURNAL BEARING—5283081
F—RETAINER, GASKET—5282858
G—VALVE, RELIEF—587310
H—JOURNAL, UNIVERSAL JOINT, ASSY—YT-2284934
J—FITTING, LUBRICATION—504268
K—NUT, SAFETY, 1/2-20—442801
L—YOKE, FLANGE (CENTER FRONT)—7350527
M—YOKE, FLANGE (CENTER REAR)—7350526
N—BOLT, SPECIAL, 1/2-20 X 1 3/8—7350528
P—YOKE, FLANGE (AT TRANSFER)—7350526
Q—YOKE, SLEEVE, ASSY—7350525
R—SLEEVE, SLEEVE YOKE—7350523

checking, and looseness. If sleeve is damaged, replace with new part. With replacer 41–R–2395–535 (fig. 153), drive new sleeve onto sleeve yoke.

(3) *Journal bearing assemblies.*—Examine bearing assembly (D) needle roller bearings for broken condition or scoring. Bearing rollers are held in place with a retainer. If retainer or rollers are damaged, replace bearing assembly.

(4) *Universal joint journal assemblies.*—Inspect bearing surfaces on journals for roughness or scoring. Make certain that lubricant passages in journals are clean and free.

(5) *Journal Bearing Caps and Lock Straps.*—Journal bearing caps (C) should be inspected for damage. As a general rule, lock straps (B) should be replaced at assembly; however, straps can be reused if one lug at each end has not been bent.

## 221. Assembly

*Note.* Key letters noted in parentheses are in figure 152. Thoroughly lubricate all parts with automotive and artillery grease (GAA).

*a.* Install relief valve (G) and lubrication fitting (J) in universal joint journal (H).

*b.* Install new gasket retainer (F) and journal bearing gasket (E) on each end of journal.

*c.* Assemble front universal joint first. Place universal joint journal assembly (H) in place in sleeve yoke assembly (Q).

*d.* Install a journal bearing assembly (D) into sleeve yoke over each end of journal cross.

*e.* Insert journal into center front flange yoke; then install journal bearing over each end of journal cross.

*f.* Install journal bearing cap (C) over each journal bearing, engaging cap lug with slot in bearing retainer.

*g.* Position lock strap (B) over each cap. Install two $5/16$–24 x $1/2$ cap screws (A) and tighten firmly. Bend one lug of lock strap against face of cap screw.

*h.* Assemble opposite universal joint in the same manner as described in *a* through *g* above.

*i.* Bolt center front and center rear flange yokes (L) and (M) together, using four $1/2$–20 x $1 3/8$ special bolts (N) and four $1/2$–20 safety nuts (K). Tighten nuts firmly.

*Note.* The front and rear universal joint assemblies must be in same plane when bolted together.

## Section IV. REBUILD OF PILLOW BLOCK ASSEMBLY

## 222. Description

*Note.* Key letters noted in parentheses are in figure 155 unless otherwise indicated.

*a.* Pillow block assembly (fig. 155) consists of a shaft (K) supported by a ball bearing assembly (G) at both ends. Bearings are pressed on the shaft and are retained in place in pillow block by shaft bearing retainers (E) which are bolted to pillow block (H). Ball bearing at front is equipped with a snap ring (F) which fits in groove of bearing and shaft bearing retainer. Double lip shaft bearing oil seal assemblies (D) are installed in shaft bearing retainers. A splined flange assembly with oil seal sleeve and deflector (B) is installed over splines at each end of shaft and held in place with shaft nut (A) and cotter pin.

*b.* The pillow block assembly is mounted on forward rear axle housing (fig. 149) and supports the two propeller shafts required to transmit power from transfer to rear rear axle.

### 223. Disassembly

*Note.* Key letters noted in parentheses are in figure 154 unless otherwise indicated.

*a.* Remove pipe plug (J), inspection hole plug and gasket (K) and (L), pipe plug (Q), lubrication and oil level plug (M), lubrication fitting elbow (N), and lubrication fitting (P) and allow lubricant to drain from housing.

*b.* Place assembly firmly in a vise. With holding tool 41-T-3215-910 over flange with deflector and sleeve assembly (B) at rear in similar manner as illustrated in figure 51, remove cotter pin at shaft nut (A). Remove shaft nut.

*c.* Remove flange with deflector and sleeve assembly (B) from end of shaft. Use brass hammer to tap flange from shaft.

*d.* Remove four $3/8$–16 x $1\frac{1}{8}$ cap screws (R) and lock washers (S) which retain the shaft bearing retainer (D) to pillow block (H). Remove shaft bearing retainer (D) with shaft bearing oil seal assembly (C), and bearing retainer gasket (E). Do not remove shaft bearing oil seal assembly (C) until inspection (par. 224*b*).

*e.* At the front end, remove shaft nut (A), flange with deflector and sleeve assembly (B), shaft bearing retainer (D), shaft bearing oil seal assembly (C), and bearing retainer gasket (E) in same manner as described in *b*, *c*, and *d* above.

*f.* With brass drift at rear of shaft (T), drive shaft through ball bearing assembly (G) at rear. Shaft (T) and ball bearing (G) at front, together with front bearing snap ring (F), will be removed as an assembly. Press ball bearing from shaft.

### 224. Cleaning, Inspection, and Repair

*Note.* Key letters noted in parentheses are in figure 154 unless otherwise indicated.

*a. Cleaning.*—Thoroughly clean parts in dry-cleaning solvent or volatile mineral spirits. Do not spin ball bearings with compressed air.

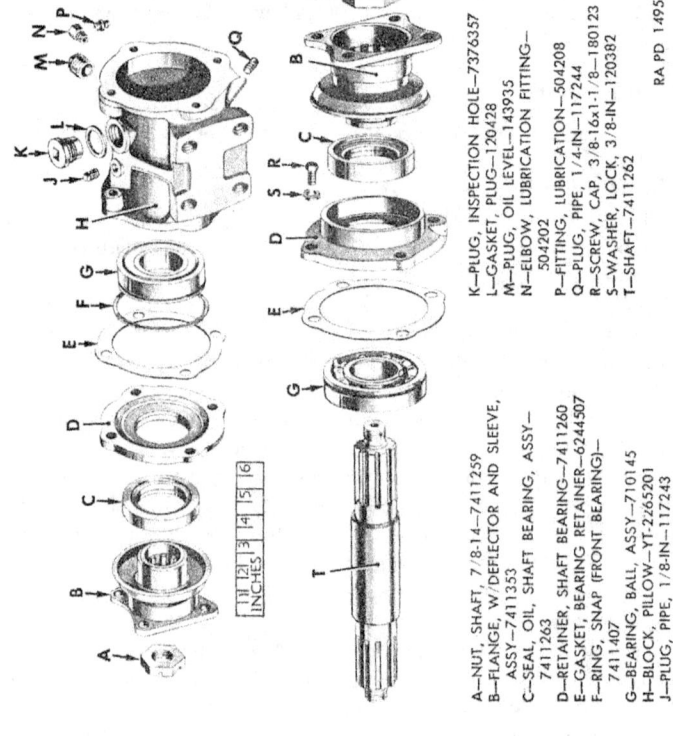

A—NUT, SHAFT, 7/8-14—7411259
B—FLANGE, W/DEFLECTOR AND SLEEVE, ASSY—7411353
C—SEAL, OIL, SHAFT BEARING, ASSY—7411263
D—RETAINER, SHAFT BEARING—7411260
E—GASKET, BEARING RETAINER—6244507
F—RING, SNAP (FRONT BEARING)—7411407
G—BEARING, BALL, ASSY—710145
H—BLOCK, PILLOW—YT-2265201
J—PLUG, PIPE, 1/8-IN—117243
K—PLUG, INSPECTION HOLE—7376357
L—GASKET, PLUG—120428
M—PLUG, OIL LEVEL—143935
N—ELBOW, LUBRICATION FITTING—504202
P—FITTING, LUBRICATION—504208
Q—PLUG, PIPE, 1/4-IN—117244
R—SCREW, CAP, 3/8-16x1-1/8—180123
S—WASHER, LOCK, 3/8-IN—120382
T—SHAFT—7411262

RA PD 149555

*Figure 154. Pillow block assembly components.*

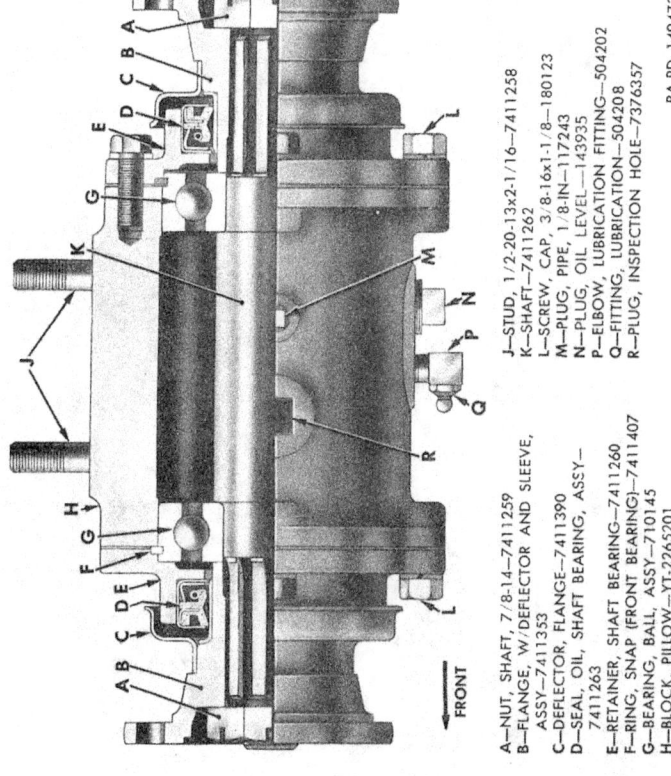

A—NUT, SHAFT, 7/8-14—7411259
B—FLANGE, W/DEFLECTOR AND SLEEVE, ASSY—7411353
C—DEFLECTOR, FLANGE—7411390
D—SEAL, OIL, SHAFT BEARING, ASSY—7411263
E—RETAINER, SHAFT BEARING—7411260
F—RING, SNAP (FRONT BEARING)—7411407
G—BEARING, BALL, ASSY—710145
H—BLOCK, PILLOW—YT-2265201
J—STUD, 1/2-20-13x2-1/16—7411258
K—SHAFT—7411262
L—SCREW, CAP, 3/8-16x1-1/8—180123
M—PLUG, PIPE, 1/8-IN—117243
N—PLUG, OIL LEVEL—143935
P—ELBOW, LUBRICATION FITTING—504202
Q—FITTING, LUBRICATION—504208
R—PLUG, INSPECTION HOLE—7376357

RA PD 149472

*Figure 155. Sectional view of pillow block assembly.*

243

*b. Inspection and Repair.*
   (1) *Pillow block.*—Thoroughly examine pillow block (H) for cracks, stripped thread, or broken studs. If threads are stripped or if housing is damaged, replace with new part. If studs are damaged, replace. Driven height of each stud

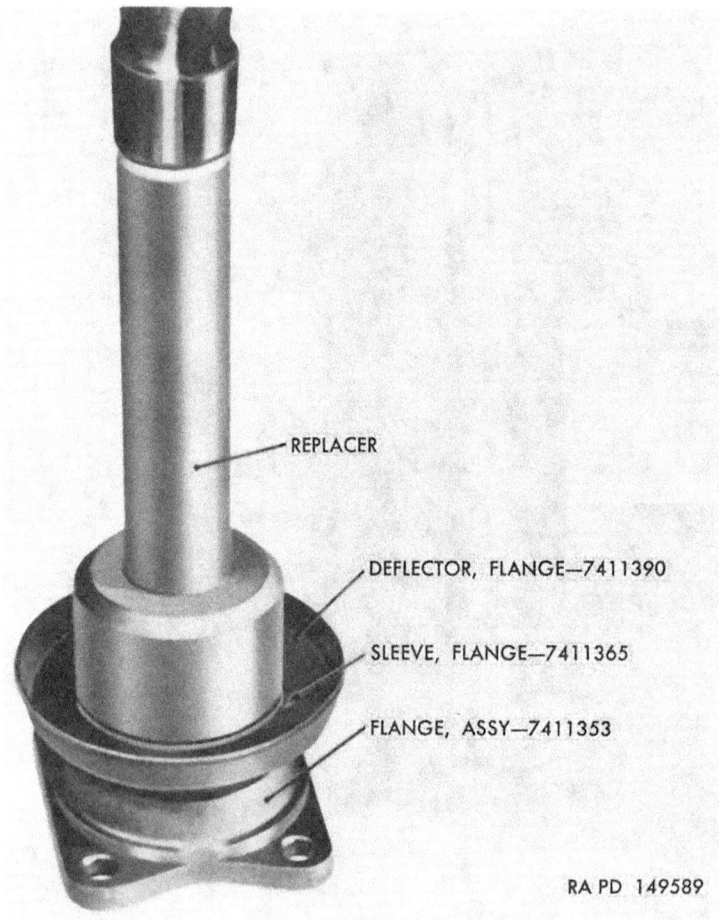

*Figure 156. Installing flange oil seal sleeve with replacer 41-R-2395-528.*

is five-sixteenths of an inch from face of pillow block. Discard plug gasket (L) and replace with new part at assembly.
   (2) *Flange with deflector and sleeve assembly.*—Inspect splines of flange with deflector and sleeve assemblies (B) for distortion and chipped or excessively worn condition. Replace

assembly if these conditions exist. Examine surface of oil seal sleeve on flange for grooves, nicks, or worn condition. If oil seal sleeve is damaged, replace as described in (*a*) below. Inspect condition of deflector on flange. If loose or damaged, replace as described in (*b*) below.

(*a*) *Oil seal sleeve replacement.*—Use ball peen hammer, and tap entire circumference of outer surface of flange sleeve (fig. 157). Peening of this surface will cause metal in sleeve to stretch until sleeve can be removed from flange. Install sleeve on flange using replacer 41–R–2395–528 (fig. 156).

(*b*) *Deflector replacement.*—Deflector is pressed on flange as shown in figure 157. The deflector is either spot welded or peened in three or four places to flange. Remove old deflector. Install new deflector in place on flange. Peen or spot weld as shown in figure 157.

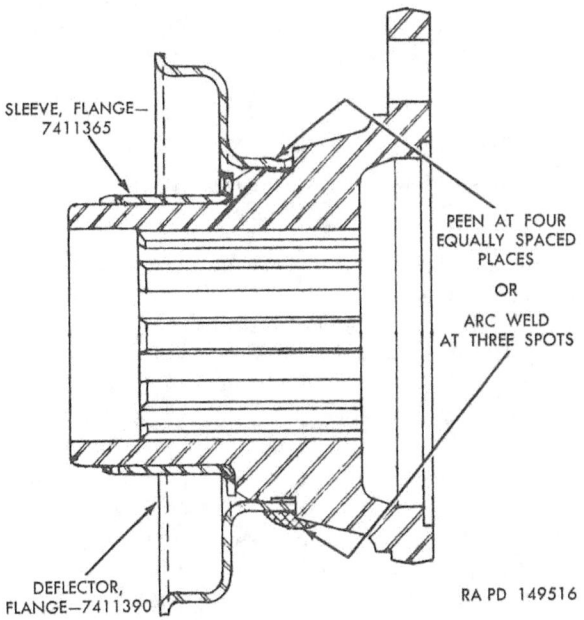

*Figure 157. Sectional view of flange assembly.*

(3) *Shaft bearing retainer and oil seal.*—Inspect shaft bearing retainers (D) for damage. If shaft bearing oil seal assemblies (C) are damaged, replace with new parts. Press old seal assemblies out of retainers. Coat outer surface of oil seal assemblies with plastic type gasket cement. Make cer-

tain that cavity in the inside circumference of seal is lubricated with automotive and artillery grease (GAA). Press oil seal assembly into retainer with mark "OUTSIDE" on seal toward outer side of retainer. When pressing new oil seal into retainer, bear against outer edge of seal only. Discard bearing retainer gaskets (E) and replace with new parts at assembly.

(4) *Ball bearing assemblies.*—Throughly inspect ball bearing assemblies (G) for rough balls, scored outer surface, or broken ball retainers. The ball bearing assembly which is used at front of pillow block assembly includes a snap ring (F). Use a new snap ring when replacing the ball bearing. The other bearing also has a snap ring groove but no ring is used.

(5) *Shaft.*—Examine splines on each end of shaft for distortion, chipped condition, or excessively worn condition. If surfaces upon which bearings are mounted are scored or scuffed, replace shaft. Check condition of threads at each end of shaft.

## 225. Assembly

*Note.* Key letters noted in parentheses are in figure 154 unless otherwise indicated. The sectional view (fig. 155) shows the component parts in their assembled position.

*a.* Thoroughly lubricate all parts with automotive and artillery grease (GAA) during assembly.

*b.* Mount pillow block housing firmly in a vise. Position ball bearing assembly (G) into rear end of pillow block with snap ring groove of bearing toward outside.

*Note.* Do not use a snap ring in rear bearing. The snap ring is used only at front bearing.

*c.* Press ball bearing assembly (G), with snap ring in place, on one end of shaft (T).

*d.* Insert shaft, with front bearing assembled, through front end of pillow block (H), and press shaft into bearing at rear. Bearings must be pressed firmly against shoulders of shaft.

*e.* Position new bearing retainer gasket (E) to front end of pillow block; then install shaft bearing retainer (D), with shaft bearing oil seal assembly (C) in place, to pillow block, using four $\frac{3}{8}$–16 x $1\frac{1}{8}$ cap screws (R) and $\frac{3}{8}$-inch lock washers (S). Tighten cap screws firmly.

*f.* Install new bearing retainer gasket (E), shaft bearing retainer (D), and shaft bearing oil seal assembly (C) to rear end of pillow block in same manner described in *e* above.

*g.* Install flange with deflector and sleeve assemblies (B) on each end of shaft. Flange assemblies must be in the same plane. Install shaft nuts (A). Tighten to a torque of 175 to 225 pound-feet. Install new cotter pins and bend over flat of nuts.

*h.* Install pipe plug (J), inspection hole plug and gasket (K) and (L), oil level plug (M), ¼-inch pipe plug (Q), and lubrication fitting elbow (N) and fitting (P).

*i.* Fill housing with automotive and artillery grease (GAA) through lubrication fitting until lubricant is level with bottom of oil level plug hole.

# CHAPTER 10

# SERVICE BRAKE SYSTEM

## Section I. DESCRIPTION AND DATA

### 226. Service Brake System Description

*a. General.*—Service brake system is a combined air-hydraulic brake system comprised primarily of a manually-operated brake pedal which is interconnected to the hydraulic master cylinder to build up the initial hydraulic pressure; an air-hydraulic power cylinder to increased the hydraulic pressure; two hydraulic wheel cylinders at each wheel to transmit the hydraulic pressure to the brake shoes; a compressed air system which maintains a supply of compressed air for operation of the air-hydraulic power cylinder; and interconnecting air lines, hydraulic lines, and fittings.

*b. Master Cylinder* (fig. 170).—Master cylinder is integral tank type, with fluid reservoir and cylinder barrel cast in one piece. The functioning components of the master cylinder are the piston with secondary cup (H), primary cup (K), secondary cup (G), and piston return spring (M). Inlet port (J) and compensating port (L) connect the fluid reservoir with the cylinder bore. The filler cap assembly, comprised of a filler cap (S), vent line fitting (T), gaskets (V), and special bolt (U), is installed on top of the filler pipe (P), which is threaded into the top of the master cylinder reservoir. The special bolt in the filler cap is drilled to permit venting the fluid reservoir into the vehicle air vent system. Piston stop (E) is retained in end of cylinder bore by a retainer (snap ring) which fits into a groove in the cylinder bore.

*c. Air-Hydraulic Power Cylinder.*—The air-hydraulic power cylinder is a hydraulically-actuated, air-operated power cylinder which utilizes the energy in compressed air for displacing hydraulic fluid under high pressure into the wheel cylinders. Air-hydraulic power unit is comprised basically of three interconnected units, the control valve, the power cylinder, and the slave cylinder. The reactionary type control valve, actuated by hydraulic fluid pressure from the master cylinder, controls the flow of compressed air into and out of the power cylinder shell. The power cylinder, comprised primarily of a cylinder shell and piston assembly, transmits the power of the compressed air to the hydraulic piston in the slave cylinder. The slave cylinder piston, when forced forward in the slave cylinder by the power cylinder piston, displaces hydraulic fluid from the slave cylinder through the hydraulic lines into the wheel cylinders at each wheel. Refer to paragraph 1 for number of technical manual which contains detailed information on the air-hydraulic power cylinder.

*d. Wheel Cylinders.*—Two dual-piston wheel cylinders (fig. 172), mounted between ends of brake shoes on each backing plate transmit the hydraulic pressure to the brake shoes through push rods. Detail parts of upper and lower wheel cylinders at each brake are identical, but the bleeder screw and the hydraulic line connector are assembled to opposite holes to provide topmost location of the bleeder screws when cylinders are installed in correct position on backing plate. Internal parts of each wheel cylinder (fig. 71) are two pistons, two piston cups, two piston cup expanders, and one piston return spring. The tapered outer edges of the piston cup expanders fit inside the bevelled lips of the piston cups and force the lips of the cups outward against the cylinder wall during brake application. The expanders also prevent lips of piston cups from collapsing during and after brake release. Rubber boot at both ends of each cylinder prevents entrance of water and dirt.

*e. Brake Shoes and Attaching Parts* (fig. 158).—Two brake shoes are mounted on each backing plate in conjunction with two anchor blocks. Anchor blocks serve as shoe stops and shoe centering points, and provide the fulcrums around which the shoes pivot during brake application. Four brake shoe return springs at each brake hold shoe ends firmly against anchors when brakes are released. Heel of each shoe anchors against steel pins which pivot in the anchor blocks; toe of each shoe anchors against adjusting screws which are threaded into the anchor blocks. Each anchor block is attached to backing plate by two cap screws threaded into tapped holes in anchor block from inner side of backing plate. Each adjusting screw has two grooves running lengthwise its entire length. Adjusting wheel, assembled in slot in anchor block, has two lugs on its inside diameter which engage the grooves in the adjusting screw. When adjusting wheel is turned by the adjusting stud and gear, it turns the adjusting screw in the anchor block, causing screw to move in or out of block, depending on the direction the adjusting wheel is turned. One-piece lining is attached to each brake shoe by 14 countersunk rivets. All brake shoes and attaching parts on one side of the vehicle, front and rear, are identical. All brake parts shown in figure 163 are interchangeable between right and left sides of the vehicle except the backing plate, anchor blocks, connecting lines, heat shields, and adjusting screws.

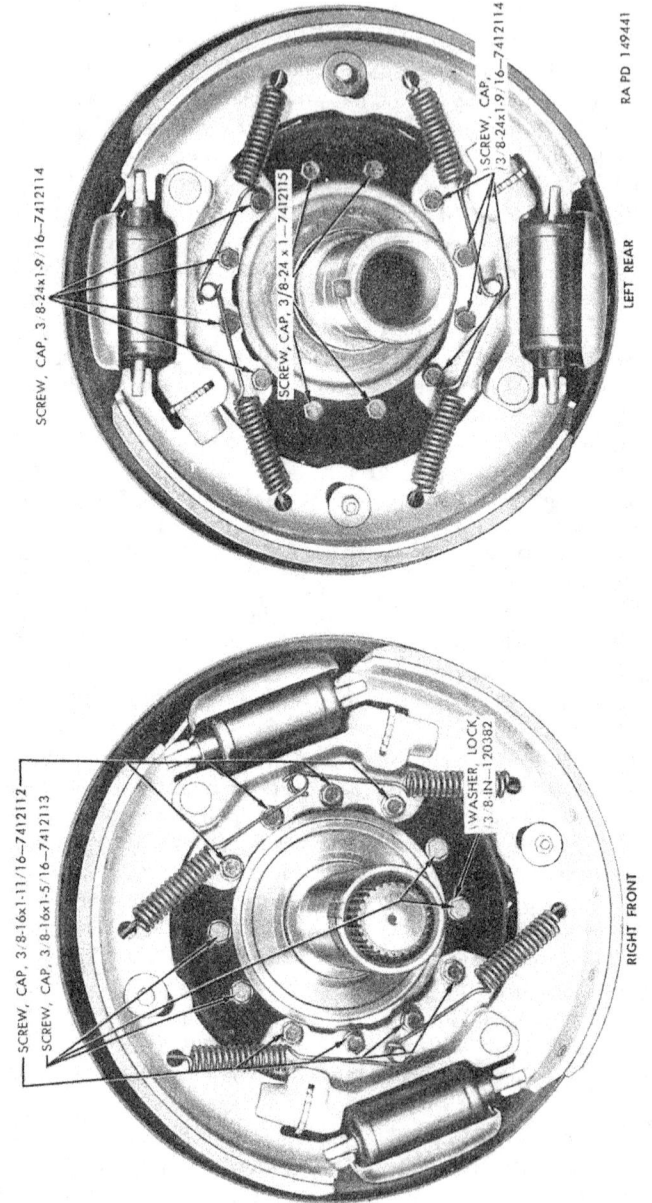

Figure 158. Front and rear brake assemblies installed on axles.

*f. Brake Drums.*—Brake drums may be either cast iron or of centrifuse construction. All brake drums on vehicle are identical. Brake drums are attached to hubs by means of brake drum backs which serve as brake drum adapters. Different drum backs are used at front and rear and for dual rear wheels to properly position brake drum over brake linings.

*g. Compressed Air System.*—Compressed air system comprises the air compressor and governor assembly, two air tanks, and interconnecting air lines and fittings. Refer to TM 9-819A and TM 9-1819AA for information on air compressor and governor assembly.

## 227. Service Brake System Data

*a. Master Cylinder.*

| | |
|---|---|
| Type | integral tank |
| Bore | 1¾ in. |
| Stroke | 1⁷⁄₁₆ in. |

*b. Air-hydraulic Power Cylinder.*

| | |
|---|---|
| Manufacturer | Bendix Products Division |
| Model | A35-15-154 |
| Cylinder shell diameter (inside) | 4½ in. |
| Power cylinder piston stroke | 3⅞ in. |
| Slave cylinder bore | 1⅛ in. |
| Slave cylinder piston stroke | 3¾ in. |
| Control valve bore | 1¹⁄₁₆ in. |

*c. Wheel Cylinders.*

| | |
|---|---|
| Type | straight bore, dual piston |
| Bore | 1¼ in. |

*d. Brake Lining.*

| | |
|---|---|
| Width | 3 in. |
| Thickness | ⅜ in. |

*e. Brake Drums.*

| | |
|---|---|
| Type | cast or centrifuse |
| Diameter | 15 in. |

## Section II. REBUILD OF BRAKE SHOES AND DRUMS

### 228. General

Removal and installation of brake shoe and lining assemblies for replacement purposes are covered in TM 9-819A; however, for purposes of a complete brake overhaul, the complete brake assembly is removed from the axle and disassembly, cleaning, inspection, repair, and assembly procedures accomplished on a bench as described herein. Except for slight differences noted in the text, procedures for removal, overhaul, and installation of right and left brake assemblies at front and rear axles are the same.

*Note.* Any painted parts from which the paint is removed during the process of cleaning and inspection must be repainted with metal primer as directed in TM 9-2851 after inspection is completed.

### 229. Removal of Brake Assembly

*a. Remove Brake Drum.*—Remove safety nuts and plain washers from 18 studs attaching brake drum to brake drum adapter. Tap brake drum to loosen from adapter; then lift drum off studs.

*b. Remove Hub and Bearings.*—Remove hub and bearings as directed in paragraph 256*b*.

*c. Disconnect Brake Line.*

 (1) *Front.*—Remove three cap screws and lock washers attaching brake line shield (fig. 159) to steering knuckle support and swing shield to one side. Disconnect axle brake line assembly from rubber brake line assembly at bracket on outer end of axle housing, remove end of rubber brake line assembly from bracket and pull line through bracket on shield; then unscrew rubber brake line assembly from distributor fitting at backing plate.

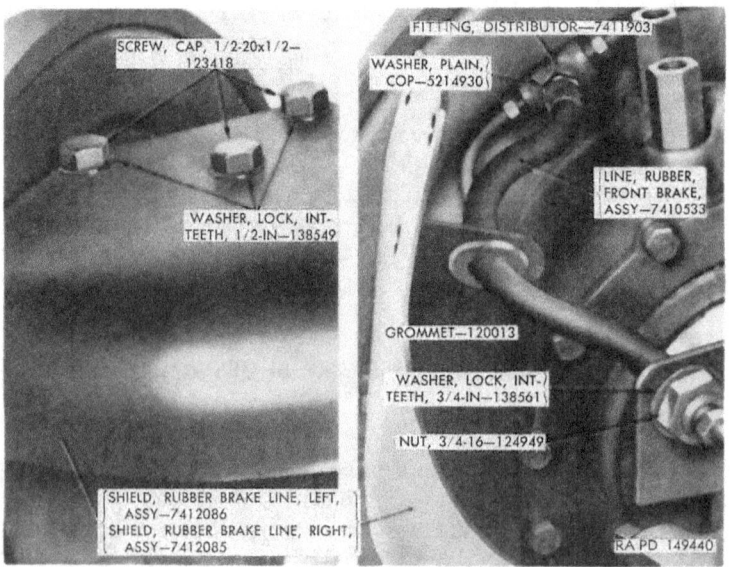

*Figure 159. Front brake rubber line installed (left side shown).*

252

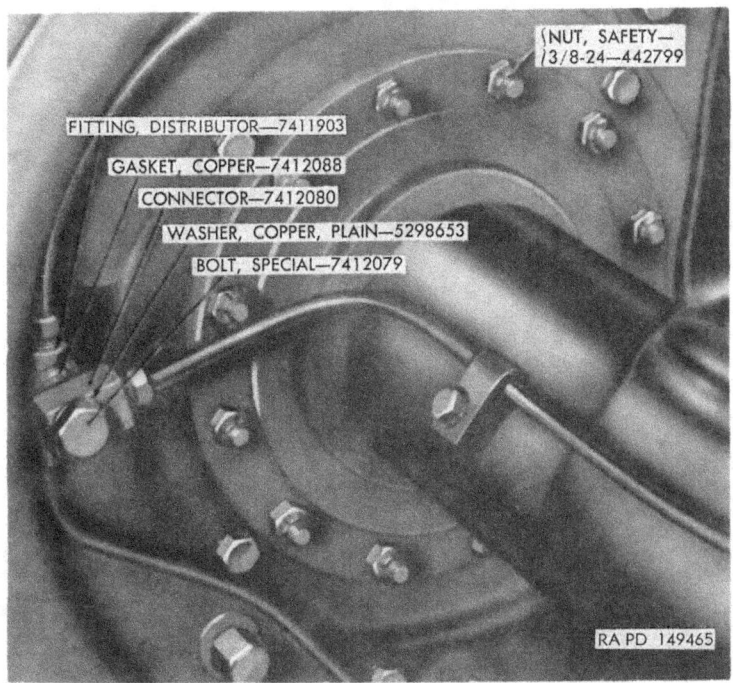

*Figure 160. Inner view of rear brake assembly installed on axle.*

(2) *Rear.*—Disconnect axle brake line assembly from distributor fitting at backing plate by removing special bolt attaching axle brake line connector to distributor fitting (fig. 160).

*d. Remove Brake Assembly.*—Remove 12 cap screws and lock washers (fig. 158) attaching brake assembly to steering knuckle support (front), or remove 12 cap screws and safety nuts attaching brake assembly to flange on axle housing (rear). Lift brake assembly (fig. 161) off axle.

### 230. Disassembly of Brake Components

*Note.* Key letters noted in parentheses are in figure 163 unless otherwise indicated.

*a. General.*—The following procedure covers removal of brake components from backing plate after the brake assembly is removed from the axle.

*b. Remove Brake Shoe Return Springs.*—Using brake spring remover and replacer 41–R–2375–20, unhook brake shoe return springs (X) from return spring pins (JJ) as shown in figure 162; then remove springs from brake shoes.

253

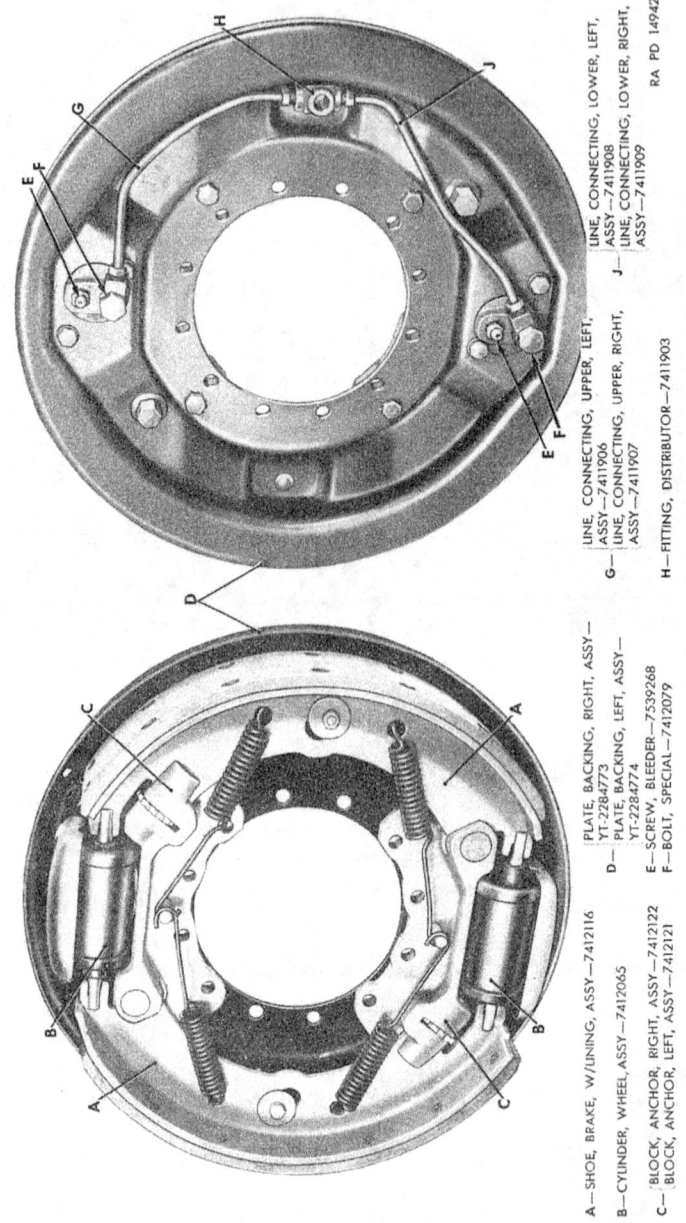

Figure 161. Brake assembly removed from axle.

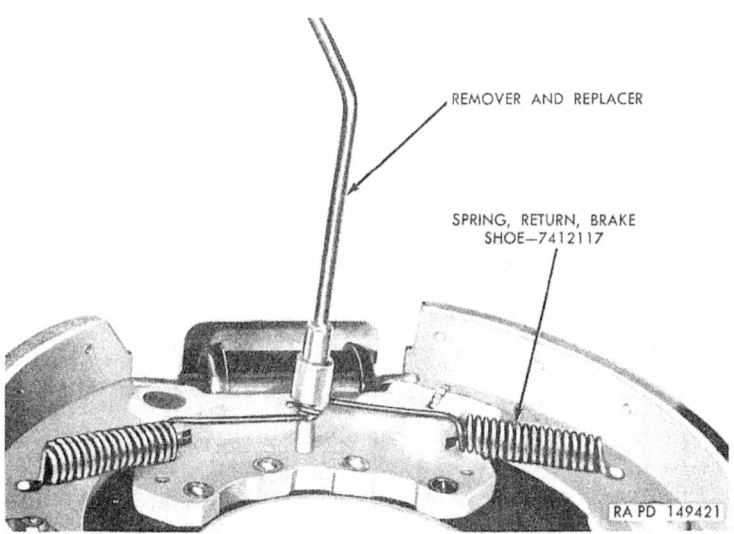

*Figure 162. Removing brake shoe return springs with remover and replacer 41–R–2375–20.*

*c. Remove Front Brake Shoe and Lining Assembly.*—Remove ¼–28 nut (**GG**), ¼-inch lock washer (**FF**), plain washer (**EE**), and sleeve (**DD**) from guide bolt (**V**) and remove bolt. Lift ends of front brake shoe and lining assembly (**LL**) out of anchor blocks (**CC**) and wheel cylinder push rods to remove.

*d. Remove Rear Brake Shoe and Lining Assembly.*—Unscrew ¼–28 x 1 cap screw (**KK**) from distributor fitting (**T**) and remove cap screw ¼-inch lock washer (**FF**), plain washer (**EE**), and sleeve (**DD**). Lift ends of rear brake shoe and lining assembly (**LL**) out of anchor blocks (**CC**) and wheel cylinder push rods to remove.

*e. Remove and Disassemble Brake Lines.*—Remove special bolts (**A**) and plain copper washers (**B**) attaching brake line fittings (**C**) to wheel cylinders. Remove upper connecting line assembly (**D**), lower connecting line assembly (**W**), fittings (**C**), and distributor fitting (**T**) as an assembly from backing plate (**L**). Unscrew tubing nuts (**E**) from fittings (**C**) and distributor fitting (**T**).

*f. Remove Wheel Cylinder Assemblies and Heat Shields.*—Remove bleeder screw (**F**) from each wheel cylinder assembly (**Y**). Remove two ⁵⁄₁₆–18 x ⅝ cap screws (**G**) and ⁵⁄₁₆-inch lock washers (**H**) attaching each wheel cylinder assembly (**Y**) to backing plate (**L**). Lift wheel cylinder assemblies and heat shields (**Z**) off backing plate. Refer to paragraphs 239 through 241 for overhaul of wheel cylinder assemblies.

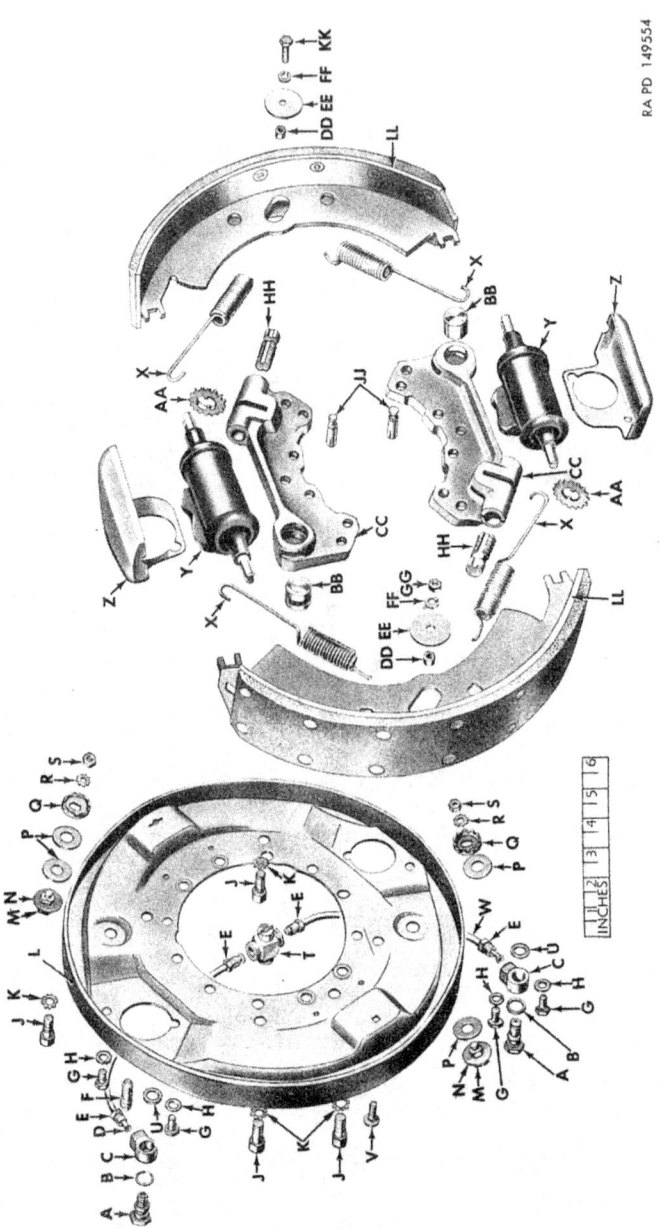

A—BOLT, SPECIAL—7412079
B—WASHER, PLAIN, COPPER—5298653
C—FITTING—5323092
D—{LINE, CONNECTING, UPPER, LEFT, ASSY—7411906
     LINE, CONNECTING, UPPER, RIGHT, ASSY—7411907
E—NUT—142432
F—SCREW, BLEEDER—7539268
G—SCREW, CAP, $5/16$-18 × $5/8$—180075
H—WASHER, LOCK, $5/16$-IN.—120214
J—SCREW, CAP, $3/8$-24 × $5/8$—181629
K—WASHER, LOCK, EXT-TEETH, $3/8$-IN.—138489
L—{PLATE, BACKING, LEFT—7412111
     PLATE, BACKING, RIGHT—7412110
M—STUD, ADJUSTING—YT-2289512
N—WASHER, SPRING—7412119
P—WASHER, PLAIN—7412120
Q—GEAR, ADJUSTING—7412104
R—WASHER, LOCK, INT-TEETH, $1/4$-IN.—138538
S—NUT, $1/4$-28—126368
T—FITTING, DISTRIBUTOR—7411903

U—GASKET, COPPER—7412088
V—BOLT, GUIDE—7411760
W—{LINE, CONNECTING, LOWER, LEFT, ASSY—7411908
     LINE, CONNECTING, LOWER, RIGHT, ASSY—7411909
X—SPRING, RETURN, BRAKE SHOE—7412117
Y—CYLINDER, WHEEL, ASSY—7412065
Z—{SHIELD, HEAT, LEFT—7412068
     SHIELD, HEAT, RIGHT—7412050
AA—WHEEL, ADJUSTING—7412123
BB—PIN, ANCHOR—7412106
CC—{BLOCK, ANCHOR, LEFT—YT-2284772
     BLOCK, ANCHOR, RIGHT—YT-2284771
DD—SLEEVE—7412103
EE—WASHER, PLAIN—5323088
FF—WASHER, LOCK, $1/4$-IN.—126380
GG—NUT, $1/4$-28—121996
HH—{SCREW, ADJUSTING, LEFT—7412108
     SCREW, ADJUSTING, RIGHT—7412109
JJ—PIN, RETURN SPRING—7412107
KK—SCREW, CAP, $1/4$-28 × 1—181568
LL—SHOE, BRAKE, W/LINING, ASSY—7412116

*Figure 163. Brake assembly components.*

*g. Remove and Disassemble Anchor Blocks.*—Remove two 3/8–24 x 5/8 cap screws (J) and 3/8-inch external-teeth lock washers (K) attaching each anchor block (CC) to backing plate (L) and remove anchor blocks. Remove anchor pins (BB) from holes in anchor blocks. Unscrew adjusting screws (HH) from anchor blocks (CC) and remove adjusting wheels (AA).

*h. Remove Adjusting Gears and Studs.*—Remove 1/4–28 nut (S), 1/4-inch internal-teeth lock washer (R), adjusting gear (Q), and plain washer (P) from each adjusting stud (M). Remove adjusting studs (M) with plain washers (P) and spring washers (N) from backing plate (L). Remove spring washers (N) from adjusting studs (M).

### 231. Cleaning and Inspection of Brake Components

*Note.* Key letters noted in parentheses are in figure 163 unless otherwise indicated.

*a. Cleaning.*—Wash all parts except brake shoe and lining assemblies in dry-cleaning solvent or volatile mineral spirits and wipe or blow dry. Wipe brake shoes clean with a cloth dampened with cleaning solution. Use a wire brush if necessary to remove rust or dirt from backing plate and brake shoes. Do not wire-brush brake shoe linings.

*b. Inspection.*

(1) *Brake shoe and lining assemblies.*—Inspect lining on brake shoe with lining assemblies (LL). Linings which have been in use will have a glazed appearance; this is a normal condition and glazed surface should not be wire brushed. If linings are scored, rough, oil-soaked, or worn down close to rivet heads, they must be replaced (par. 232*a*). All brake shoes on an axle must be relined at the same time. Examine brake shoes for evidence of distortion, and for evidence of roughness or wear on ends of shoe webs where they contact the wheel cylinder push rods, adjusting screws, and anchor pins. Make sure shoe web is tight in table of shoe.

(2) *Anchor blocks, anchor pins, and adjusting screws.*—Examine all threaded holes in anchor blocks (CC) for damaged threads. Check action of adjusting screws (HH) in anchor blocks. Screws must turn freely in anchor blocks. (Adjusting screws and anchor blocks used on left side of vehicle have left-hand threads, and right-hand threads on right side of vehicle.) Inspect anchor pins (BB) and anchor pin bore in anchor blocks for evidence of wear (par. 348) or corrosion. Bore and anchor pin must be clean and smooth. Any worn or damaged parts must be replaced. Check return spring pins (JJ) which are pressed into anchor blocks. If damaged or loose, pins must be replaced (par. 232*b*).

(3) *Backing plates.*—Examine backing plates (L) for distortion. Check flange at outer edge of backing plate for runout. Flange must run true within 0.010 inch total dial indicator reading. Raised rib at each guide bolt hole must be clean and smooth to permit free action of brake shoes. Replace backing plate if any damage is evident.

(4) *Adjusting gears and wheels.*—Examine teeth on adjusting gears (Q) and adjusting wheels (AA) for wear or damage. If wear is evident or if damaged in any way, replace with new parts.

(5) *Brake connecting line assemblies.*—Examine brake connecting line assemblies (D) and (W) for dents, sharp bends, or flattened condition. Blow through lines from both ends to make sure they are unobstructed. Make sure flared ends are smooth and well formed. Inspect nuts (E) on each line assembly for damaged threads. If any damage is evident, replace with new connecting line assembly.

(6) *Brake shoe return springs.*—Check brake shoe return springs (X) for free length and loaded length (par. 348). Measurements must be taken at points shown on figure 164.

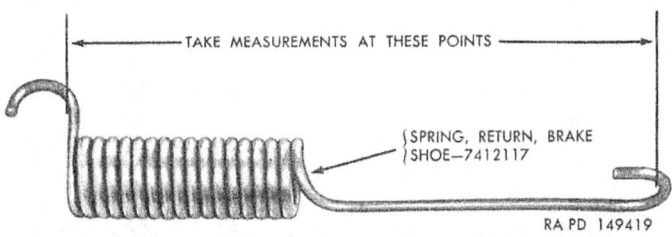

*Figure 164. Brake shoe return spring showing points of measurement.*

(7) *Miscellaneous parts.*—Examine all bolts and nuts for damaged threads and replace with new parts as necessary. Make sure drilled passages through special bolts (A) are open. Discard all lock washers (H), (K), (R), and (FF), plain copper washers (B), and copper gaskets (U) and obtain new parts for assembly.

## 232. Repair of Brake Components

*a. Brake Shoe Relining.*

*Note.* When brake drums are refinished, shim stock must be installed between lining and brake shoe. Refer to paragraph 235.

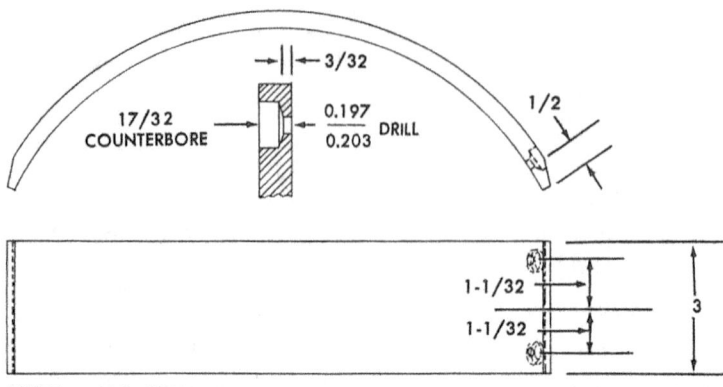

*Figure 165. Brake shoe lining dimensions and location of end rivet holes.*

(1) Remove rivets, using deliver punch on a conventional brake relining machine, and remove lining from brake shoe.

(2) Remove all dirt and corrosion from brake shoe with a wire brush; then wash shoe in dry-cleaning solvent or volatile mineral spirits and dry thoroughly.

(3) Locate two points at one end of lining as shown in figure 165, and drill through lining with a 0.197 to 0.203-inch (No. 8) drill.

(4) Position lining on shoe with two drilled holes alined with two end holes in table of shoe. Insert two rivets through lining and shoe to hold lining in position, but do not upset rivets. Clamp lining securely to shoe with a conventional lining applier; then remove rivets used to aline lining and shoe.

(5) Using brake relining machine, countersink the two drilled holes and drill and countersink two end holes at opposite end to dimensions shown in figure 165. Install $\frac{3}{16}$ x $\frac{7}{16}$ rivets in two holes at each end and upset rivets.

(6) Remove lining applier; then drill and countersink balance of holes and install rivets. Make sure lining fits firmly against shoe and that rivets are properly upset.

*b. Return Spring Pin Replacement.*—Press return spring pin (JJ) out of anchor block (CC) from inner side. Press new pin into anchor block to position shown in figure 166 and stake in place at three points.

### 233. Assembly of Brake Components

*Note.* Key letters noted in parentheses are in figure 163 unless otherwise indicated.

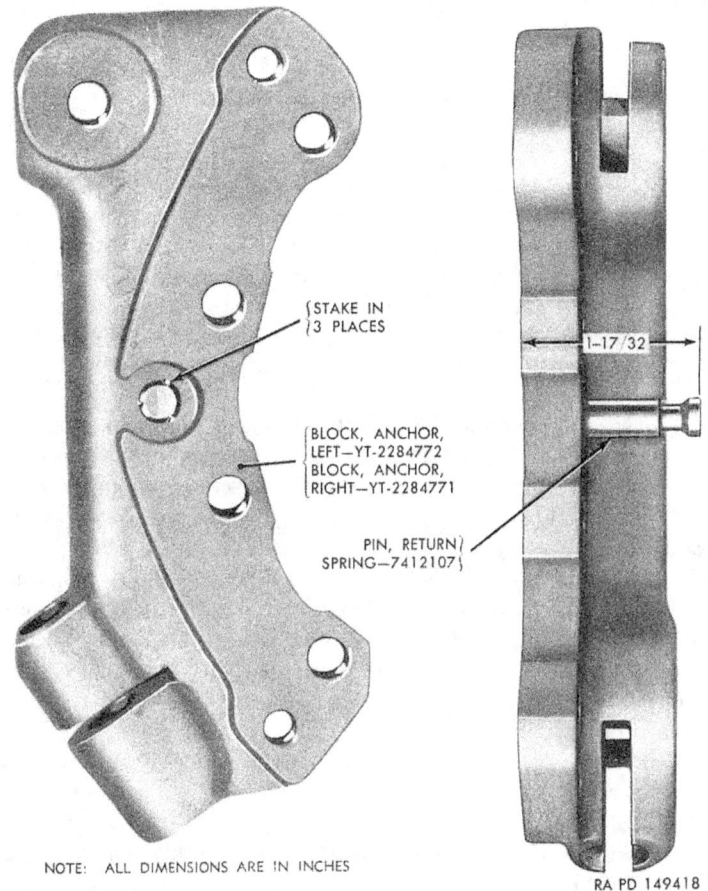

*Figure 166. Return spring pin installed in anchor block.*

  *a. General.*—The following procedure covers assembly of brake components to the backing plate with the backing plate removed from the axle.

  *b. Install Adjusting Studs and Gears.*—Press spring washer (N) onto shoulder on adjusting stud (M) with dished side toward threaded end of stud. Place one plain washer (P) over adjusting stud and insert adjusting stud through backing plate from inner side. At outer side of backing plate, place plain washer (P) and adjusting gear (Q) on adjusting stud and secure in place with a ¼-inch internal-teeth lock washer (R) and ¼–28 nut (S). Tighten nut firmly against shoulder on stud. Install other adjusting stud and gear in same manner.

 *c. Assemble and Install Anchor Blocks.*

  (1) Adjusting screws and anchor blocks used on right side of vehicle have right-hand threads and left-hand threads on left side of vehicle; parts must be assembled and installed accordingly.

  (2) Thread adjusting screw (HH) into anchor block (CC). As end of adjusting screw approaches slot in anchor block, position adjusting wheel (AA) in slot and engage lugs on inside of adjusting wheel in grooves in adjusting screw. Thread adjusting screw in until two threads on screw are exposed at threaded end of anchor block, or until plain end of adjusting screw is below unthreaded end of anchor block to a depth of approximately 1 inch.

  (3) Position assembled anchor blocks on backing plate in positions shown in figure 161, and attach each anchor block to backing plate with two 3/8–24 x 5/8 cap screws (J) and 3/8-inch external-teeth lock washers (K). Tighten cap screws firmly.

 *d. Install Wheel Cylinders and Heat Shields.*—Position heat shield (Z) over mounting base on each wheel cylinder assembly (Y) and position wheel cylinders on backing plate. Both wheel cylinders must be positioned with long end toward adjusting screw in anchor block, and right or left heat shields must be used, depending upon which side of vehicle brake assembly is to be used. Attach each wheel cylinder to backing plate with two 5/16–18 x 5/8 cap screws (G) and 5/16-inch lock washers (H). Tighten cap screws firmly. Install bleeder screw (F) in upper opening in each wheel cylinder.

 *e. Install Brake Lines.*

  (1) Refer to legend for figure 163 for identification of brake connecting line assemblies (D) and (W) to be used on right and left brakes. Assemble connecting line assemblies (D) and (W) to fittings (C) and distributor fitting (T), but do not tighten tube nuts (E).

  (2) Position assembled lines on inner side of backing plate with fittings (C) at lower openings in wheel cylinders. Place plain copper washer (B) over each special bolt (A), insert bolts through fittings (C), then place copper gasket (U) over each bolt and thread bolts into lower openings in wheel cylinders. Tighten bolts firmly.

  (3) Position lugs on distributor fitting (T) in slots in backing plate (L); then tighten four nuts (E) firmly, using care not to overtighten.

*f. Install Rear Brake Shoe and Lining Assembly.*

*Note.* Large portion of brake shoe web which engages slots in anchor blocks is curved at one end and flat at the other end. Shoes must be installed with curved end at adjusting screw and flat end engaging groove in anchor pin anchor block. Front and rear brake shoe and lining assemblies are identical and are interchangeable.

Place anchor pin (BB) in lower anchor block (CC) with slot in pin alined with slot in anchor block. Position rear brake shoe with lining assembly (LL) at backing plate, with shoe web engaging slots in wheel cylinder push rods, anchor blocks, and anchor pins. Place ¼-inch lock washer (FF), plain washer (EE), and sleeve (DD) on ¼-28 x 1 cap screw (KK), insert cap screw through shoe web, and thread into distributor fitting (T). Tighten cap screw firmly, making sure lugs on distributor fitting are in place in slots in backing plate.

*g. Install Front Brake Shoe and Lining Assembly.*—Refer to NOTE under *f* above. Place anchor pin (BB) in upper anchor block (CC) with slot in pin alined with slot in anchor block. Position front brake shoe and lining assembly (LL) at backing plate with shoe web engaging slots in wheel cylinder push rods, anchor block, and anchor pin. Insert guide bolt (V) through backing plate and brake shoe web, with square shoulder on bolt engaging square hole in backing plate. Install sleeve (DD), plain washer (EE), and ¼-inch lock washer (FF) on guide bolt; then install ¼-28 nut (GG) on guide bolt and tighten firmly.

*h. Install Brake Shoe Return Springs.*—Install brake shoe return springs (X), installing end of each spring in brake shoe web; then hooking springs onto return spring pins (JJ), using brake spring remover and replacer 41-R-2375-20 as shown in figure 167.

*i. Adjust Brake Shoes.*—Using a conventional ring type brake shoe gage, adjust brake shoes to a diameter of 14.870 to 14.885 inches before installing on axle. Adjustment is made by turning adjusting studs (M) at inner side of backing plate.

## 234. Installation of Brake Assembly

*a. Install Brake Assembly.*

(1) *Front.*—Make sure steering knuckle is properly positioned against steering knuckle support with milled oil drain slot in flange at bottom. Position front brake assembly at steering knuckle support with wheel cylinders positioned in relation to vertical centerline as shown in figure 158. Attach brake assembly to steering knuckle support with 12 cap screws and lock washers as follows: Install eight ⅜-16 x 1¹¹⁄₁₆ cap screws with ⅜-inch lock washers through anchor blocks and thread into steering knuckle support, and install four ⅜-16 x 1⁵⁄₁₆ cap screws with lock washers through backing

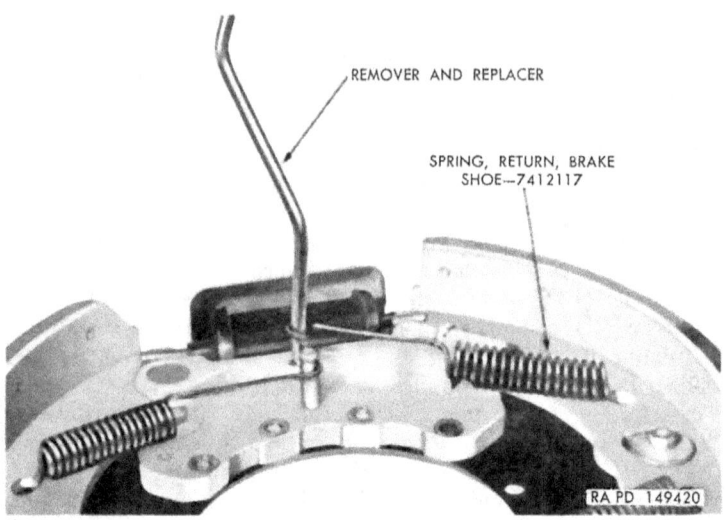

*Figure 167. Installing brake shoe return springs with remover and replacer 41–R–2375–20.*

plate between anchor locks and thread into steering knuckle support. Tighten cap screws evenly and firmly.

(2) *Rear.*—Position rear brake assembly on flange on axle housing with wheel cylinders positioned at top and bottom as shown in figure 158. Attach brake assembly to flange on axle housing with 12 cap screws and safety nuts as follows: Install eight $3/8$–24 x $1 9/16$ cap screws through anchor blocks and flange on axle housing, and install four $3/8$–24 x 1 cap screws through backing plate and flange on axle housing between anchor blocks. Install twelve $3/8$–24 safety nuts on cap screws and tighten evenly and firmly.

b. *Connect Brake Line.*

(1) *Front* (fig. 159).—Place plain copper washer on rubber brake line fitting, thread fitting into distributor fitting and tighten firmly. Insert fitting at other end of rubber brake line assembly through bracket on brake line shield and into bracket at axle housing, install $3/4$-inch internal-teeth lock washer and $3/4$–16 nut on fitting, and tighten firmly. Make sure grommet is properly positioned around rubber brake line in bracket on brake line shield. Thread axle brake line nut into rubber brake line fitting at axle bracket and tighten firmly, using care not to overtighten. Position brake line shield on three spacer nuts on top of steering knuckle support

and secure in place with three ½–20 x ½ cap screws and ½-inch internal-teeth lock washers.

 (2) *Rear* (fig. 160).—Place plain copper washer over special bolt, insert bolt through axle brake line connector, place copper gasket on bolt, then thread bolt into distributor fitting and tighten firmly.

 *c. Install Hub and Bearings.*—Install hub and bearings and adjust bearings as directed in paragraph 256 *e* and *f*.

 *d. Install Brake Drum.*—Install brake drum on studs in brake drum adapter and attach with 18 plain washers and ⅜–24 safety nuts. Tighten nuts to torque of 20 to 27 pound-feet.

## 235. Brake Drums

 *a. Inspection.*—Inspect brake drums for cracks, distortions, or scored braking surface. Place drum in brake drum lathe and check runout of braking surface. If drum is cracked or badly distorted, replace with new part. If braking surface is scored or out-of-round more than 0.005 inch, brake drum can be refinished (*b* below).

 *b. Refinishing Brake Drums.*—Refinish braking surface of brake drum, using a conventional brake drum lathe. The maximum allowable removal of metal is 0.125 inch on the inside diameter. If necessary to go beyond this limit to obtain a true braking surface, replace with new part. After grinding or cutting drum, the braking surface must be honed to remove all traces of tool ridges. When drums are machined over 0.030 inch on the inside diameter, shim stock of a thickness corresponding to the amount of metal removed from drum must be installed between the brake lining and brake shoe (par. 232*a*) to maintain full lining-to-drum contact.

## Section III. REBUILD OF MASTER CYLINDER

## 236. Disassembly of Master Cylinder

 *Note.* Key letters noted in parentheses are in figure 168 unless otherwise indicated.

 *a. General.*—Refer to figure 170 for sectional view of master cylinder assembly showing assembled position of parts.

 *b. Remove Filler Cap Components.*—Unscrew special bolt (N) attaching vent line fitting (M) and filler cap (K) to filler pipe (G). Remove filler cap C-type washer (H) from special bolt (N), then remove filler cap (K), vent line fitting (M), and gaskets (L) from special bolt. Remove filler cap gasket (J) from filler cap (K).

 *c. Remove Filler Pipe.*—Remove filler pipe (G) from top of master cylinder reservoir.

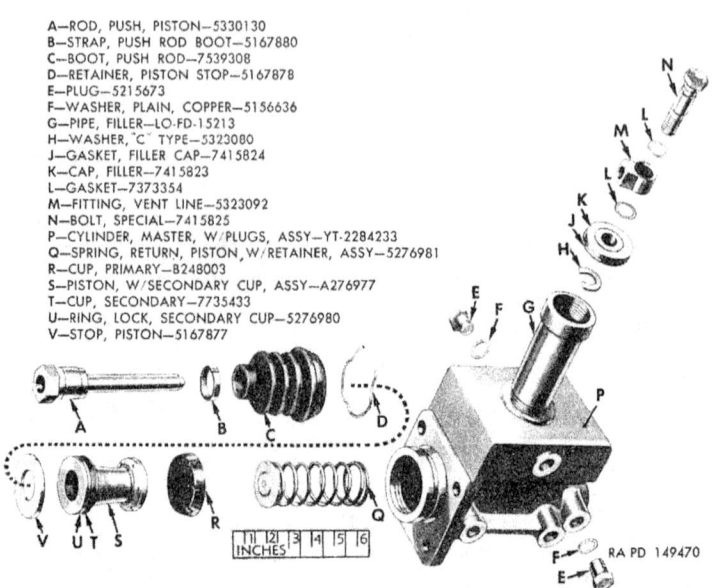

*Figure 168. Master cylinder components.*

*d. Remove Piston Push Rod and Boot.*—Pull push rod boot (C) out of groove on end of cylinder, then remove piston push rod (A) and boot (C). Pry push rod boot strap (B) off boot and push rod, then remove boot from push rod.

*e. Remove Piston Stop, Piston, Primary Cup, and Return Spring.*—Remove piston stop retainer (D) from groove in cylinder bore, then remove piston stop (V) from cylinder. Remove piston with secondary cup assembly (S), primary cup (R), and piston return spring with retainer assembly (Q) from cylinder bore.

*f. Disassemble Piston.*—Remove piston secondary cup lock ring (U) by working ring off over lip of secondary cup (T) (fig. 169).

*Figure 169. Removal and installation of secondary cup lock ring.*

Remove secondary cup (T) from piston; then remove lock ring from piston.

*g. Remove Plugs.*—Remove plug (E) and plain copper washer (F) from each side of cylinder to permit thorough cleaning of interior.

## 237. Cleaning, Inspection, and Repair of Master Cylinder

*Note.* Key letters noted in parentheses are in figure 168 unless otherwise indicated.

*a. Cleaning.*—Wash all parts of master cylinder in denatured alcohol to remove hydraulic brake fluid. Never use cleaning solutions which contain petroleum products for cleaning hydraulic brake parts. Blow inside of cylinder bore and reservoir dry with compressed air, and wipe small parts dry.

*b. Inspection.*
  (2) *Master cylinder body.*—Carefully examine cylinder bore for scored or rusted condition. If either of these conditions are evident, the cylinder bore must be reconditioned by honing (*c* below). Check for damaged threads in tapped holes. If any damage is evident, a new master cylinder body must be used. Make sure inlet port and compensating port (fig. 170) connecting reservoir to cylinder bore (fig. 170) are open.
  (2) *Piston.*—Examine piston for burs or cracks, and for loose protector which is riveted to end of piston. If any damage is evident, replace with new piston with secondary cup assembly (S).
  (3) *Plugs, special bolt, vent line fitting, and filler pipe.*—Examine plugs (E), special bolt (N), vent line fitting (M), and filler pipe (G) for damaged threads and replace with new parts as necessary.
  (4) *Piston return spring and retainer assembly.*—Check piston return spring with retainer assembly (Q) for proper free length and compression (par. 349). Replace with new assembly if not within specifications. Make sure retainer is secure on end of spring.
  (5) *Piston push rod.*—Inspect piston push rod (A) for damaged threads in outer end for roughness at rounded inner end. Roughness on inner end may be honed off. If threads are damaged, replace push rod.
  (6) *Piston primary and secondary cups (R and T), plain copper washers (F), piston stop retainer (D), gaskets (J and L), and push rod boot (C).*—Always discard these parts when rebuilding master cylinder and obtain new parts for assembly.

*c. Repair.*
  (1) *Hone cylinder.*—Select hone of proper size and install in chuck of electric drill clamped in a vise. Slide cylinder

back and forth over revolving hone a few times; then inspect cylinder walls to see if they are cleaned up. Do not hone away any more than is required to remove scores and clean up cylinder. Use burring tool to remove burs which form around inlet and compensating ports. Do not remove more than 0.004 inch from original bore diameter. Refer to repair and rebuild standards (ch. 21) for bore diameters.

(2) *Check piston fit in cylinder bore.*—Insert piston, with cups removed, into cylinder bore and check clearance between piston and cylinder wall with a feeler gage. If clearance is in excess of 0.005 inch, new master cylinder body must be used.

### 238. Assembly of Master Cylinder

*Note.* Key letters noted in parentheses are in figure 168 unless otherwise indicated.

*a. General.*—Refer to figure 170 for sectional view of master cylinder assembly showing assembled position of parts. Dip all parts in hydraulic brake fluid before assembling.

*b. Install Plug.*—Install plug (E) with plain copper washer (F) in opening in each side of cylinder and tighten firmly.

*c. Assemble Piston.*—Place piston secondary cup lock ring (U) over piston, install secondary cup (T) on piston; then install lock ring over secondary cup, working lock ring over lip of cup as shown in figure 169.

*Caution:* Do not use a tool with sharp edges, and use extreme care not to cut or damage lip of cup while installing lock ring.

*d. Install Piston Return Spring, Primary Cup, Piston, and Piston Stop.*—Insert piston return spring with retainer assembly (Q) in cylinder bore with retainer toward open end of cylinder. Install primary cup (R) in cylinder bore with lip of cup over the return spring retainer. Insert piston with secondary cup assembly (S) into cylinder bore with open end of piston toward open end of cylinder, carefully guiding lip of secondary cup into the cylinder. Push piston into cylinder, compressing piston return spring; then install piston stop (V) and secure in place with piston stop retainer (D). Make sure retainer is fully seated in groove in cylinder bore.

*e. Install Piston Push Rod and Boot.*—Install push rod boot (C) on piston push rod (A), with bead at small end of boot seated in groove next to hex end of push rod. Secure boot to push rod with push rod boot strap (B). Insert rounded end of piston push rod through piston stop into piston, and place bead at large end of boot into groove at end of cylinder.

*f. Install Filler Pipe.*—Thread filler pipe (G) into top of master cylinder reservoir and tighten to a torque of 75 to 80 pound-feet.

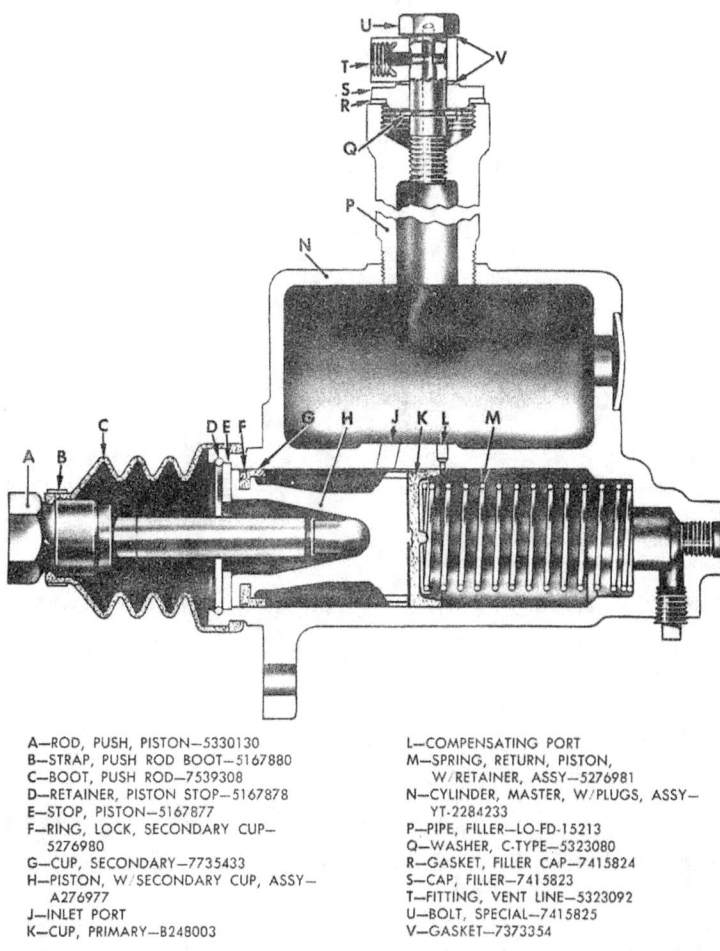

A—ROD, PUSH, PISTON—5330130
B—STRAP, PUSH ROD BOOT—5167880
C—BOOT, PUSH ROD—7539308
D—RETAINER, PISTON STOP—5167878
E—STOP, PISTON—5167877
F—RING, LOCK, SECONDARY CUP—5276980
G—CUP, SECONDARY—7735433
H—PISTON, W/SECONDARY CUP, ASSY—A276977
J—INLET PORT
K—CUP, PRIMARY—B248003
L—COMPENSATING PORT
M—SPRING, RETURN, PISTON, W/RETAINER, ASSY—5276981
N—CYLINDER, MASTER, W/PLUGS, ASSY—YT-2284233
P—PIPE, FILLER—LO-FD-15213
Q—WASHER, C-TYPE—5323080
R—GASKET, FILLER CAP—7415824
S—CAP, FILLER—7415823
T—FITTING, VENT LINE—5323092
U—BOLT, SPECIAL—7415825
V—GASKET—7373354

RA PD 149576

*Figure 170. Sectional view of master cylinder assembly.*

*g. Install Filler Cap Components.*—Install new filler cap gasket (J) in groove in filler cap (K). Place gasket (L), vent line fitting (M), and another gasket (L) over special bolt (N), insert bolt through filler cap (K); then install filler cap C-type washer (H) in groove in bolt. Place filler cap on top of filler pipe, thread special bolt into threads in filler pipe, and tighten to a maximum torque of 30 pound-feet.

## Section IV. REBUILD OF WHEEL CYLINDERS

### 239. Disassembly of Wheel Cylinder

*Note.* Key letters noted in parentheses are in figure 171.

*a.* Pull piston rod (A) out of boot (B) at each end of cylinder body (G). Pull bead on large end of each boot (B) out of groove in ends of body (G) to remove boots.

*b.* Push pistons (C), piston cups (D), piston cup expanders (E), and piston return spring (F) out of body (G).

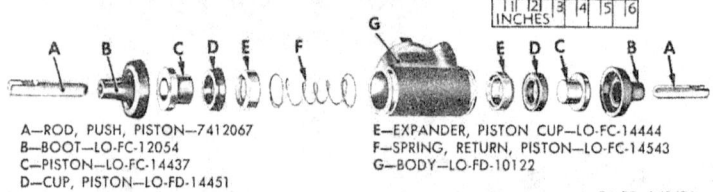

A—ROD, PUSH, PISTON—7412067
B—BOOT—LO-FC-12054
C—PISTON—LO-FC-14437
D—CUP, PISTON—LO-FD-14451
E—EXPANDER, PISTON CUP—LO-FC-14444
F—SPRING, RETURN, PISTON—LO-FC-14543
G—BODY—LO-FD-10122

RA PD 149471

*Figure 171. Wheel cylinder components.*

### 240. Cleaning, Inspection, and Repair of Wheel Cylinder

*Note.* Key letters noted in parentheses are in figure 171.

*a. Cleaning.*—Wash all parts of wheel cylinder in denatured alcohol to remove hydraulic brake fluid. Never use cleaning solutions which contain petroleum products for cleaning hydraulic brake parts. Wipe parts dry with a clean cloth.

*b. Inspection.*—Carefully examine cylinder bore for scored or rusted condition. If either of these conditions is evident, the cylinder bore must be reconditioned by honing (*c* below). Examine pistons (C) and piston cup expanders (E) for burs or cracks. Install new parts if damaged in any way. Check piston return spring (F) for free length and compression (par. 350); replace with new spring if not within specified limits. Always install new piston cups (D) and boots (B) when rebuilding wheel cylinders.

*c. Repair.*—Recondition wheel cylinder bore in same manner described for master cylinder (par. 237c).

### 241. Assembly of Wheel Cylinder

*Note.* Key letters noted in parentheses are in figure 171. Refer to figure 172 for sectional view of wheel cylinder showing assembled position of parts.

*a.* Dip all parts in hydraulic brake fluid before assembling. Install piston cup (D) and piston cup expander (E) on each piston (C). Insert one piston assembly into body (G) so that the expander is facing

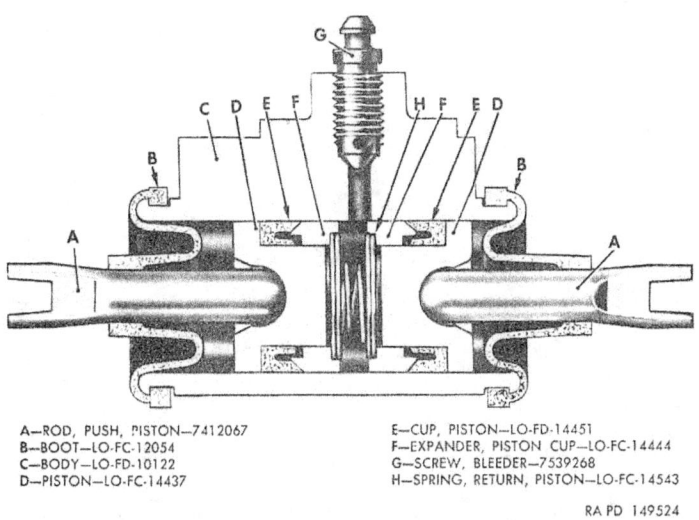

A—ROD, PUSH, PISTON—7412067
B—BOOT—LO-FC-12054
C—BODY—LO-FD-10122
D—PISTON—LO-FC-14437
E—CUP, PISTON—LO-FD-14451
F—EXPANDER, PISTON CUP—LO-FC-14444
G—SCREW, BLEEDER—7539268
H—SPRING, RETURN, PISTON—LO-FC-14543

RA PD 149524

*Figure 172. Sectional view of wheel cylinder assembly.*

inward, install piston return spring (F) in body; then insert other piston assembly into other end of body. Make sure ends of piston return spring are seated in counterbore in piston cup expanders, and use care when guiding lips of cups into body not to damage cups.

*b.* Install boot (B) on each end of body (G), seating bead on large end of each boot in groove in each end of body. Insert rounded end of piston push rod (A) through small end of each boot until ends of rods extend approximately seven-eighths of an inch out of ends of boots. Plug tapped openings in body to prevent entrance of dirt prior to installation.

# CHAPTER 11

# PARKING BRAKE SYSTEM

## Section I. DESCRIPTION AND DATA

### 242. Description

*a. Mechanical Parking Brake.*—Mechanical parking brake system comprises an external-contracting one-piece band type brake located at rear of transfer assembly, operated by a parking brake lever located in cab and connected to brake band through rods and a relay lever. The lever bracket and relay lever are equipped with replaceable bushings. A pilot pressed into center of brake drum web fits into counterbore in transfer drive flange to locate drum. Brake lining is secured to brake band by 26 tubular rivets.

*b. Temporary (Electric) Parking Brake.*—Temporary (electric) parking brake system comprises a solenoid valve connected into the master cylinder hydraulic outlet line, operated by a two-position switch on instrument panel. This parking brake is for emergency use only in the event of failure of the mechanical parking brake, and should not be depended upon to hold the vehicle for extended periods.

### 243. Data

 *a. Mechanical Parking Brake.*

| | |
|---|---|
| Type | external-contracting band |
| Location | rear of transfer |
| Brake drum diameter | 9½ in. |
| Brake lining width | 3 in. |
| Brake lining thickness | 5/16 in. |

 *b. Temporary (Electric) Parking Brake.*

| | |
|---|---|
| Switch: | |
|  Manufacturer | Delco-Remy |
|  Model | 1997889 |
| Solenoid valve: | |
|  Manufacturer | Wagner Electric Corp |
|  Model | FD–15018H |

## Section II. REBUILD OF MECHANICAL PARKING BRAKE

### 244. General

Replacement of mechanical parking brake components is covered in TM 9-819A. Procedures which follow cover inspection and repair operations which are beyond the scope of the using organization.

### 245. Inspection of Mechanical Parking Brake Components

*Note.* Key letters noted in parentheses are in figure 173.

*a. General.*—Wash all parts except brake band and lining assembly in dry-cleaning solvent or volatile mineral spirits and dry thoroughly. Examine all small parts for distortion, evidence of wear, and damaged threads and replace with new parts as necessary. Check release spring (LL) and tension spring (PP) for free length and compression (par. 351). If not within specified limits, replace with new parts. Inspect major components of mechanical parking brake as directed in *b* through *f* below.

*b. Lever Bracket and Bushings.*—Examine lever bracket assembly (M) for cracks or visible distortion. Replace with new part if damaged in any way. Check bushings in bracket for wear (par. 351). If worn beyond specified limits, replace bushings (par. 246a).

*c. Relay Lever and Bushings.*—Examine relay lever assembly (W) for cracks or visible distortion. Make sure arms are securely welded to lever hub. If damaged in any way, replace with new part. Check bushings in hub for wear (par. 351). If worn beyond specified limits, replace bushings (par. 246b).

*d. Brake Drum and Pilot.*—Examine parking brake drum assembly (HH) for visible cracks or distortion and for rough or scored braking surface. Check runout of braking surface, which should not exceed 0.005-inch total dial indicator reading. If brake drum is cracked, or if runout is above specified maximum by an excessive amount, replace drum assembly. If braking surface is rough or scored or if runout is slight, drum can be refinished (par 246c). Check pilot in center of brake drum web for damage or looseness, and replace if necessary (par 246d).

*e. Brake Band and Lining.*—Inspect parking brake band with lining assembly (JJ) for distorted band and for loose or broken band anchor brackets. Replace if damaged in any way. Examine band lining for roughness, wear, or loose rivets. If lining is worn down close to rivet heads or if damaged in any way, replace lining (par. 246e).

*f. Special Pin and Relay Lever Shaft.*—Check special pin (N) and relay lever shaft (U) for wear at bearing contact area (par. 351). Replace with new parts if not within specified limits.

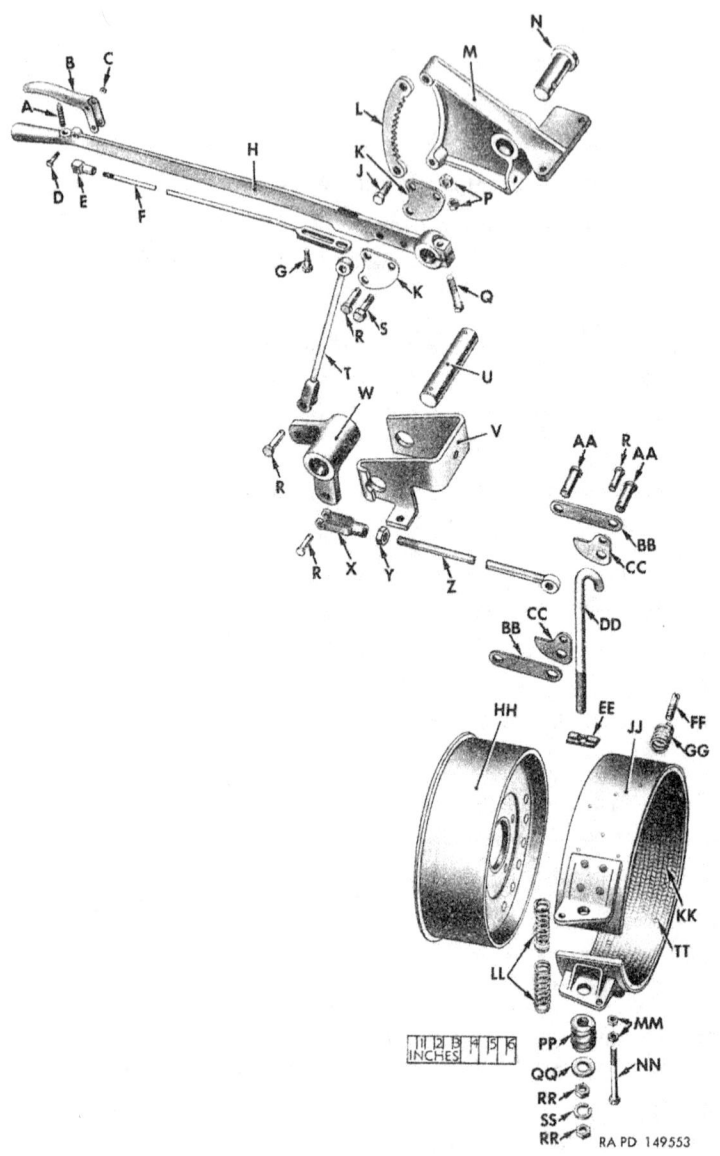

*Figure 173. Mechanical parking brake components.*

A—SPRING, LATCH—A291252
B—LATCH, GRIP—7350481
C—NUT, MACH-SCREW, NO. 10—110924
D—SCREW, HEX-HD, NO. 10-24 X 7/8—7415983
E—END, LEVER ROD—7350480
F—ROD, LEVER—7350485
G—SCREW, SPECIAL, 1/4-28 X 7/8—A291224
H—LEVER, PARKING BRAKE, ASSY—7350482
J—BOLT, 3/8-24 X 2—181648
K—ADAPTER, LEVER—7350476
L—SECTOR—7413006
M—BRACKET, LEVER, ASSY—YT-2265707
N—PIN, SPECIAL—7350483
P—NUT, SAFETY, 3/8-24—442799
Q—BOLT, CLAMP, 3/8-24 X 1 3/4—181646
R—PIN, CLEVIS, 3/8 X 15/16—138084
S—BOLT, 3/8-24 X 1 3/8—181641
T—ROD, PARKING-BRAKE-LEVER-TO-RELAY-LEVER, ASSY—7350486
U—SHAFT, RELAY LEVER—7350512
V—BRACKET, RELAY LEVER—7350477
W—LEVER, RELAY, ASSY—7350484
X—YOKE—144243
Y—NUT, 3/8-24—120369
Z—ROD, RELAY-LEVER-TO-CAM-LEVER, ASSY—7350510
AA—PIN, CLEVIS, 1/2 X 1 13/64—138086
BB—LINK, SPACER—7735388
CC—LEVER, CAM—7411344
DD—BOLT, ADJUSTING—6245684
EE—SHOE, OPERATING, CAM LEVER—6245683
FF—SCREW, ANCHOR—6245840
GG—SPRING, ANCHOR—6245694
HH—DRUM, PARKING BRAKE, ASSY—7411357
JJ—BAND, PARKING BRAKE, W/LINING, ASSY—6248820
KK—LINING, BRAKE BAND—YT-2196088
LL—SPRING, RELEASE—5284008
MM—NUT, 1/4-20—120375
NN—BOLT, LOCATING, 1/4-20 X 3 1/4—187068
PP—SPRING, TENSION—6245692
QQ—WASHER, PLAIN,—120396
RR—NUT, 7/16-14—124834
SS—WASHER, LOCK, 7/16-IN.—120383
TT—RIVET, TUBE, 9/64 X 3/8—136497

*Figure 173.*—Continued.

## 246. Repair of Mechanical Parking Brake Components

*a. Lever Bracket Bushing Replacement* (fig. 174). Press old bushings out of bracket. Press new bushing into each end of bracket until outer end of each bushing is flush with face of bracket. After pressing bushings into place burnish to diameter shown in figure 174.

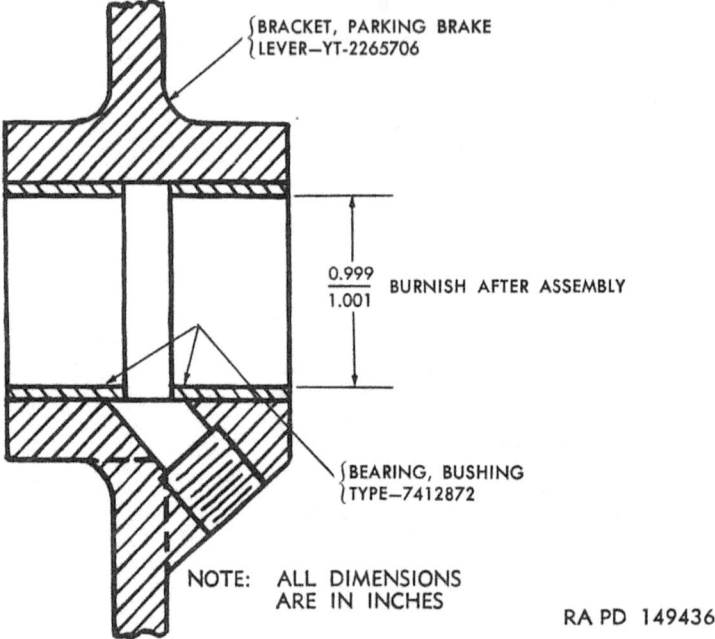

*Figure 174. Sectional view of lever bracket and bushings.*

*b. Relay Lever Bushing Replacement* (fig. 175).—Press old bushings out of lever hub. Press new bushing into each end of hub until outer edge of each bushing is at bottom of chamfer in hub bore. After pressing bushings into place, line ream to diameter shown in figure 175.

*c. Refinishing Brake Drum.*—When refinishing brake drum, remove only enough metal to smooth up braking surface or to correct runout. A $\frac{1}{16}$-inch radius must be maintained at brake drum flange.

*d. Brake Drum Pilot Replacement* (fig. 176).—Press pilot out of brake drum web. Press new pilot into brake drum web until it extends through the front (flanged) side to dimension shown in figure 176.

*e. Brake Band Lining Replacement.*
  (1) Remove rivets, using deliner punch in a conventional brake relining machine. Remove lining from brake band.

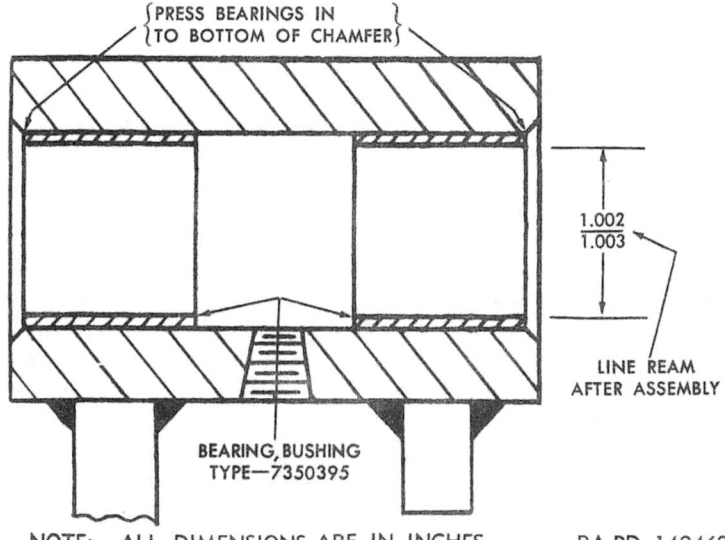

*Figure 175. Sectional view of relay lever and bushings.*

(2) Thoroughly clean brake band, using a wire brush if necessary to remove corrosion.

(3) Position new lining in brake band with two end holes alined, install two 9/64 x 3/8 tubular rivets, and upset rivets with a conventional brake relining machine. Install and upset balance of 9/64 x 3/8 tubular rivets (total of 26). Make sure lining fits firmly against band at all points and that rivets are properly upset.

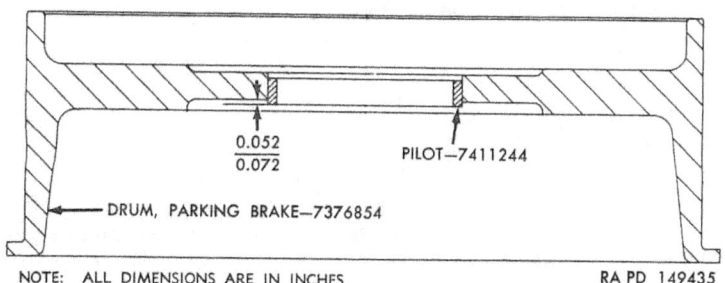

*Figure 176. Sectional view of brake drum and pilot.*

277

## Section III. REBUILD OF TEMPORARY (ELECTRIC) PARKING BRAKE VALVE

### 247. General

Temporary (electric) parking brake valve (fig. 177) is used for emergency parking by retaining hydraulic pressure in the brake system by means of a solenoid-actuated seal which is released by breaking the circuit supplying the solenoid. Valve consists primarily of a plunger and seal cup contained in a pressuretight tube surrounded by a solenoid coil, which in turn is contained in a watertight case. A check valve cup and spring are also contained in outlet end of valve to allow hydraulic brake fluid to be introduced into the brake system when the solenoid is energized. Waterproof cables, equipped with bayonet type connectors and female connector shells, are provided for connecting the solenoid coil into the vehicle electrical system. A two-position switch, mounted on instrument panel in cab, is included in the circuit to energize and deenergize the solenoid as required.

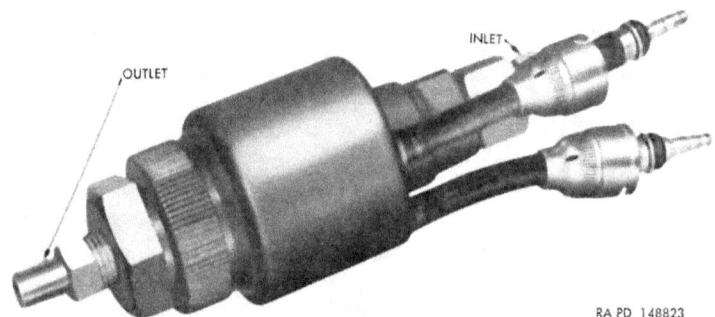

*Figure 177. Overall view of temporary (electric) parking brake valve.*

### 248. Operation of Parking Brake Valve

*Note.* Key letters noted in parentheses are in figure 178.

*a.* Solenoid coil can be energized (switch turned "ON") either before or after applying the brakes, but must be energized before brakes are released to hold brakes applied.

*b.* When brakes are applied with solenoid coil (H) deenergized (switch turned "OFF"), hydraulic brake fluid from the master cylinder enters at the inlet plug (M), passes around the hexagon-shaped plunger (J), through the center hole in plunger stop (E), and check valve cup (P), and out the outlet plug (A) to the air-hydraulic power cylinder, applying the brakes. If solenoid coil (H) is now energized by turning switch "ON," magnetic field causes plunger (J) to seat against plunger stop (E) and plunger cup (F) seats around center hole in plunger stop, preventing brake fluid from returning to master cylinder, holding brakes applied.

*c.* When brakes are applied with solenoid coil (H) energized (switch turned "ON"), hydraulic brake fluid from the master cylinder enters at the inlet plug (M) and passes around the hexagon-shaped plunger (J). Since the plunger cup (F) is sealing center hole in plunger stop (E), brake fluid passes through the small holes near edge of plunger stop and forces the check valve cup (P) away from plunger stop, then passes through center hole in cup and out the outlet plug (A) to the air-hydraulic power cylinder applying the brakes. As

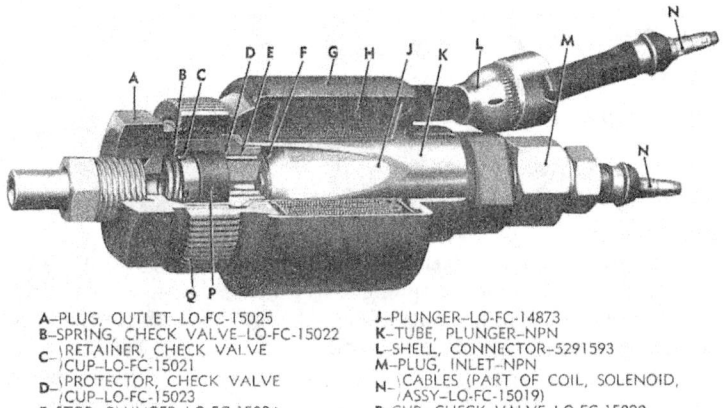

A—PLUG, OUTLET–LO-FC-15025
B—SPRING, CHECK VALVE–LO-FC-15022
C—\RETAINER, CHECK VALVE
   /CUP–LO-FC-15021
D—\PROTECTOR, CHECK VALVE
   /CUP–LO-FC-15023
E—STOP, PLUNGER–LO-FC-15024
F—CUP, PLUNGER–LO-FC-14974
G—CASE, SOLENOID–LO-FC-15086
H—\COIL, SOLENOID, ASSY–
   /LO-FC-15019
J—PLUNGER–LO-FC-14873
K—TUBE, PLUNGER–NPN
L—SHELL, CONNECTOR–5291593
M—PLUG, INLET–NPN
N—\CABLES (PART OF COIL, SOLENOID,
   /ASSY–LO-FC-15019)
P—CUP, CHECK VALVE–LO-FC-15020
Q—\END, KNURLED, SOLENOID–
   /LO-Q-15027

RA PD 148822

*Figure 178. Cut-away section of temporary (electric) parking brake valve*

soon as pressure is removed from brake fluid passing through valve (brake pedal released), check valve spring (B) seats check valve cup (P) against plunger stop (E), sealing the holes in edge of plunger stop against the return of brake fluid and holding the brakes applied.

*d.* When solenoid coils (H) are deenergized by turning switch "OFF," magnetic field around plunger collapses, releasing plunger. Hydraulic brake fluid pressure on outlet end forces plunger away from plunger stop (E) and returns through center holes in check valve cup (P) and plunger stop (E) through the inlet plug (M) to the master cylinder, permitting brakes to release.

**Warning:** Temporary (electric) parking brake valve is not designed to hold high pressures for an indefinite period, and its use should therefore be restricted to EMERGENCY parking only, and for not over 1 hour at a time. Use the mechanical parking brake whenever possible.

## 249. Testing Parking Brake Valve

Connect inlet end of valve to source of hydraulic brake fluid pressure, and connect a hydraulic pressure gage to outlet end. Connect cables to source of 24-volt current. With 24 volts applied to the solenoid coil, valve must trap 1,100 p. s. i. hydraulic pressure which must not drop below 1,000 p. s. i. within 10 seconds. If pressure drops below 1,000 p. s. i. within 10 seconds, valve assembly must be rebuilt or replaced.

## 250. Disassembly of Parking Brake Valve

*Note.* Key letters in text refer to figure 178.

*a.* Suitable radius blocks with serrations on inside diameter to fit knurled end of solenoid should be used when clamping the parking brake valve in vise for disassembly.

*b.* With solenoid knurled end (Q) clamped in vise, remove outlet plug (A).

*c.* Remove check valve cup (P), check valve cup retainer (C), check valve spring (B), and plunger stop (E). Remove check valve cup protector (D) from plunger stop (E) by inserting a small wire through hole near edge of plunger stop.

*d.* Remove plunger (J) from plunger tube (K); then remove plunger cup (F) from plunger.

*e.* Do not attempt to further disassemble parking brake valve. If inspection (par. 251*b*) reveals any defects in solenoid case (G), solenoid coil (H), or plunger tube (K), the complete parking brake valve assembly must be replaced.

## 251. Cleaning and Inspection of Parking Brake Valve

*Note.* Key letters noted in parenthesis are in figure 178.

*a. Cleaning.*—Wash all parts except solenoid, case, and tube assembly in denatured alcohol. Never use cleaning solutions containing petroleum products for cleaning hydraulic brake parts. Wipe or blow hydraulic brake fluid out of inside of plunger tube (K).

*b. Inspection.*
  (1) *Plunger and plunger cup.*—Examine face of plunger (J) around cup bore; surface must be smooth, and edge of cup bore must be sharp and free of nicks and burs. Face of plunger cup (F) must be smooth and without transverse grooves; concentric grooves, if small, are not harmful. Install cup in plunger; cup must extend at least 0.012 inch and

not more than 0.018 inch from end of plunger. Best operation is obtained with cup extending 0.012 to 0.014 inch. Make sure small bleed hole in plunger at bottom of cup bore is open to permit air or fluid behind cup to escape so cup will seat in bottom of bore.

(2) *Plunger stop.*—All surfaces of plunger stop (E) must be free of burs and scratches. Make sure eight small holes near edge of stop are clean and open. Make sure tapered shoulder is perfectly smooth and will form a tight seal against seat at end of plunger tube (K).

(3) *Check valve cup.*—Inspect check valve cup (P) for nicks or scratches at the sealing edges or other imperfections. Replace cup if any damage is evident.

(4) *Check valve spring.*—Inspect check valve spring (B) for free length and compression (par. 352). Replace spring if not within specified limits.

(5) *Outlet end plug.* Inspect outlet plug (A) for nicks or scratches on tapered tube seat and for damaged threads. Bottom of large bore in end plug must have concentric grooves without nicks or scratches, to provide a good seat for outer end of plunger stop when assembled.

(6) *Solenoid, case, and tube assembly.*—Interior of brass plunger tube (K) must be smooth, and the tapered sealing face toward threaded opening must not be marred or scratched. Check circuit continuity through solenoid coil (H) and check coil for ground, using 24-volt battery current and test lamp. If coil tests satisfactory for circuit continuity and ground, test resistance of coil with ohmmeter; resistance must be between 75 and 90 ohms. Check condition of cables, terminals, and connector shells. Grommet, bushing, and shell can be replaced if damaged (par. 293). If any other damage is evident, replace complete parking brake valve assembly.

### 252. Assembly of Parking Brake Valve

*Note.* Key letters noted in parenthesis are in figure 178.

*a.* Coat all internal parts with hydraulic brake fluid before assembling.

*b.* Place plunger stop (E) in solenoid end, with check valve cup bore facing open end.

*c.* Drop check valve cup protector (D) into check valve cup bore in plunger stop; then place check valve cup (P) in plunger stop with open end of cup facing outward.

*d.* Place check valve cup retainer (C) inside of check valve cup, install check valve spring (B) inside of retainer; then position outlet plug (A) over spring and plunger stop and thread into solenoid knurled end (Q).

*e.* Clamp solenoid knurled end in vise, using radius blocks (par. 250*a*) to prevent distorting unit or damaging serrations. Tighten outlet plug to a torque of 145 to 155 pound-feet torque.

# CHAPTER 12

# WHEELS AND HUBS

## Section I. DESCRIPTION AND DATA

### 253. Description

a. *Wheels.*
   (1) Single front wheels are used on all truck models covered by this manual; single rear wheels are used on some models and dual rear wheels are used on other models. Wheels used on vehicles equipped with single rear wheels are not interchangeable with wheels used on models equipped with dual rear wheels; however, all wheels used on any vehicle are interchangeable with each other.
   (2) Each wheel consists of a riveted disk and rim assembly. One tire bead seat is formed by the rim flange and the removable tire retaining ring forms the other tire bead seat. Wheels do not pilot on hub; taper on wheel nuts engage chamfered holes in wheel, positioning wheel concentric with hub. When dual rear wheels are used, two nuts are used on each wheel stud to attach wheels to hub. The inner nuts, which have internal and external threads, secure inner wheel to hub. The outer wheel is installed over the inner nuts, and the outer nuts are threaded onto the external threads of the inner nuts. Front wheel installation is illustrated in figure 179, and both single and dual rear wheel installations are shown in figure 180.

b. *Hubs.*—Front and rear hubs are identical on all vehicles; however, different wheel studs and brake drum backs are used and drive flange studs must be installed in correct end of hub for right side, left side, front, single rear, and dual rear wheel application. Flange on hub is nearer one end than the other. When used at front on all vehicles and at rear when single rear wheels are used, the drive flange studs are installed in the short end of the hub. When dual rear wheels are used, the drive flange studs are installed in the long end of the hub. Refer to figure 182 for identification of brake drum back to

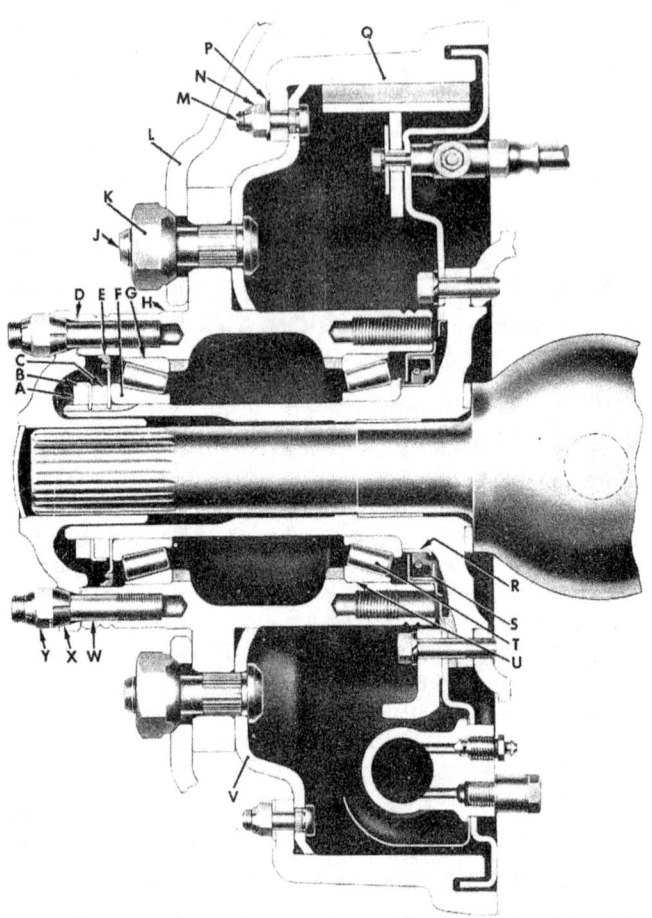

A—NUT, ADJUSTING, WHEEL BEARING—7411379
B—LOCK, ADJUSTING NUT—7411378
C—NUT, ADJUSTING, WHEEL BEARING—7411379
D—GASKET—7411265
E—SEAL, OIL, OUTER, ASSY—7411430
F—CONE, ROLLER BEARING, OUTER—712868
G—CUP, ROLLER BEARING, OUTER—712869
H—HUB, W/BEARING CUPS, ASSY—7411427
J—{STUD, WHEEL (RIGHT)—501245 / STUD, WHEEL (LEFT)—501246
K—{NUT, WHEEL STUD (RIGHT)—537803 / NUT, WHEEL STUD (LEFT)—537804
L—{WHEEL, ASSY (SINGLE WHEELS)—7389617 / WHEEL, ASSY (DUAL WHEELS)—7389620
M—STUD—7411426
N—NUT, SAFETY, 3/8-24—442799
P—WASHER, PLAIN—120394
Q—DRUM, BRAKE—7411425
R—SLEEVE, INNER OIL SEAL—7411433
S—SEAL, OIL, INNER, ASSY—7411429
T—CONE, ROLLER BEARING, INNER—712868
U—CUP, ROLLER BEARING, INNER—712869
V—BACK, BRAKE DRUM—7413231
W—STUD, 2-7/8-IN LONG—7411269
X—WEDGE, DOWEL, TAPERED—7411264
Y—NUT, SAFETY, 1/2-20—442801

RA PD 149593

*Figure 179. Sectional view of front hub installed on axle.*

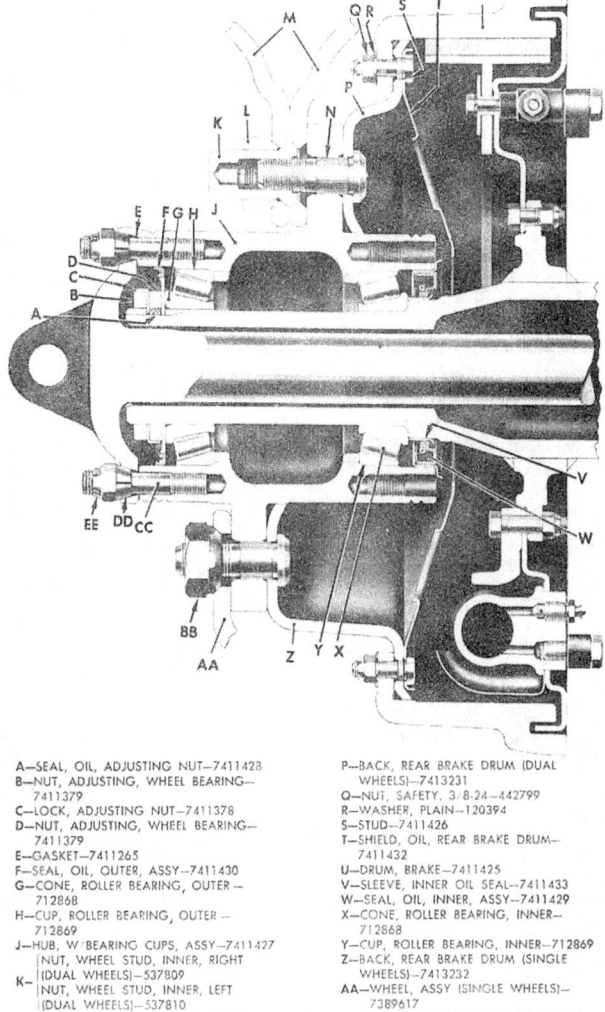

A—SEAL, OIL, ADJUSTING NUT—7411428
B—NUT, ADJUSTING, WHEEL BEARING—7411379
C—LOCK, ADJUSTING NUT—7411378
D—NUT, ADJUSTING, WHEEL BEARING—7411379
E—GASKET—7411265
F—SEAL, OIL, OUTER, ASSY—7411430
G—CONE, ROLLER BEARING, OUTER—712868
H—CUP, ROLLER BEARING, OUTER—712869
J—HUB, W/BEARING CUPS, ASSY—7411427
K—{ NUT, WHEEL STUD, INNER, RIGHT (DUAL WHEELS)—537809
      NUT, WHEEL STUD, INNER, LEFT (DUAL WHEELS)—537810
L—{ NUT, WHEEL STUD, OUTER, RIGHT (DUAL WHEELS)—537805
     NUT, WHEEL STUD, OUTER, LEFT (DUAL WHEELS)—537808
M—WHEEL, ASSY (DUAL WHEELS)—7389620
N—{ STUD, WHEEL (RIGHT)—501245
      STUD, WHEEL (LEFT)—501246

P—BACK, REAR BRAKE DRUM (DUAL WHEELS)—7413231
Q—NUT, SAFETY, 3/8-24—442799
R—WASHER, PLAIN—120394
S—STUD—7411426
T—SHIELD, OIL, REAR BRAKE DRUM—7411432
U—DRUM, BRAKE—7411425
V—SLEEVE, INNER OIL SEAL—7411433
W—SEAL, OIL, INNER, ASSY—7411429
X—CONE, ROLLER BEARING, INNER—712868
Y—CUP, ROLLER BEARING, INNER—712869
Z—BACK, REAR BRAKE DRUM (SINGLE WHEELS)—7413232
AA—WHEEL, ASSY (SINGLE WHEELS)—7389617
BB—{ NUT, WHEEL STUD, RIGHT (SINGLE WHEELS)—537803
      NUT, WHEEL STUD, LEFT (SINGLE WHEELS)—537804
CC—STUD, 2-7/8-IN LONG—7411269
DD—WEDGE, DOWEL, TAPERED—7411264
EE—NUT, SAFETY, 1/2-20—442801

RA PD 149594

*Figure 180. Sectional view of rear hub installed on axle.*

be used for a specific application, and for wheel studs to be used on right and left sides. Brake drum backs are secured to the hub flange by the fluted-shoulder wheel studs which are pressed into the hub flange.

## 254. Data

*a. Wheels.*

```
Ordnance number:
    Single front and rear_____ 7389617
    Single front and dual rear_____ 7389620
Rim size_____ 20 x 7.50
Offset:
    Single front and rear_____ 5⅛ in.
    Single front and dual rear_____ 6³⁄₁₆ in.
Bolt circle diameter_____ 8¾ in.
Wheel bore diameter_____ 6.469 to 6.473 in.
```

*b. Hubs.*

```
Wheel stud circle diameter_____ 8¾ in.
Drive flange stud holes:
    Diameter of stud circle_____ 5⁵⁄₁₆ in.
    Hole threads_____ ½-13
Drive flange studs:
    Thread size_____ ½-13-20
    Stud length_____ 2⅞ in.
    Stud driven height (when tightened to 50 to 60
        pound-feet)_____ 1⁵⁄₁₆ to 1⁷⁄₁₆ in.
```

## Section II. REBUILD OF WHEELS AND HUBS

## 255. Wheels

*a. Inspection.*—Inspect wheel (disk and rim assembly) for visible distortion, loose rivets, or other damage. Mount wheel in a suitable

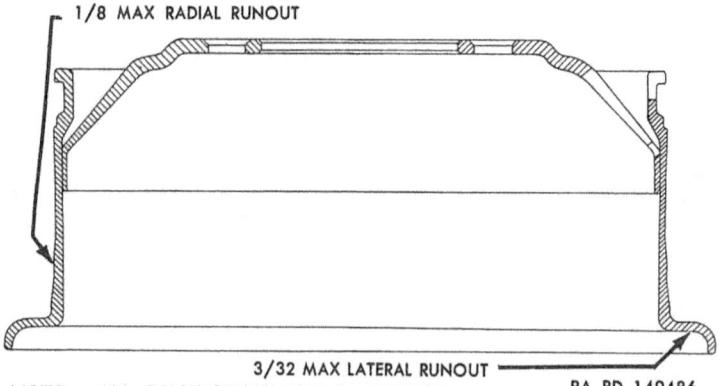

NOTE: ALL DIMENSIONS ARE IN INCHES     RA PD 149486

*Figure 181. Wheel (disk and rim assembly) showing checking points.*

fixture and check for lateral runout (wobble) and for radial runout (out-of-round) at points shown in figure 181. If wheel is only slightly distorted or if rivets are loose, wheel can be repaired (*b* below). If badly damaged, replace with new part.

*b. Repair.*—If adequate equipment is available, wheels that are slightly distorted can be straightened. Lateral and radial runout must not exceed maximum limits shown in figure 181. Tighten or replace any loose rivets attaching disk to rim.

*a. General.*—The procedures which follow cover removal, inspection, and repair of the hub, cups, studs, and brake drum back assembly, and installation and adjustment of the hub and bearings. Sectional views of these parts installed on the axle are shown in figures 179 and 180. Key letters in text refer to figure 182.

## 256. Hubs and Bearings

*b. Hub and Bearing Removal.*
  (1) Remove brake drum (par. 229*a*).
  (2) Remove eight nuts from drive flange studs. Strike hub drive flange (front axle) or axle shaft (rear axle) a sharp blow with a soft metal hammer to loosen tapered dowel wedges; then remove dowel wedges from studs. Remove hub drive flange (front axle) or axle shaft (rear axle). On front axles, two ½-20 puller screws may be used in tapped holes in hub drive flange if necessary to remove hub drive flange.
  (3) Bend tangs of adjusting nut lock (C) away from wheel bearing adjusting nuts (B and D). Remove wheel bearing adjusting nut (B), adjusting nut lock (C), wheel bearing adjusting nut (D), and outer oil seal assembly (F) from steering knuckle (front axle) or axle housing (rear axle). Pull hub and bearing assembly straight off steering knuckle or axle housing.
  (4) Lift outer roller bearing cone (G) out of outer end of hub. Using a suitable driver through outer end of hub to exert force on inner roller bearing cone, force inner roller bearing cone (L) and inner oil seal assembly (M) out of inner end of hub.

*c. Inspection.*
  (1) Inspect brake drum back (P) and brake drum oil shield (S) (used at rear only) for distortion and replace if this condition is found (*d*(2) below). Examine studs (R) which are pressed into brake drum back for damaged threads. If damaged, replace studs (*d*(1) below).
  (2) Examine right and left wheel studs (Q) for damaged threads. If damaged, replace studs (*d*(2) below).

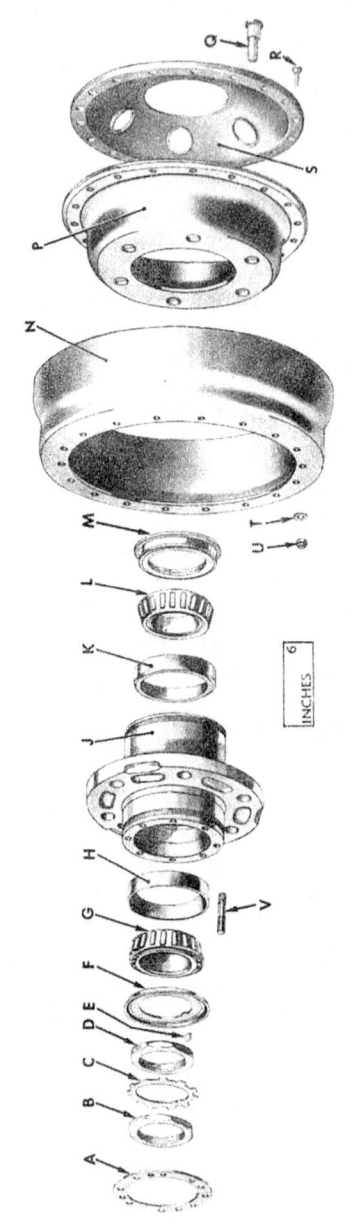

A—GASKET—7411265.
B—NUT, ADJUSTING, WHEEL BEARING—7411379
C—LOCK, ADJUSTING NUT—7411378
D—NUT, ADJUSTING, WHEEL BEARING—7411379
E—SEAL, OIL, ADJUSTING NUT (REAR ONLY)—7411428
F—SEAL, OIL, OUTER, ASS'Y—7411430
G—CONE, ROLLER BEARING (OUTER)—712868
H—CUP, ROLLER BEARING (OUTER)—712869
J—HUB, W/ BEARING CUPS, ASS'Y—7411427
K—CUP, ROLLER BEARING (INNER)—712869
L—CONE, ROLLER BEARING (INNER)—712868
M—SEAL, OIL, INNER, ASS'Y—7411429
N—DRUM, BRAKE—7411425
 BACK, FRONT BRAKE DRUM—7413231
 BACK, REAR BRAKE DRUM (SINGLE WHEELS)—7413232
P—BACK, REAR BRAKE DRUM (DUAL WHEELS)—7413231
Q—{STUD, WHEEL (RIGHT)—501245
 {STUD, WHEEL (LEFT)—501246
R—STUD—7411326
S—{SHIELD, OIL, BRAKE DRUM (REAR ONLY)—7411143
T—WASHER, PLAIN—120394
U—NUT, SAFETY, 3 8-24—442799
V—STUD, 2-7-8-IN LONG—7411269

RA PD 149433

*Figure 182. Hub, bearing, drum back, and stud components.*

(3) Inspect drive flange studs (V) for damaged threads or bent condition. If bent or if threads are damaged, replace studs ($d(3)$ below).

(4) Examine outer and inner roller bearing cups (H and K) for pitting, cracks, or evidence of wear. If any of these conditions are evident, replace bearing cups ($d(4)$ below).

d. *Repair.*

(1) *Drum back stud replacement.*—Press studs (R) out of brake drum back, supporting drum back near studs to prevent distorting drum back. Brake drum oil shield (S) (used at rear only) will be removed when studs are pressed out.

> *Note.* Before installing studs and oil shield on rear hub, right and left wheel studs (Q) and brake drum back (P) must be replaced if necessary as indicated by inspection (*b* above).

Heads of wheel studs are not accessible after oil shield is installed. On rear hub only, position brake drum oil shield on drum back. Support drum back near outer edge to prevent distortion. Press studs into drum back until heads of studs seat firmly against drum back or oil shield.

(2) *Brake drum back or wheel stud replacement.*—To remove brake drum back (P) or right and left wheel studs (Q), press wheel studs out of hub flange and brake drum back. Position drum back on hub flange, referring to figure 182 for identification of drum back to be used for front, single rear, or dual rear wheels and for wheel studs to be used on right and left sides. At rear only, brake drum oil shield (S) and studs (R) must be removed from drum back ((1) above) before wheel studs can be installed. Insert studs through brake drum back and hub flange as far as possible by hand; then support hub flange and press studs in until heads of studs bottom against brake drum back. At rear only, install brake drum oil shield (S) and studs (R) ((1) above).

(3) *Drive flange stud replacement.*—Damaged studs (V) can be removed from end of hub with a stud remover and replacer. If all studs are removed, or when installing studs in a new hub and cup assembly, determine end of hub in which studs are to be installed (par. 253*b*). Make sure threads in tapped holes are not damaged and that holes are not partially filled with grease or dirt which would prevent driving studs to proper height. Tighten studs into hub, using a stud remover and replacer, to a torque of 50 to 60 pound-feet; then check stud height. Studs should extend from end of hub $1\frac{5}{16}$ to $1\frac{7}{16}$ inches after being tightened to above torque.

(4) *Bearing cup replacement.*—Using a brass drift through opposite end of hub, drive out each bearing cup. Four knockout notches are provided in each bearing flange on inside of hub. Alternately drive on opposite sides of bearing cup to prevent cocking bearing cup and damaging machined bore in hub. Install new bearing cups in hub, driving them into place with replacer 41–R–2392–635 as shown in figure 183. Make sure cups are driven in squarely and are fully seated against flanges on inside of hub.

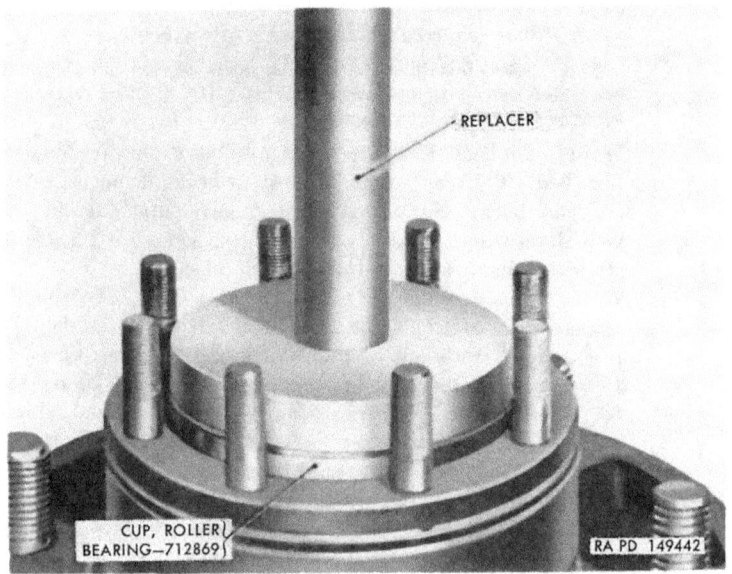

*Figure 183. Installing bearing cups in hub with replacer 41–R–2292–635.*

e. *Hub and Bearing Installation.*
  (1) Clean, inspect, and lubricate outer and inner roller bearing cones (G and L), inside of hub with bearing cups assembly (J), and steering knuckle or axle housing as directed in TM 9–819A.
  (2) Place inner roller bearing cone (L) in inner end of hub. Position new inner oil seal assembly (M) on inner end of hub and drive into place with replacer 41–R–2392–635 (fig. 184). Oil seal flange must seat against inner end of hub.
  (3) Install hub assembly on steering knuckle (front axle) or axle housing (rear axle), using care not to damage inner oil seal. Place outer rolling bearing cone (G) on steering knuckle or axle housing and press into outer end of hub with

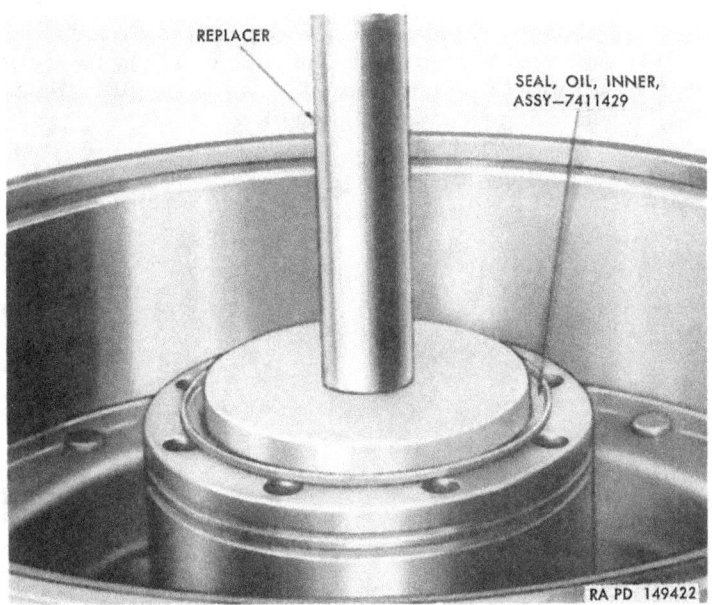

*Figure 184. Installing inner oil seal with replacer 41-R-2392-635.*

fingers. Install outer oil seal assembly (F) on steering knuckle or axle housing and press into hub against bearing. On rear axles only, install adjusting nut oil seal (E) in groove in axle housing.

(4) Install wheel bearing adjusting nut (D) on steering knuckle or axle housing, using care not to dislodge adjusting nut oil seal (E) at rear axle. Adjust bearings, and complete the installation as directed in *f* below.

*f. Bearing Adjustment.*

(1) Using wrench 41-W-3825-66 at front or 7950690 at rear with torque wrench 41-W-3634, tighten wheel bearing adjusting nut (D) to a torque of 60 to 75 pound-feet; then back nut off three-eighths of a turn. Lock adjusting nut in this position by installing adjusting nut lock (C) and bending two tangs of nut lock over flats of adjusting nut.

(2) Install another wheel bearing adjusting nut (B) and, using same tools mentioned in (1) above, tighten nut to a torque of 100 to 150 pound-feet. Secure nut by bending two tangs of nut lock over flats of nut.

(3) Place new gasket (A) on drive flange studs in hub. Install axle shaft (rear axle) or hub drive flange (front axle) and

**291**

attach to hub with eight tapered dowel wedges and eight ½-20 nuts. Tighten nuts to a torque of 55 to 65 pound-feet.

(4) Install brake drum on studs in brake drum adapter and attach with 18 plain washers and ⅜-24 safety nuts. Tighten nuts to a torque of 20 to 27 pound-feet.

# CHAPTER 13

# STEERING GEAR AND DRAG LINK

## Section I. DESCRIPTION AND DATA

### 257. Description

*Note.* Key letters noted in parentheses are in figure 185 unless otherwise indicated.

*a. General.*—The steering system consists of a recirculating-ball type steering gear assembly, mounted on left frame side rail, interconnected from Pitman arm to front axle left steering arm with a drag link. Movement of the steering wheel is transmitted through steering gear mechanism and drag link to the axle steering arm. Both wheels are turned by means of a tie rod interconnecting the front axle right and left steering knuckles. Components of the tie rod and front axle are covered in chapter 7.

*b. Steering Gear Assembly.*
  (1) *Construction.*
    (*a*) *Steering shaft.*—A worm is integrally welded to steering shaft assembly (G). The worm portion of the shaft is mounted between two roller bearings. The shaft lower bearing (J) is adjustable toward shaft upper bearing (C) by means of the worm bearing adjuster (K) for purpose of eliminating end play and maintaining suitable bearing preload. The steering shaft assembly extends through the column jacket with steering wheel mounted on upper end of shaft. A ball-type bearing assembly, mounted in upper end of jacket, takes the radial load on the upper end of the shaft.
    (*b*) *Worm ball nut.*—The worm ball nut (F) fits over the worm portion of the steering shaft. The bore of the ball nut is threaded with helical grooves corresponding with groove in shaft worm. Within the length of the ball nut, helical grooves are filled with steel worm balls (E). There are two separate circuits in the ball nut. To complete each circuit and to keep balls from running out at ends, ball

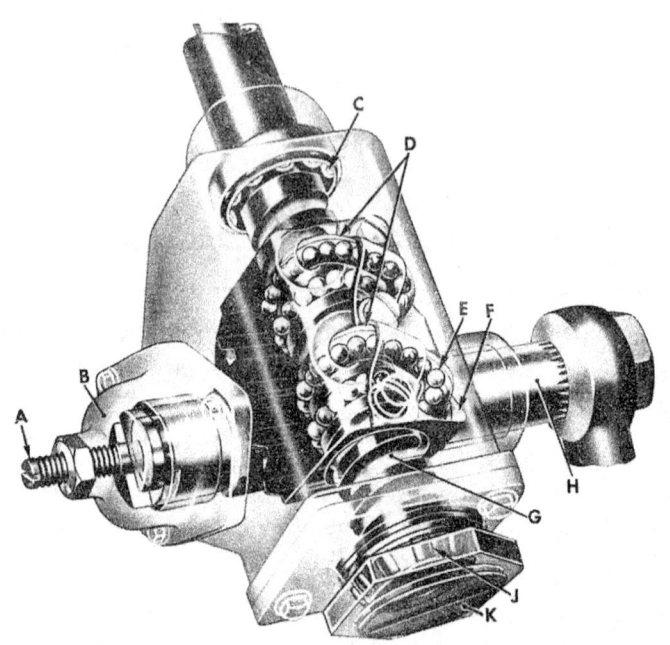

A—PITMAN ARM SHAFT ADJUSTING SCREW
B—SIDE COVER ASSEMBLY
C—SHAFT UPPER BEARING
D—BALL RETURN GUIDE
E—WORM BALLS
F—WORM BALL NUT
G—STEERING SHAFT ASSEMBLY
H—PITMAN ARM SHAFT
J—SHAFT LOWER BEARING
K—WORM BEARING ADJUSTER

RA PD 149424

*Figure 185. Phantom view of steering gear mechanism.*

nut is fitted with ball return guides (D). Each guide deflects the balls from their helical path when they reach the end of the ball nut, thus returning the balls to the helical path in the ball nut at start of circuit.

(c) *Pitman arm shaft.*—The Pitman arm shaft (H) which is integral with sector gear, is mounted in bushings in steering gear housing and side cover assembly (B). The side cover supports the Pitman arm shaft adjusting screw (A) which permits lash adjustment by shifting Pitman arm shaft along its axis. The teeth on the sector gear mesh with similar teeth on worm ball nut.

(d) *Horn connections.*—The horn button contact cable assembly (M, fig. 198) is mounted on steering shaft assembly as shown on figure 197. The cable extends inside the shaft to the horn button contact assembly (X, fig. 198) in the steering wheel center. The horn button (V, fig. 198) in center of steering wheel is depressed to make contact. The cable connector (K, fig. 198) is threaded to column jacket, and is sealed with cable connector seal (L, fig. 198) between connector nut and column jacket.

(2) *Operation* (fig. 185).—As steering wheel turns steering shaft with integral worm, worm ball nut travels on worm as with an ordinary screw thread. Balls, circulating in helical grooves of ball nut and shaft worm, travel within their separate circuits. A screw action, with rolling instead of sliding contact, is obtained. The movement of worm ball nut teeth in contact with the sector teeth on Pitman arm shaft turns the Pitman arm shaft. The movement of the Pitman arm, attached to the arm shaft, turns front wheels by means of the interconnecting drag link.

*c. Drag Link.*—A tubular type drag link (fig. 204) is connected to Pitman arm and left steering arm of front axle with tapered ball studs and nuts. Ball studs are mounted in drag link ends in special bearing material which permits movement of the ball studs but requires no lubrication. A rubber dust seal is placed around taper of each stud to prevent dirt and water entering ball stud bearings.

## 258. Data

*a. Steering Gear Assembly.*

| | |
|---|---|
| Manufacturer | Saginaw Steering Gear Div. |
| Model number | 552–D–6 |
| Ratio | 28.14 to 1 |
| Steering wheel diameter | 20 in. |

*b. Drag Link.*

| | |
|---|---|
| Manufacturer | Saginaw Steering Gear Div |

## Section II. REBUILD OF STEERING GEAR ASSEMBLY

### 259. Preliminary Procedures

*Note.* Key letters noted in parentheses are in figure 198 unless otherwise indicated.

**Caution:** Whenever necessary to turn the steering mechanism to extreme right or left, approach either extreme gently to avoid damage to ball return guides on worm ball nut.

*a. Cleaning Assembly.*—Clean exterior of entire steering gear assembly (fig. 186) with dry-cleaning solvent or volatile mineral spirits before attempting to disassemble the unit.

*b. Facilities.*—Mount steering gear assembly in a vise or holding fixture in a manner that will permit access to both the end and side

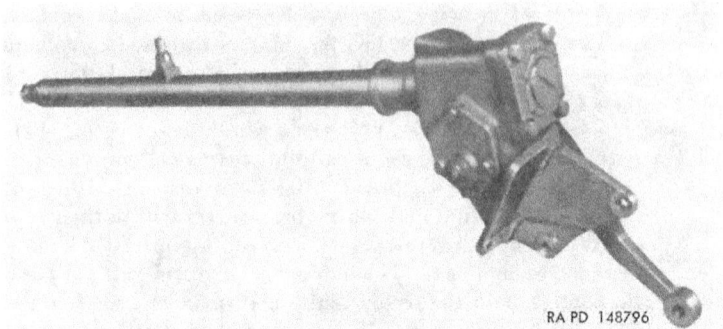

*Figure 186. External view of steering gear assembly.*

covers. Do not grip the steering gear housing tightly in vise. Grip the mounting flange with the vise. Disassemble steering gear on a clean bench and use clean receptacles to hold the parts.

  *c. Steering Wheel Removal.*—Steering wheel may remain with steering gear when assembly is removed from vehicle. If so, remove steering wheel in following manner.

   (1) Remove four No. 10 x ⅞ screws (U) which attach horn button retaining ring (T) to steering wheel assembly. Re-

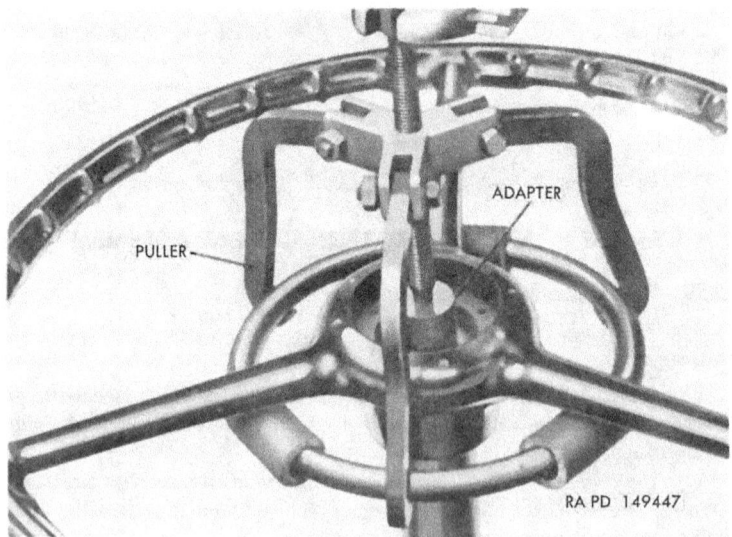

*Figure 187. Removing steering wheel using adapter 41-A-27-430 with puller 41-P-2954.*

move retaining ring, horn button (V), and horn button contact assembly (X).

(2) Remove steering wheel nut (S).

(3) Install adapter 41–A–27–430 with puller 41–P–2954 at steering wheel in manner shown in figure 187. Pull steering wheel from shaft.

(4) Remove shaft bearing spring (Q) and shaft bearing spring seat (P).

## 260. Disassembly Procedures

*Note.* Key letters noted in parentheses are in figure 188 unless otherwise indicated. The sectional view (fig. 198) illustrates parts in their respective positions.

*a.* Remove cable connector (G) from column jacket (F). Discard cable connector seal (H).

*b.* If Pitman arm has not been removed from Pitman arm shaft (JJ), remove Pitman arm nut (L) and $7/8$-inch lock washer (M); then install puller 41–P–2952 in manner shown in figure 189. Remove Pitman arm from shaft.

*c.* Remove housing drain plug (HH) and gasket (K) to drain lubricant from housing. Remove housing filler plug (J) and gasket (K).

*d.* Remove $7/16$–20 jam nut (RR) and $15/32$ ID plain washer (SS) from Pitman arm shaft adjusting screw (LL). Turn adjusting screw a few turns counterclockwise.

*e.* Remove three side cover bolts (TT) and $3/8$-inch lock washers (UU). Pry side cover assembly (PP) away from steering gear housing (VV). After lubricant has drained from housing, remove side cover assembly (PP) by turning Pitman arm shaft adjusting screw (LL) clockwise through side cover. Remove side cover packing (QQ) and side cover gasket (NN).

*f.* Turn steering shaft until sector gear on Pitman arm shaft will pass through housing opening; then remove Pitman arm shaft (JJ). Pitman arm shaft adjusting screw (LL) and adjusting screw shim (MM) can then be removed. Remove Pitman arm shaft packing (P) and packing retainer (N) from steering gear housing (VV).

*g.* Mount assembly horizontally on bench to prevent worm ball nut from running down to end of shaft. Remove worm bearing adjuster nut (EE), end cover packing (DD), and worm bearing adjuster (CC) from housing end cover (BB). Remove four $7/16$–14 x $1 1/8$ cap screws (FF) and $7/16$-inch lock washers (GG) which attach end cover to housing. Remove end cover. Shaft lower bearing (Z) may remain with cover or on end of shaft. Remove bearing and end cover gasket (AA).

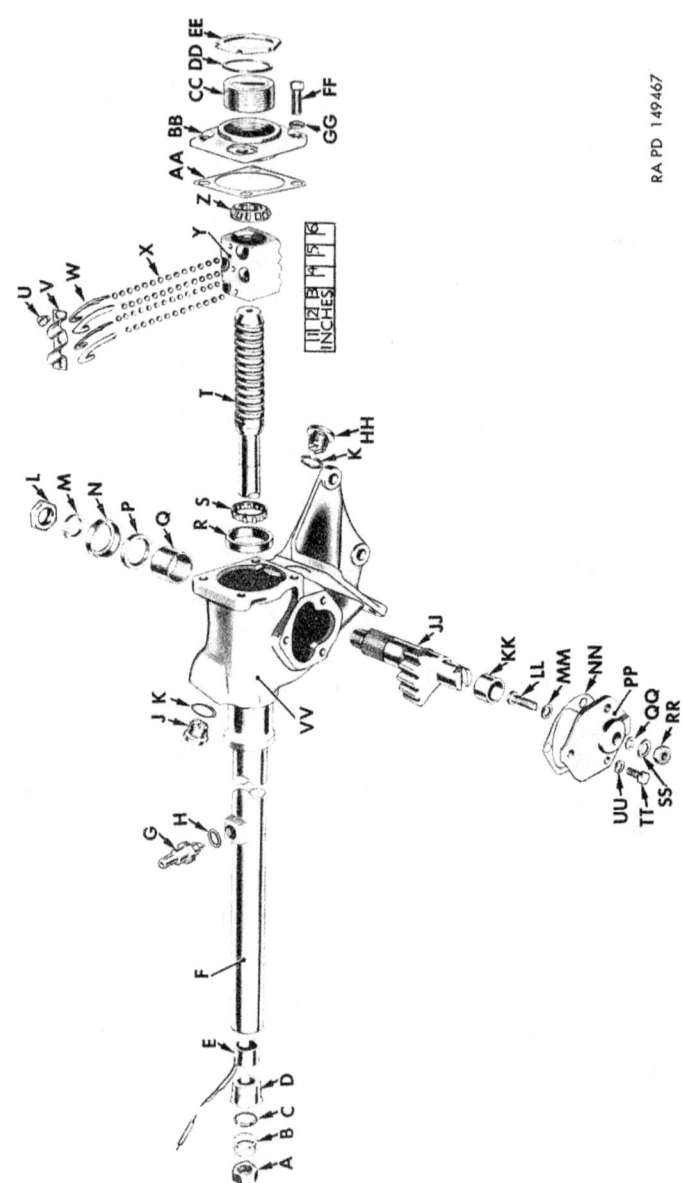

A—NUT (STEERING WHEEL)—7376315
B—SPRING, SHAFT BEARING—SSG-262250
C—SEAT, SHAFT BEARING SPRING—7377383
D—BEARING, ASSY—SSG-262251
E—CABLE, HORN BUTTON CONTACT, ASSY—SSG-267898
F—JACKET, COLUMN—7376355
G—CONNECTOR, CABLE—SSG-5662809
H—SEAL, CABLE CONNECTOR—7412926
J—PLUG, HOUSING FILLER—7376357
K—GASKET—120428
L—NUT, PITMAN ARM—7000667
M—WASHER, LOCK, ⅞-IN.—131047
N—RETAINER, PACKING—7000668
P—PACKING, PITMAN ARM SHAFT—7696444
Q—BEARING, BUSHING TYPE, PITMAN SHAFT (HOUSING)—7000406
R—RACE, ROLLER BEARING, OUTER—708617
S—BEARING, SHAFT, UPPER—707690
T—SHAFT, STEERING, ASSY—7376441
U—SCREW, ½-28 x ⁵⁄₁₆—187375
V—CLAMP, BALL RETURN GUIDE—7000670
W—GUIDE, BALL RETURN—7000671
X—BALL, WORM, ⁹⁄₃₂-IN.—SSG-266800
Y—NUT, WORM BALL—SSG-267610
Z—BEARING, SHAFT, LOWER—707690

AA—GASKET, END COVER—7696441
BB—COVER, HOUSING END—7696439
CC—ADJUSTER, WORM BEARING—7376974
DD—PACKING, END COVER—7696437
EE—NUT, WORM BEARING ADJUSTER—SSG-267100
FF—SCREW, CAP, ⁷⁄₁₆-14 x 1⅛—180146
GG—WASHER, LOCK, ⁷⁄₁₆-IN.—120382
HH—PLUG, HOUSING DRAIN—7376357
JJ—SHAFT, PITMAN ARM—7376442
KK—BEARING, BUSHING TYPE, PITMAN SHAFT (SIDE COVER)—7373552
LL—SCREW, ADJUSTING, PITMAN ARM SHAFT—7376973
MM ⎰ SHIM, ADJUSTING SCREW—0.063 THK—SSG-266903
   ⎱ SHIM, ADJUSTING SCREW—0.065 THK—SSG-266905
     SHIM, ADJUSTING SCREW—0.067 THK—SSG-266907
     SHIM, ADJUSTING SCREW—0.069 THK—SSG-266909
NN—GASKET, SIDE COVER—7376353
PP—COVER, SIDE, ASSY—7376352
QQ—PACKING, SIDE COVER—7696438
RR—NUT, JAM, ⁷⁄₁₆-20—124929
SS—WASHER, PLAIN, ¹⁵⁄₃₂ ID—120395
TT—BOLT, SIDE COVER—7000401
UU—WASHER, LOCK, ⅜-IN.—120382
VV—HOUSING, STEERING GEAR—SSG-5662812

*Figure 188. Steering gear assembly components.*

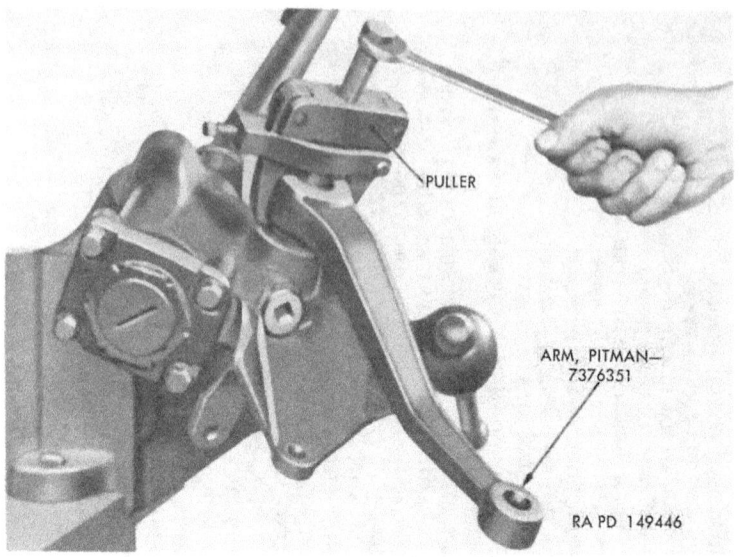

*Figure 189. Removing Pitman arm with puller 41–P–2952.*

*h.* Carefully withdraw steering shaft assembly (T) and worm ball nut (Y) as an assembly from housing and column jacket.

**Caution:** If the shaft with the ball nut is held in a vertical position, the ball nut will travel by its own weight to the end of the shaft. If ball nut sharply strikes either end of shaft worm, ball return guides will be damaged. Lay the assembly flat on a bench. Tape each end of shaft worm to prevent ball nut rotating to ends if the ball nut does not need to be disassembled. Make inspection of action of ball nut (par. 262) before disassembling.

*i.* To disassemble ball nut, remove three $1/4$–28 x $5/16$ screws (U) which attach ball return guide clamp (V) to worm ball nut (Y). Remove clamp. Pull ball return guides (W) out of ball nut in manner shown in figure 190. Turn ball nut upside down, and rotate shaft back and forth until all $9/32$-inch worm balls (X) have dropped out of ball nut into a clean pan. With the balls removed, pull ball nut endwise from shaft worm.

*j.* Remove bearing assembly (D) from upper end of column jacket in manner illustrated in figure 191.

*k.* Further disassembly of the steering shaft assembly, steering gear housing and column jacket, and side cover assembly may be deferred until inspection of parts (par. 262).

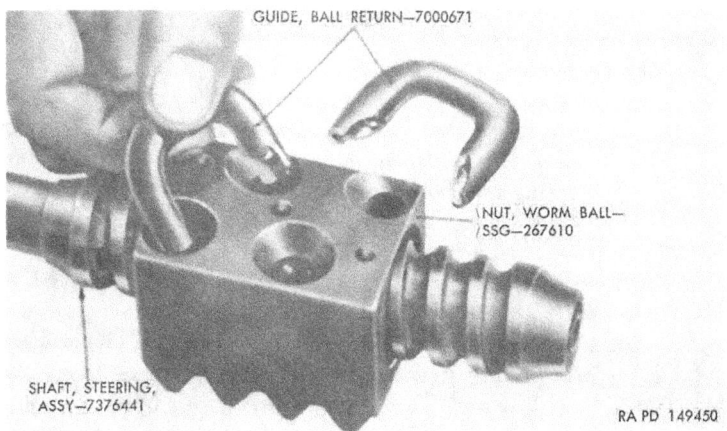

*Figure 190. Removing or installing ball return guides.*

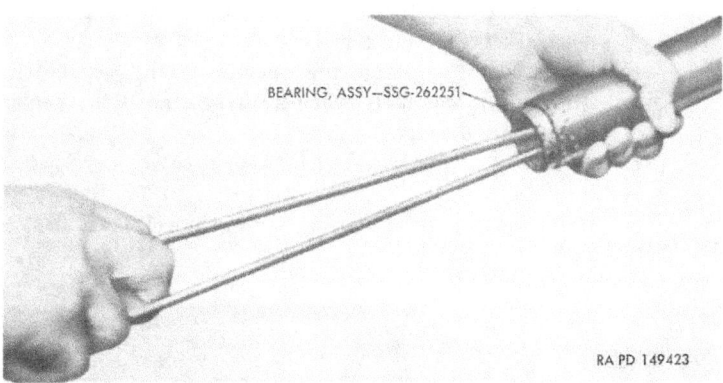

*Figure 191. Removing bearing assembly from column jacket.*

## 261. Cleaning Parts

All parts should be thoroughly cleaned with dry-cleaning solvent or volatile mineral spirits before inspection of the parts is made. Particular care should be taken when bearings are cleaned to make sure that all particles of grit and old lubricant have been removed. Dry all parts thoroughly; then place them in clean containers for inspection.

## 262. Inspection and Repair of Parts

*Note.* Key letters noted in parentheses are in figure 188 unless otherwise indicated.

*a. Steering Gear Housing and Column Jacket.*
(1) *Inspection.*
 (*a*) Examine steering gear housing (VV) for cracks, and stripped threads in cover mounting surfaces. If column jacket (F) is damaged, replace both column jacket and steering gear housing (VV).
 (*b*) Check clearance (par. 353) between Pitman arm shaft (JJ) and Pitman shaft bushing type bearing (Q) (in steering gear housing). If bushing is worn or scored, replace as described in (2) below.
 (*c*) Check condition of outer roller bearing race (R) which is pressed into steering gear housing. If race is cracked, scored, or worn excessively, replace race. When installing race, seat race firmly against end of column jacket in steering gear housing.
 (*d*) Check condition of housing filler plug (J) and housing drain plug (HH). Replace if plugs are damaged. Use new gaskets (K) at assembly.
 (e) Check condition of bearing assembly (D), shaft bearing spring (B), and shaft bearing spring seat (C). Replace if parts are damaged.
(2) *Replacement of Pitman shaft bushing type bearing in steering gear housing.*
 (a) Drive bearing from housing with remover 41–R–2370 as shown in figure 192.

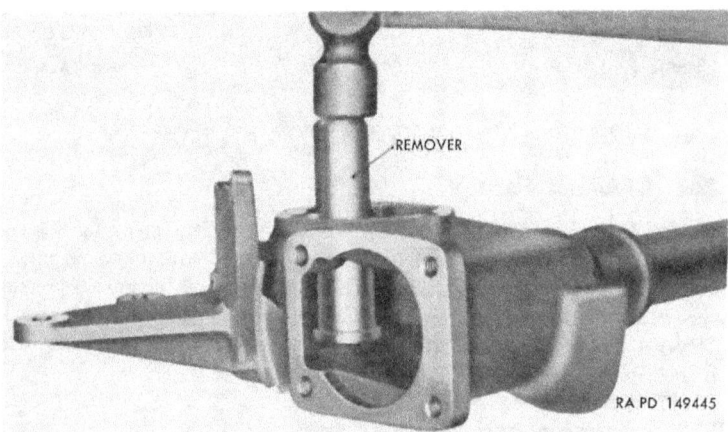

*Figure 192. Remove Pitman arm shaft bearings from steering gear housing with remover 41–R–2370.*

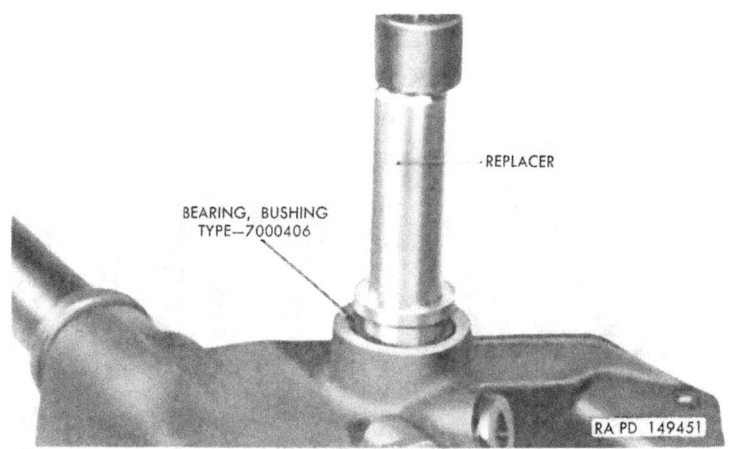

*Figure 193. Installing Pitman arm shaft bearing into steering gear housing with replacer 41–R–2388–730.*

    (*b*) Install bearing with replacer 41–R–2388–730 as shown in figure 193. Tool will properly locate bearing in housing as shown in figure 194.

    (*c*) Ream bearing to size shown in figure 194.

*b. Side Cover Assembly.*

    (1) *Inspection.*—Examine side cover assembly (PP) for cracks or damage. Check clearance (par. 353) between Pitman arm

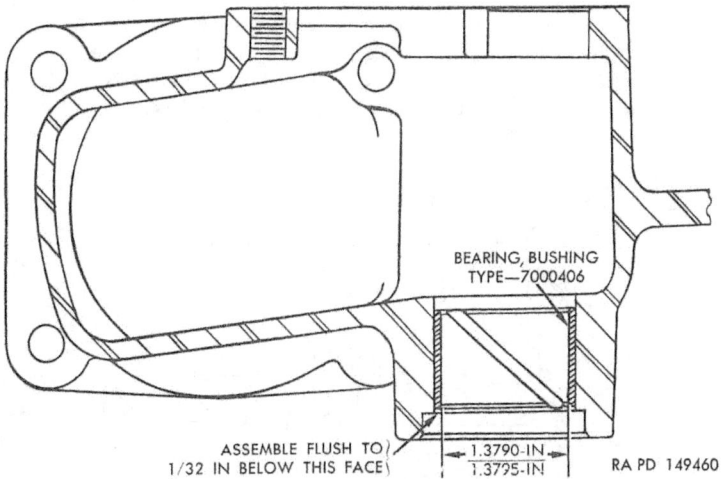

*Figure 194. Sectional view of steering gear housing showing Pitman shaft bearing installed.*

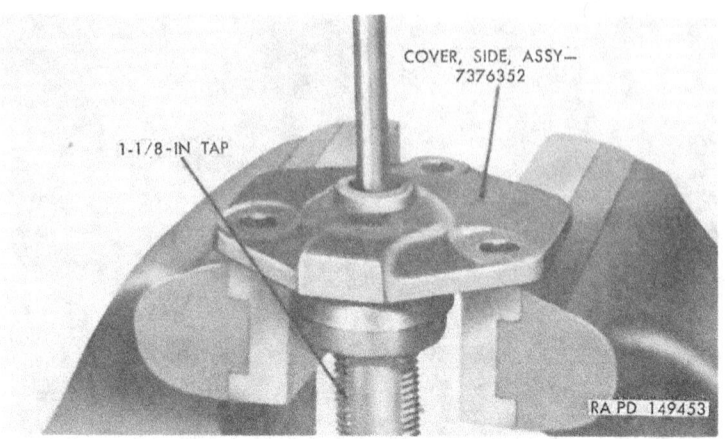

*Figure 195. Method of removing bearing from side cover.*

shaft (**JJ**) and Pitman shaft bushing type bearing (**KK**) in side cover. If clearance is not correct or if bearing is scored or otherwise damaged, replace bearing as described in (2) below. Side cover gasket (**NN**) must be replaced with new part at assembly.

(2) *Replacement of bearing in side cover.*—Thread a 1⅛-inch tap into bearing the full length of bearing. Place cover in a press. With suitable driver or rod inserted through adjusting screw hole in cover against tap, press out bearing (fig. 195). Press new bearing into place as shown in figure 196. Finish bearing to dimensions shown in figure 196.

  *c. End Cover.*—Examine housing end cover (**BB**) for cracks or distortion. Check condition of worm bearing adjuster threads in

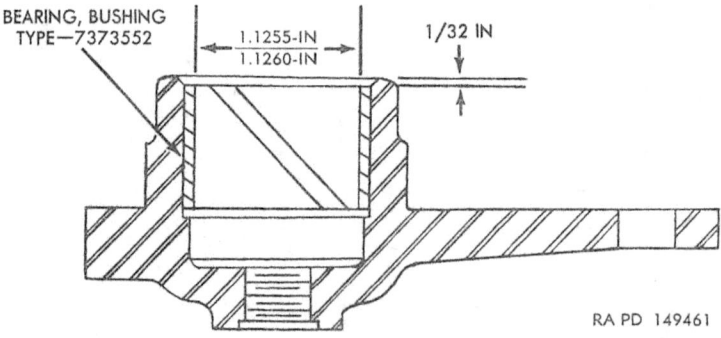

*Figure 196. Sectional view of side cover assembly.*

housing. Replace end cover if damage is apparent. End cover gasket (AA) must be replaced with new part at assembly.

*d. Worm Bearing Adjuster.*—Check condition of threads in worm bearing adjuster (CC). Examine inner end of adjuster for scores or roughness. The inner end of adjuster serves as an outer race for shaft lower bearing (Z). If adjuster is damaged as explained, replace with new part. Check condition of worm bearing adjuster nut (EE), and if damaged, replace. End cover packing (DD) must be replaced with new part at assembly.

*e. Pitman Arm Shaft.*
  (1) Inspect Pitman arm shaft (JJ) for damaged serrations and threads at Pitman arm end. Replace shaft if damaged.
  (2) Examine sector teeth on shaft for signs of scuffing, scoring, and other damage. Replace shaft if damaged.
  (3) Check outside diameter of shaft at housing and side cover ends. If wear is evident (par. 353), replace shaft.
  (4) Pitman arm shaft packing (P) and packing retainer (N) must be replaced at assembly. Check condition of Pitman arm nut (L) and 7/8-inch lock washer (M). Replace if parts are damaged.

*f. Steering Shaft Assembly.*
  (1) *Inspection.*
    (*a*) Thoroughly inspect worm at end of steering shaft assembly for evidence of scoring and wear. Check action of assembled worm ball nut on worm. If worm ball nut rotates smoothly without evidence of binding or roughness, do not disassemble worm ball nut. Refer to paragraph 260*j*, if necessary, to disassemble worm ball nut. If shaft worm is damaged as explained, replace.
    (*b*) Check condition of shaft upper bearing (S) and shaft lower bearing (Z). If bearings are worn or damaged, replace.
    (*c*) If shaft upper bearing (S) is damaged and requires replacement, replace as described in (2) below.
  (2) *Replacement of shaft upper bearing.*
    (*a*) Unsolder cable terminal (fig. 197) at upper end of shaft. Remove terminal and fiber insulator.

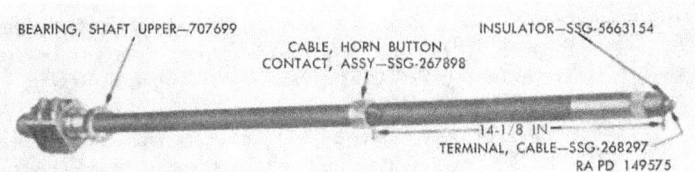

*Figure 197. Construction of steering shaft assembly.*

(b) Pull horn button contact cable assembly (E) from shaft. Remove horn button contact cable assembly from shaft. Shaft upper bearing may then be removed from shaft.

(c) Install new shaft upper bearing on shaft.

(d) Install horn button contact cable assembly on shaft to dimension shown in figure 197. Insert wire through shaft; then install fiber insulator. Solder terminal to end of cable. Remove all surplus solder.

g. *Worm Ball Nut Parts.*

(1) Examine rack teeth of worm ball nut (Y) for scuffing, scoring, or wear. Check holes and passages for obstructions. Thoroughly examine worm ball nut for external and internal damage. Replace worm ball nut if damage explained exists.

(2) Check all of the $9/32$-inch worm balls (X) for flat spots, checking, wearing, or damage. Balls should be the same size within 0.0001 inch.

(3) Examine ball return guides (W) for distortion. Place two halves together and try action of balls in the two halves. Replace guides if any restriction exists. Replace ball return guide clamp (V) if distorted.

h. *Pitman Arm.*—Examine Pitman arm serrations and threads for damage. If Pitman arm is bent, replace. Do not attempt to straighten.

i. *Steering Wheel.*—Inspect wheel spokes and rim for damage or distortion. Replace wheel if these conditions exist.

*Note.* Do not attempt to repair any part of the steering gear assembly by welding or machining.

## 263. Assembly of Steering Gear

*Note.* Key letters noted in parentheses are in figure 88 unless otherwise indicated. Reference should also be made to the sectional view (fig. 198) which shows relative position of the assembled parts.

a. *General.*—One of the most important phases of assembling steering gear components is cleanliness. All parts must be kept clean. Any bits of abrasive material which may get inside of the housing during assembly will quickly damage the gear mechanism. Grease and oil used at assembly must be free from dirt. When handling parts, make certain that hands are clean, and that clean cloths are used. Prelubricate all bearings and moving parts (except bearing assembly (D)) at assembly with universal gear lubricant (GO).

b. *Assembly of Worm Ball Nut.*

(1) Place steering shaft assembly (T) flat on bench. Place worm ball nut (Y) over shaft worm with ball return guide holes in ball nut in upper surface. Aline grooves in worm and ball nut by sighting through bottom of ball return guide holes.

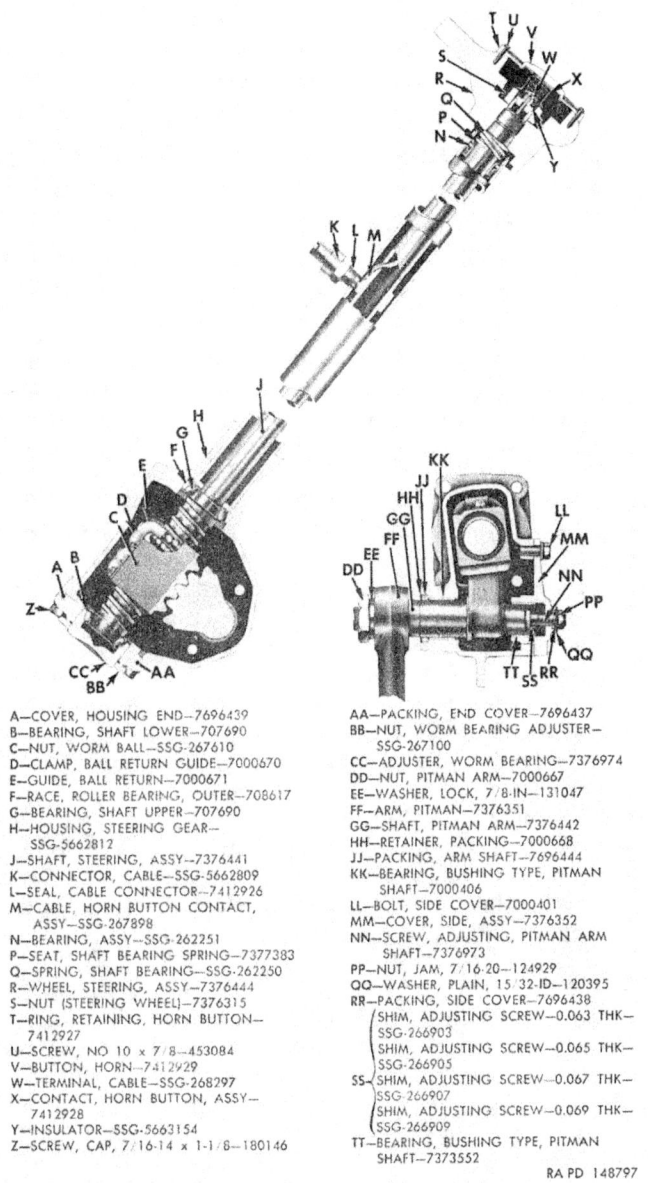

A—COVER, HOUSING END—7696439
B—BEARING, SHAFT LOWER—707690
C—NUT, WORM BALL—SSG-267610
D—CLAMP, BALL RETURN GUIDE—7000670
E—GUIDE, BALL RETURN—7000671
F—RACE, ROLLER BEARING, OUTER—708617
G—BEARING, SHAFT UPPER—707690
H—HOUSING, STEERING GEAR— SSG-5662812
J—SHAFT, STEERING, ASSY—7376441
K—CONNECTOR, CABLE—SSG-5662809
L—SEAL, CABLE CONNECTOR—7412926
M—CABLE, HORN BUTTON CONTACT, ASSY—SSG-267898
N—BEARING, ASSY—SSG-262251
P—SEAT, SHAFT BEARING SPRING—7377383
Q—SPRING, SHAFT BEARING—SSG-262250
R—WHEEL, STEERING, ASSY—7376444
S—NUT (STEERING WHEEL)—7376315
T—RING, RETAINING, HORN BUTTON— 7412927
U—SCREW, NO 10 x 7/8—453084
V—BUTTON, HORN—7412929
W—TERMINAL, CABLE—SSG-268297
X—CONTACT, HORN BUTTON, ASSY— 7412928
Y—INSULATOR—SSG-5663154
Z—SCREW, CAP, 7/16-14 x 1-1/8—180146

AA—PACKING, END COVER—7696437
BB—NUT, WORM BEARING ADJUSTER— SSG-267100
CC—ADJUSTER, WORM BEARING—7376974
DD—NUT, PITMAN ARM—7000667
EE—WASHER, LOCK, 7/8-IN—131047
FF—ARM, PITMAN—7376351
GG—SHAFT, PITMAN ARM—7376442
HH—RETAINER, PACKING—7000668
JJ—PACKING, ARM SHAFT—7696444
KK—BEARING, BUSHING TYPE, PITMAN SHAFT—7000406
LL—BOLT, SIDE COVER—7000401
MM—COVER, SIDE, ASSY—7376352
NN—SCREW, ADJUSTING, PITMAN ARM SHAFT—7376973
PP—NUT, JAM, 7/16-20—124929
QQ—WASHER, PLAIN, 15/32-ID—120395
RR—PACKING, SIDE COVER—7696438
SS—{SHIM, ADJUSTING SCREW—0.063 THK— SSG-266903
SHIM, ADJUSTING SCREW—0.065 THK— SSG-266905
SHIM, ADJUSTING SCREW—0.067 THK— SSG-266907
SHIM, ADJUSTING SCREW—0.069 THK— SSG-266909}
TT—BEARING, BUSHING TYPE, PITMAN SHAFT—7373552

RA PD 148797

*Figure 198. Sectional view of steering gear assembly.*

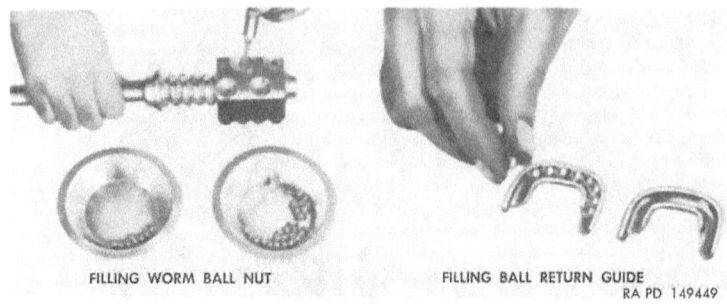

*Figure 199. Installing worm balls into worm ball nut and return guides.*

(2) Count 53 or one-half of the total quantity of balls (106 total) into a clean container. This is the proper number of balls for one circuit. Drop balls into one of the ball return guide holes in the upper circuit (fig. 199). Gradually turn shaft away from that hole while inserting balls. Continue until the circuit is filled from bottom of one hole to the bottom of the other, or until stopped by reaching the end of the shaft worm.

(3) In the event balls are stopped by reaching end of shaft worm, hold down the balls, already inserted, with a rod or punch in manner shown in figure 199. Turn shaft in the reverse direction a few turns. Filling of the circuit can then be continued. It may be necessary to work shaft back and forth, holding balls down first in one hole then in the other. This will close up spaces between balls, filling the circuit completely and solidly.

(4) Lay one-half of ball return guide (W) with groove up, on bench. Place the remaining balls of the 53 selected into groove of guide. Close this half of guide with the other half. Hold the two halves together; then plug each open end with heavy chassis grease to prevent balls from dropping out.

(5) Push ball return guide completely into holes in worm ball nut (fig. 190). This completes one circuit of balls.

(6) Fill lower circuit in worm ball nut (fig. 199) in same manner described in (2) through (5) above).

(7) Install ball return guide clamp (V) to ball nut using three $\frac{1}{4}$–28 x $\frac{5}{16}$ screws (U). Tighten screws firmly.

(8) Make certain that ball nut and balls are thoroughly lubricated. Test assembly by rotating ball nut on steering shaft worm. Do not rotate ball nut to end of worm threads. Assembly must move freely. If there is any bind in motion of

worm ball nut, remove and check condition of ball return guides. These guides must not be bent to restrict movement of balls. Tape shaft at both ends of ball nut until ready to install assembly into steering gear housing.

*c. Roller Bearing Outer Race Installation.*—With a suitable driver, install roller bearing outer race (R) into housing. Make certain that bearing seat is clean, permitting bearing to bottom squarely in housing.

*d. Steering Shaft Assembly Installation.*—Remove tape previously installed at each end of ball nut ($b(8)$ above). Grasp steering shaft worm below and above ball nut to prevent nut from running to extreme ends. With shaft upper bearing (S) in place on shaft, insert steering shaft assembly through end cover opening in housing. Guide shaft carefully through housing and column jacket until shaft upper bearing (S) contacts roller bearing outer race (R) in housing.

*e. End Cover Installation.*
  (1) Place shaft lower bearing (Z) into place in housing end cover (BB).
  (2) Swab threads of worm bearing adjuster (CC) with plastic type gasket cement. Thread adjuster a few turns into housing end cover (BB). Place new end cover gasket (AA) over end cover. Install end cover to housing with four $7/16$-$14 \times 1\frac{1}{8}$ cap screws (FF) and $7/16$-inch lock washers (GG), dipping threads of cap screws in plastic type gasket cement before installing. Tighten cap screws securely.
  (3) Temporarily tighten worm bearing adjuster (CC) until all worm bearing end play is removed.
  (4) Install new end cover packing (DD); then install worm bearing adjuster nut (EE). Do not tighten at this time.

*f. Adjusting Screw to Pitman Arm Shaft Clearance.*
  (1) Place original adjusting screw shim (MM) on Pitman arm shaft adjusting screw (LL). Insert adjusting screw and shim into slotted end of Pitman arm shaft (JJ).
  (2) Check clearance between adjusting screw head and shaft in manner shown in figure 200. Clearance must not exceed specified amount (par. 353). If clearance is greater, select another thicker adjusting screw shim (MM). Four sizes of shims are available.

*g. Pitman Arm Shaft Installation.*
  (1) With Pitman arm shaft adjusting screw (LL) and correct shim in place in end of Pitman arm shaft, start side cover assembly (PP) over end of shaft.
  (2) Insert screw driver into hole in side cover to engage slot of adjusting screw. Turn adjusting screw into side cover to pull side cover and Pitman shaft bushing type bearing (KK) over shaft end.

(3) Turn steering shaft until worm ball nut is in approximate center of steering shaft worm. Center tooth of sector gear on Pitman arm shaft must enter center rack teeth of worm ball nut.

(4) With new side cover gasket (NN) in place on side cover, insert Pitman arm shaft into steering gear housing (VV), meshing center tooth of gear on shaft with center tooth of worm ball nut. Turn Pitman arm shaft adjusting screw (LL) to pull cover over end of shaft. Back off adjusting screw to permit backlash between sector gear on shaft and worm ball nut.

(5) Install three side cover bolts (TT) with 3/8-inch lock washers (UU). Tighten bolts firmly.

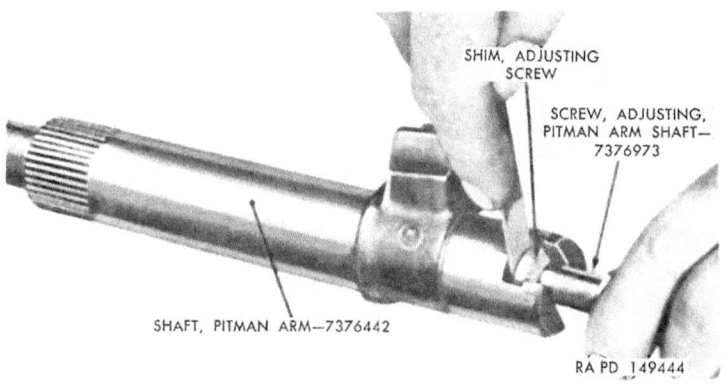

*Figure 200. Method of adjusting Pitman arm shaft adjusting screw.*

(6) Install side cover packing (QQ) over Pitman arm shaft adjusting screw (LL). Install 15/32 ID plain washers (SS) and 7/16–20 jam nut (RR) on adjusting screw.

(7) Coat outside of packing retainer (N) with plastic type gasket cement. Install Pitman arm shaft packing (P) inside of retainer. Install retainer and packing in steering gear housing with suitable tool. Seat packing firmly against Pitman shaft bushing type bearing (Q) in housing.

*h. Column Jacket Bearing Installation.*—Pack recesses of bearing assembly (N, fig. 198) with automotive and artillery grease (GAA). Squarely install bearing assembly over shaft into column jacket. Install shaft bearing spring seat (C) and shaft bearing spring (B) on shaft.

*i. Filler and Drain Plug Installation.*—Install housing drain plug (HH) with new gasket (K). Tighten plug firmly. Fill housing with universal gear lubricant (GO) until lubricant is at level of bottom of filler plug hole (with unit in operating position). Install housing filler plug (J) with new gasket (K). Tighten plug firmly.

*j. Cable Connector Installation.*—Place new cable connector seal (H) into recess of cable connector fitting on column jacket. Install cable connector (G) to column jacket. Tighten connector firmly. Place protective tape over the connector until unit is installed in vehicle.

*k. Steering Wheel and Horn Button Installation.*

*Note.* Installation of the steering wheel and horn button parts may be deferred until steering gear assembly is installed in vehicle. Steering wheel must be installed temporarily to accomplish adjustments as explained in paragraph 264. Key letters noted in parentheses are in figure 198.

(1) Install steering wheel on shaft. Install nut (S), and tighten to a torque of 40 to 55 pound-feet.

(2) Install horn button contact assembly (X) over terminal. Install horn button (V). Install horn button retaining ring (T) and attach with four No. 10 x 7/8 screws (U). Tighten screws firmly.

*l. Pitman Arm Installation.*—Installation of the Pitman arm may be deferred until adjustments (par. 264) are made. Position Pitman arm (FF) on Pitman arm shaft (GG), matching blank serration on shaft with blank serration in arm. Install 7/8-inch lock washer (EE) and Pitman arm nut (DD). Tighten nut to torque of 115 to 155 pound-feet.

## 264. Steering Gear Adjustments

*a. General.*—The steering gear assembly is designed to provide for adjustments to compensate for normal wear at steering shaft bearings and at Pitman arm shaft and mating parts. These adjustments must be accomplished before assembly is installed in the vehicle, and further checked and adjusted in the same manner after unit is installed. The following sequence of adjustments must be followed. The unit must be mounted in stand or vise in operating position.

*b. Preliminary Procedures.*—Determine straightahead position of the steering mechanism by turning steering wheel to extreme right.

**Caution:** Approach extreme end cautiously. Worm ball nut must not strike end with any degree of force.

Turn wheel to opposite extreme in same manner, counting the number of wheel turns between extreme ends. Turn wheel back one-half number of wheel turns. Mark wheel with respect to column so that center position may readily be found during adjustment procedures.

*c. Steering Shaft Bearing Adjustment.*

(1) Check tightness of housing end cover bolts. Back out Pitman arm shaft adjusting screw a few turns counterclockwise to

provide clearance between Pitman arm shaft gear and worm ball nut teeth.

(2) Turn steering wheel gently until mechanism has reached one extreme. Turn wheel back one full turn. With scale 41-S-503 on spoke of steering wheel (fig. 201), measure pull required to keep wheel moving. Pull on scale should be made at right angle to wheel spoke.

(3) If pull is not within the limits of 1½ to 2 pounds, shaft bearings must be adjusted.

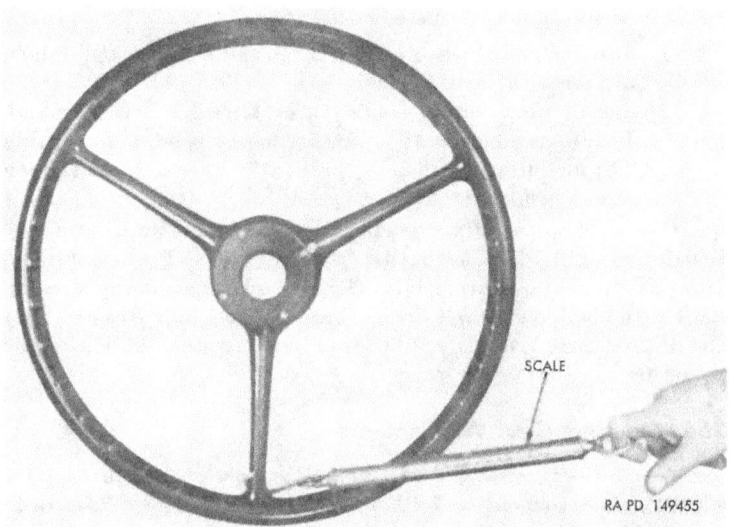

*Figure 201. Use of scale 41-S-503 on steering wheel to check pull.*

(4) Loosen worm bearing adjuster nut (fig. 202); then turn worm bearing adjuster in or out to obtain correct pull. Tighten adjuster nut and recheck pull.

(5) After shaft bearings are adjusted, Pitman arm shaft lash adjustment must be made (*d* below).

*d. Pitman Arm Shaft Lash Adjustment* (fig. 203).
  (1) Check tightness of side cover bolts.
  (2) Center steering mechanism as previously marked in *b* above.
  (3) Turn Pitman arm shaft adjusting screw in to remove all lash between gear teeth. The amount of backlash can be determined by pushing backward and forward on Pitman arm. When all backlash has been removed, tighten adjusting screw jam nut.

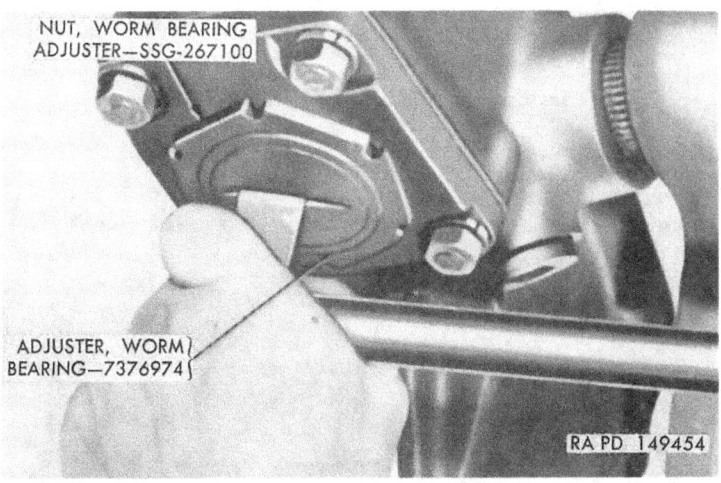

*Figure 202. Adjusting worm bearings.*

(4) Check pull of steering wheel with scale ($c(2)$ above). Measure pull as wheel is pulled through center position.

(5) If pull is not within 2½ to 3 pounds, turn shaft adjusting screw in or out to obtain proper pull. Tighten jam nut and check pull again.

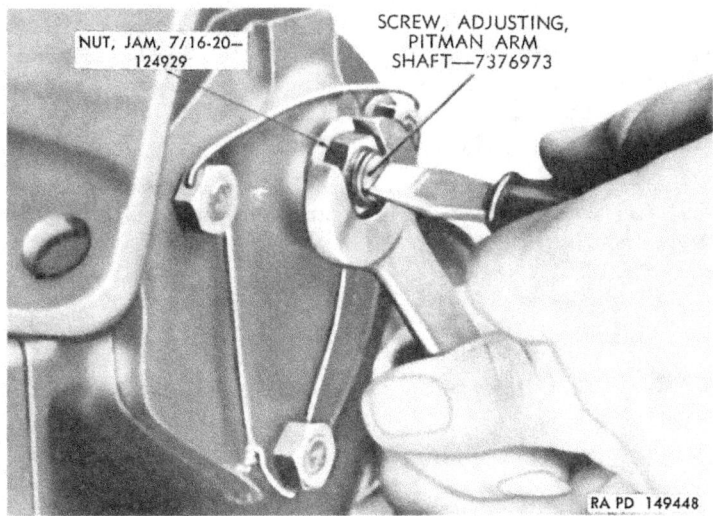

*Figure 203. Adjusting Pitman arm shaft lash.*

(6) After adjustment is made, install lock wire through heads of side cover bolts as shown in figure 203.

## Section III. REBUILD OF DRAG LINK

### 265. Description

Drag link assembly (fig. 204) is tubular-type with ball studs mounted in special bearing material. A rubber dust seal is placed around taper of each ball stud.

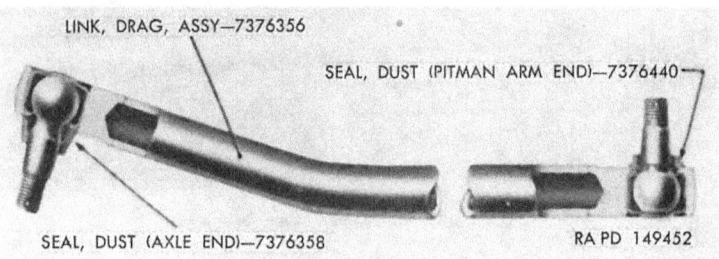

*Figure 204. Sectional view of drag link.*

### 266. Inspection and Repair

*a. Dust Seal.*—Inspect dust seal at axle end and Pitman arm end. If seals show signs of damage, replace with new parts.

*b. Ball Studs.*—Examine ball stud threads for damage. If movement of ball studs in drag link ends indicate that bearing material is damaged, replaced entired drag link. The end bearings and ball studs cannot be replaced in drag link ends.

*Note.* Do not lubricate ball stud bearings.

*c. Drag Link.*—Examine drag link for bent condition. Compare drag link with new part. If drag link is distorted in any manner, replace with new part.

# CHAPTER 14

# FRONT SPRING SUSPENSION

## Section I. DESCRIPTION AND DATA

### 267. Description

*a. General.*—Front spring suspension system components comprise the front springs, spring shackles and brackets, torque rods, and shock absorbers. Front spring suspension components are illustrated in figure 205.

*b. Front Springs, Shackles, and Brackets.*—Semielliptic front springs carry only vertical and lateral loads and are attached to frame brackets at both ends through shackles. Shackles and spring eyes are equipped with replaceable bushing type bearings. Shackle bolts and pins are drilled and equipped with lubrication fittings for lubricating spring eye and shackle bearings. Front spring front brackets and upper torque rod brackets are riveted to frame side rails.

*c. Torque Rods.*—Three torque rods, two lower and one upper, transmit front axle driving and braking forces to frame. Lower torque rod frame brackets are integral with front spring rear brackets. Upper torque rod bracket is mounted inside right frame side rail ahead of front spring rear bracket. Frame ends of torque rods are equipped with bushing type bearings mounted in material which requires no lubrication. Axle ends of torque rods are equipped with tapered end pins which are mounted in material requiring no lubrication; end pin and bearing must be replaced as an assembly.

*d. Shock Absorbers.*—Shock absorbers, used at front axle, are hydraulic, double-acting, opposed-cylinder type. Filler plug is at top front end of shock absorber body. Shock absorber arms are connected to bumper blocks at axle through links which have tapered studs mounted in rubber at each end. Construction and operation of shock absorbers are described in paragraphs 273 and 274.

## 268. Data

*a. Front Springs.*

| | |
|---|---|
| Length (center-to-center of spring eyes) | 50 in. |
| Width | 2½ in. |
| Number of leaves | 11 |
| Thickness of leaves: | |
|     1 @ | 0.360 in. |
|     4 @ | 0.323 in. |
|     6 @ | 0.291 in. |
|         Total thickness | 3.398 in. |

*b. Shock Absorbers.*

| | |
|---|---|
| Manufacturer | Delco Products |
| Model: | |
|     Right side | 2009-H |
|     Left side | 2009-G |
| Valve code: | |
|     Compression valve | G2 |
|     Rebound valve | 2N |

## Section II. REBUILD OF FRONT SPRING SUSPENSION COMPONENTS

### 269. Inspection of Front Spring Suspension Components

*Note.* Key letters noted in parentheses are in figure 205.

*a. General.*—It is not necessary to disassemble front spring suspension components further than shown in figure 205 unless replacement of parts is necessary as indicated in the inspection procedures which follow. Clean all parts in dry-cleaning solvent or volatile mineral spirits to remove all dirt to facilitate inspection.

*b. Springs.*—Examine front spring assembly (Z) for broken leaves, broken center bolt, or broken rebound clips, and check bushing type bearings (W) in spring eyes for wear (par. 354). Nos. 1 and 2 spring leaves can be replaced if broken, and bearings in No. 1 spring eyes can be replaced (par. 270*b*) if worn beyond specified limits. If rebound clips or spring leaves other than Nos. 1 and 2 are broken, the complete spring assembly must be replaced.

*c. Spring Shackles.*—Examine spring shackle assemblies (T) for visible cracks or distortion. Check bushing type bearings (S) in shackles for wear (par. 354) and replace (par. 271) if worn beyond specified limits.

*d. Shackle Bolts and Pins.*—Examine front and rear shackle bolts (P and GG) for damaged threads and serrations, and check for wear at bearing contact surfaces (par. 354). If threads or serrations are damaged, or if worn beyond specified limits, replace with new parts. Check shackle pins (X) for wear at bearing contact surfaces (par.

354). Replace with new parts if worn beyond specified limits. Make sure drilled lubricant passages in bolts and pins are unobstructed.

*e. Spring "U" Bolts.*—Examine left and right inner and outer front spring "U" bolts (Y) for damaged threads, cracks, or visible distortion. Replace with new parts if any damage is evident.

*f. Spring Bumper.*—Examine bumper assembly (K) for evidence of deterioration. Inspect stud which is bonded into bumper for damaged threads and for looseness in bumper. If any damage is evident or if stud is loose, replace with new part.

*g. Torque Rods.*—Examine upper and lower torque rod assemblies (F and CC) for distortion, damaged threads on end pins, and for loose or damaged pin and bearing assemblies at axle ends and loose or damaged bushing type bearings at frame ends. If end pins or bearings are damaged or loose, replace (par. 272). On upper torque rod assembly (F), make sure brake and vent line shield (G) is securely welded to torque rod.

*h. Torque Rod Bolts.*—Examine bolts 1–14 x $5\frac{7}{8}$ (FF) and 14 x $7\frac{5}{16}$ bolts (HH) which attach torque rods to front spring rear brackets for damaged threads and serrations. Replace with new parts if damaged.

### 270. Repair of Spring Assembly

*Note.* Key letters noted in parentheses are in figure 205.

*a. General.*—If only the bushing type bearings in spring eyes require replacement, replacement can be made (*b* below) without disassembling the spring assembly.

*b. Spring Eye Bearing Replacement.*—Press old bushing type bearing (W) out of spring eye, using a suitable bearing driver and arbor press. Make sure spring eye is clean; then press new bearing into place and burnish to a diameter of 0.749 to 0.754 inch.

*c. Disassembly of Spring.*—Remove nuts from three rebound clip bolts, then remove bolts and spacers. Clamp spring leaves firmly together, using one "C" clamp on each side of center bolt or using an arbor press. Remove nut from center bolt and remove bolt. Release "C" clamps or arbor press slowly to avoid personal injury.

*d. Assembly of Spring.*

(1) Clean all dirt and corrosion from spring leaves, using a wire brush if necessary, then wash in dry-cleaning solvent or volatile mineral spirits. Coat each spring leaf with a thin film of soft graphited grease (GG); grease must cover entire contact area.

(2) Stack spring leaves in correct order with center bolt holes alined, then compress spring leaves using "C" clamps or arbor press. Install center bolt and nut with nut at top and tighten

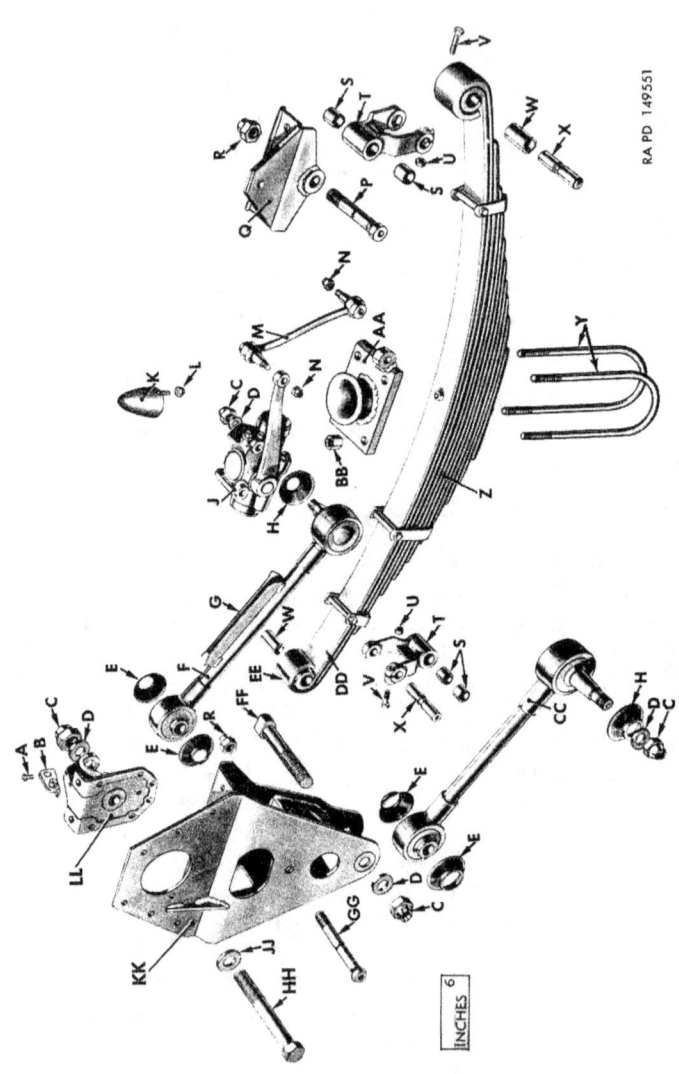

A—SCREW, CAP, ¼-20 X ⅝—180018
B—BRACKET, SUPPORT, FRONT BRAKE RUBBER LINE, ASSY—7410802
C—NUT, 1-14—454334
D—WASHER, PLAIN—7412849
E—SEAL, DUST—7411873
F—ROD, TORQUE, UPPER, ASSY—7411369
G—SHIELD, BRAKE AND VENT LINE—YT-2265451
H—SEAL, DUST—7411374
J—{ABSORBER, SHOCK, LEFT, ASSY—7350540
   {ABSORBER, SHOCK, RIGHT, ASSY—7350539
K—BUMPER, ASSY—7412911
L—NUT, ⅜-24—442799
M—LINK, ASSY—7350541
N—NUT, ½-20—442801
P—BOLT, FRONT SHACKLE—7350406
Q—BRACKET, FRONT SHACKLE, ASSY—7410567
R—NUT, ¾-16—442804
S—BEARING, BUSHING TYPE—7410889
T—SHACKLE, ASSY—7350458
U—NUT, 9/16-24—442798
V—SCREW, CAP, 9/16-24 X 1⅞—181619
W—BEARING, BUSHING TYPE—7412851
X—PIN, SHACKLE—7350457
Z—SPRING, FRONT, ASSY—7350460
Y—{BOLT, "U", LEFT FRONT SPRING—7410657
   {BOLT, "U", RIGHT FRONT SPRING INNER—7410656
   {BOLT, "U", RIGHT FRONT SPRING OUTER—7350461
AA—{BLOCK, BUMPER, LEFT—YT-2286276
    {BLOCK, BUMPER, RIGHT—YT-2286275
BB—NUT, ⅝-18—7348692
CC—ROD, TORQUE, LOWER, ASSY—YT-2277031
DD—LEAF, NO. 1, ASSY—7350455
EE—LEAF, NO. 2—7350456
FF—BOLT, 1-14 X ⅝—7411370
GG—BOLT, REAR SHACKLE—7350407
HH—BOLT, 1-14 X 7/16—7411371
JJ—WASHER—7412849
KK—BRACKET, FRONT SPRING. REAR ASSY—YT-2278214
LL—BRACKET, UPPER TORQUE ROD—YT-2283514

*Figure 205. Front spring suspension components.*

firmly. Install rebound clip spacers and 3/8–16 x 3 3/8 bolts and secure with 3/8–16 nuts. Tighten nuts firmly, pulling ends of rebound clips against ends of spacers. Remove "C" clamps or remove spring assembly from arbor press.

### 271. Repair of Spring Shackles

*a. Bearing Removal.*—Press old bushing type bearings (S) out of shackle assembly (T), using a suitable bearing driver and arbor press.

*b. Bearing Installation.*—Press a new bushing type bearing into each side of shackle until outer ends of bearings are at bottom of chamfer in shackle. After pressing both bearings into place, burnish to dimension shown in figure 206.

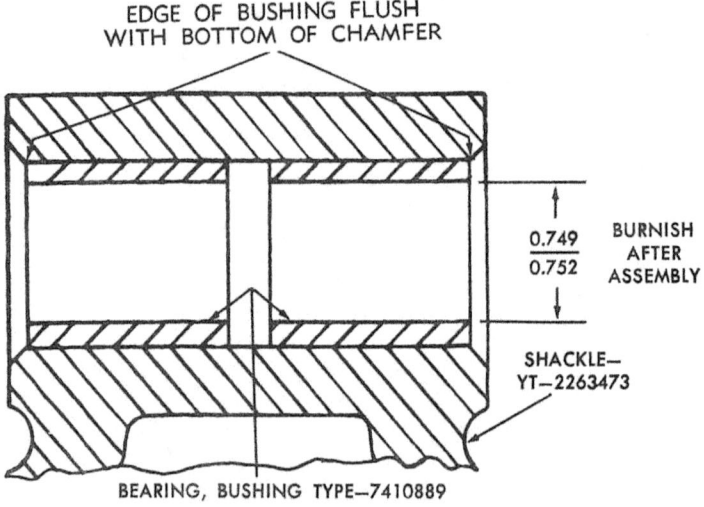

*Figure 206. Bushing type bearings installed in spring shackle.*

### 272. Repair of Torque Rods

*a. Bearing Removal.*—Using a driver to exert force on rounded edge of bearing case at point shown in figure 207, press bushing type bearing assembly out of torque rod end.

*b. Bearing Installation.*—Press new bushing type bearing assembly into torque rod end, using a driver which will exert force on rounded edge of bearing case at point shown in figure 207. Press bearing in until it is centered in torque rod end.

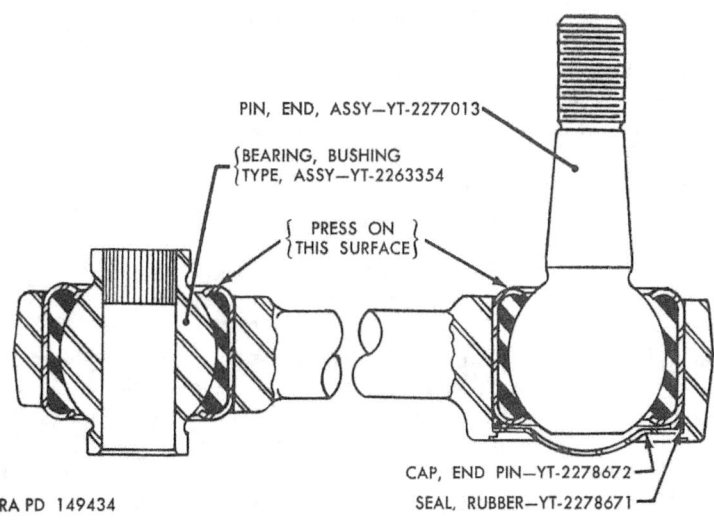

Figure 207. Sectional views of front torque rod ends.

c. *End Pin Removal.*—Using a small sharp chisel, cut off eight staked lugs which secure end pin cap in torque rod end. Using a driver over end pin which will exert force on rounded edge of bearing case at point shown in figure 207, press end pin assembly, rubber seal, and end pin dust cap out of torque rod end. File stake marks back flush with inner circumference of torque rod end.

d. *End Pin Installation.*—End pin assembly (fig. 207) must be installed in torque rod end with tapered end pin opposite the side having the stake marks. Press new end pin assembly into torque rod end, using a driver over end pin which will exert force on rounded edge of bearing case at point shown in figure 207. Press in to approximate position shown in figure 207; then install rubber seal and end pin cap and secure in place by staking at eight places. Press end pin assembly in further to compress rubber seal between bearing case and end pin cap, using care not to force cap past the stake marks.

## Section III. REBUILD OF SHOCK ABSORBERS

### 273. Construction

a. Shock absorber shaft is mounted in body on two bushing type bearings. Shaft bore at one end is sealed with an expansion plug and gasket. Shaft extends out of body on one side and is sealed by means of a packing washer and packing gland. A cam, which extends down into piston assembly, is welded to shaft inside of body. Shock absorber arm is pressed onto end of shaft which extends out of body.

Due to the welded construction of the cam and shaft, these parts cannot be disassembled; any wear or damage to the shaft, body, bearings, packing gland and washer, and arm necessitates replacing the complete shock absorber assembly.

*b.* The piston assembly consists of two piston halves, held together by two screws and springs. Intake valve and spring is secured in outer end of each piston half by a snap ring type retainer. Intake valves may be replaced individually, but the piston is serviced only as a complete assembly.

*c.* The fluid reservoir consists of the space around the shaft and cam, with the filler plug located at the top of this chamber. An end cap, with fibre gasket, is threaded into each end of body; these end caps must be removed for replacing intake valves or piston assembly.

*d.* Rebound and compression valves, which control the flow of fluid in the shock absorber, are installed in drilled passages in the shock absorber body and retained by nuts with soft aluminum gaskets. Valve identification code numbers are stamped on nuts and on valve washers as shown in figure 213. Valves having the same code number as stamped on the retaining nuts must always be used.

### 274. Operation of Shock Absorbers

*a. General.*—The flow of fluid within the shock absorber during operation is schematically illustrated in figure 208. Two separate actions take place, compression and rebound. Each phase of operation is described separately under *b* and *c* below.

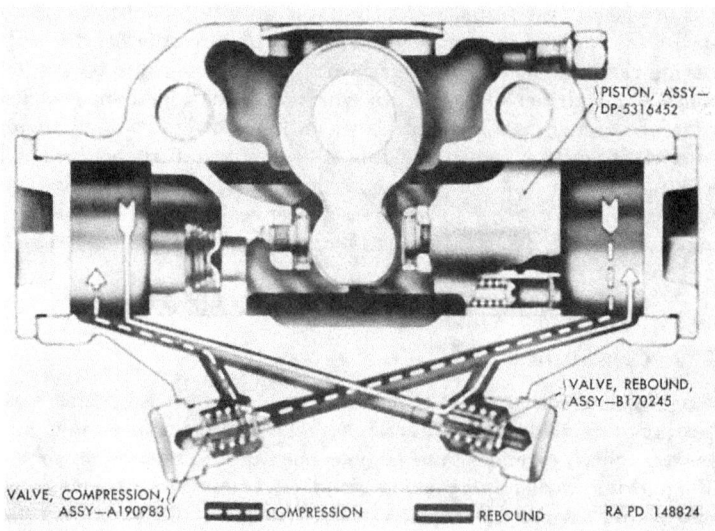

*Figure 208. Schematic sectional view of shock absorber showing operation.*

*b. Compression.*—As the shock absorber arm moves upward (compression stroke), the cam forces the piston toward the arm end of the cylinder, displacing fluid from the compression end of the cylinder. On very slight or slow axle movements, the fluid flows only through the small opening provided by compressing the small inner spring of the compression valve and into the rebound end of the cylinder. On rapid movements, the pressure overcomes the large outer spring, lifting the valve further from its seat, permitting more rapid flow of fluid into the rebound end of the cylinder. At the same time, the intake valve in the rebound end of the piston opens, permitting fluid to flow from the reservoir into the rebound end of the cylinder, compensating for any loss of fluid between compression end of piston and cylinder wall.

*c. Rebound.*—During the rebound stroke, or as the arm moves downward, the direction of flow is reversed. The cam forces the piston away from the arm end of the cylinder, displacing fluid from the rebound end of the cylinder. For slow action, the fluid flows only through the orifice in the rebound valve into the compression end of the cylinder. During rapid action, the rebound valve is lifted from its seat and the fluid passes, at a pressure controlled by the rebound valve spring, into the compression end of the cylinder. At the same time, the intake valve in the compression end of the piston opens, allowing fluid to pass from the reservoir into the compression end of the cylinder, compensating for any loss of fluid between rebound end of piston and cylinder wall.

### 275. Disassembly of Shock Absorber

*Note.* Key letters noted in parentheses are in figure 210.

*a. Removal of Rebound and Compression Valves.*

(1) Remove filler plug and gasket (G and H) and drain fluid from shock absorber. Work arm up and down to expel most of the fluid.

(2) Mount shock absorber assembly on shock absorber rebuilding stand 41-S-4977-5. Stand must be secured to bench or clamped in a vise.

(3) Remove rebound and compression valve nuts (N), using an offset or "L" shaped screw driver; then remove rebound valve assembly (R) and compression valve assembly (P) from shock absorber body, using a piece of wire with hooked end to lift valves out.

*b. Removal of Intake Valves.*

(1) Mount shock absorber on shock absorber rebuilding stand 41-S-4977-5 with the cylinder in a vertical position. Using a short length of $1\frac{1}{16}$-inch hex tool stock as an adapter for

a box end wrench, remove end plug (A) and end plug gasket (B) from upper end.

(2) Move shock absorber arm to position piston at top of cylinder. Pry intake valve spring retainer (M) out of groove in piston; then remove intake valve spring (L) and intake valve (K) from piston.

(3) Turn shock absorber over to position other end upward and repeat (1) and (2) above.

*c. Removal of Piston Assembly.*

(1) With shock absorber mounted in vertical position on shock absorber rebuilding stand 41-S-4977-5 (fig. 209), and with piston at top of cylinder use a sharp pointed punch to pierce expansion plug (C) and pry plug out of piston. Remove piston screw (D) and piston screw spring (E) from upper end of piston.

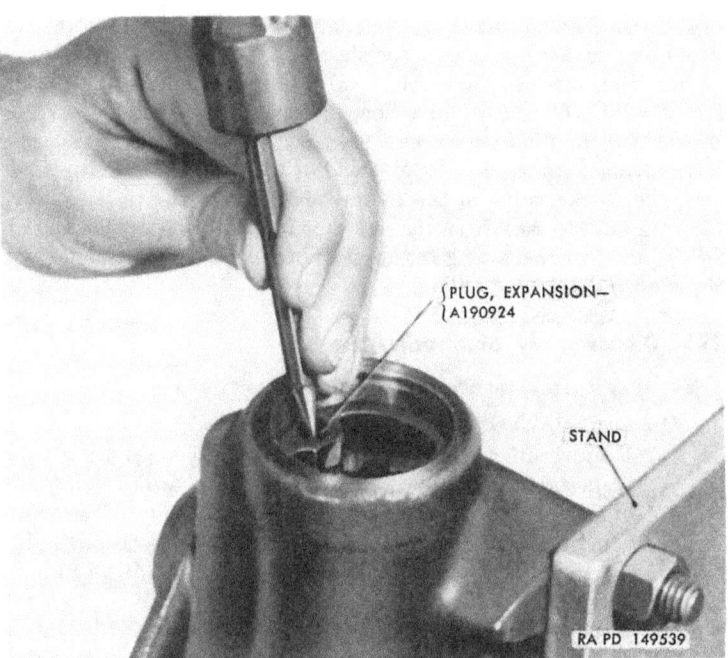

*Figure 209. Removing piston screw expansion plug with shock absorber mounted in stand 41-S-4977-5.*

(2) Turn the assembly over to position other end upward and remove expansion plug (C), piston screw (D), and piston screw spring (E) in same manner as in (1) above.

(3) Remove half of piston from each end of cylinder.

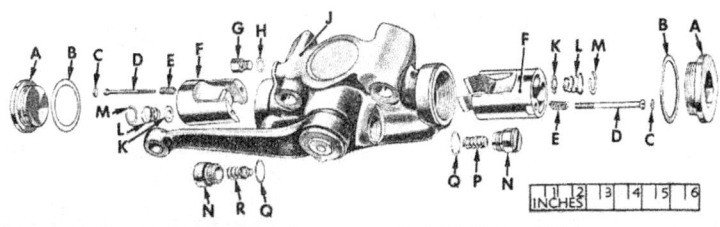

A—PLUG, END—A190913
B—GASKET, END PLUG—A190914
C—PLUG, EXPANSION—A190924
D—SCREW, PISTON—A190922
E—SPRING, PISTON SCREW—A190923
F—PISTON, ASSY—DP-5316452
G—PLUG, FILLER—A290939
H—GASKET, FILLER PLUG—5190920
J— {ABSORBER, SHOCK, LEFT, ASSY—7350540
ABSORBER, SHOCK, RIGHT, ASSY—7350539
K—VALVE, INTAKE—7350155
L—SPRING, INTAKE VALVE—A190990
M—RETAINER, INTAKE VALVE SPRING—DP-046812
N—NUT, REBOUND AND COMPRESSION VALVE—A190917
P—VALVE, COMPRESSION, ASSY—A190983
Q—GASKET, VALVE NUT—7372338
R—VALVE, REBOUND, ASSY—B170245

RA PD 148832

*Figure 210. Shock absorber components.*

## 276. Cleaning and Inspection of Shock Absorber

*a. Cleaning.*—Wash all shock absorber parts in dry-cleaning solvent or volatile mineral spirits. Wipe small parts dry, and blow inside of cylinder and internal passages in body dry with compressed air. Make sure internal passages are thoroughly cleaned. Shock absorber components are shown in figure 210.

*b. Inspection.*

(1) *Body.*—Check shaft and bushings in body for wear by moving shock absorber arm sideways. Inspect cam on shaft for galled or worn condition. Examine cylinder bore for roughness or scoring. If any of the above conditions are evident, replace the complete shock absorber assembly.

(2) *Piston.*—Insert each piston half in cylinder in operating position (cam clearance at top) and check clearance between piston and cylinder wall. If clearance is not within limits specified (par. 354), a new piston assembly must be selected which will provide clearance within these limits. Examine piston for evidence of scoring or other damage. Check thrust button at inner end of each piston half for roughness or evidence of wear. Make sure antirotation spring installed under thrust button on rebound half of piston is not bent. If any wear or damage is evident, replace piston assembly.

(3) *Rebound and compression valves.*—The only inspection that can be made on rebound and compression valves is to make a visual inspection to see that valves are securely assembled and appear to be in good condition. If there is any doubt as to whether valve springs have become weakened, replace with new valve assemblies. When obtaining new valves, make

sure valve code stamped on outer spring retainer washer on each valve is the same as the code stamped on the valve nuts (fig. 213).

### 277. Assembly of Shock Absorber

*Note.* Key letters noted in parentheses are in figure 210 unless otherwise indicated.

*a. General.*—Coat all internal parts with petroleum base hydraulic oil (OHA) before assembling shock absorber.

*b. Installation of Piston Assembly.*
   (1) Mount shock absorber body on shock absorber rebuilding stand 41–S–4977–5. File a $1/16$-inch chamfer on inner edges of compression half of piston at points indicated in figure 211 to permit ends of antirotation spring to enter piston when assembling.
   (2) Insert piston halves in cylinder, with the half having antirotation spring opposite the arm end of the shock absorber, and with cam clearance side of piston halves toward top of shock absorber.
   (3) Install piston screws (D) and piston screw springs (E) in piston. Tighten screws; then back off each screw 1 to $1\frac{1}{2}$ turns to prevent binding between cam and thrust buttons. Install new expansion plug (C) in screw hole in outer end of each piston half.

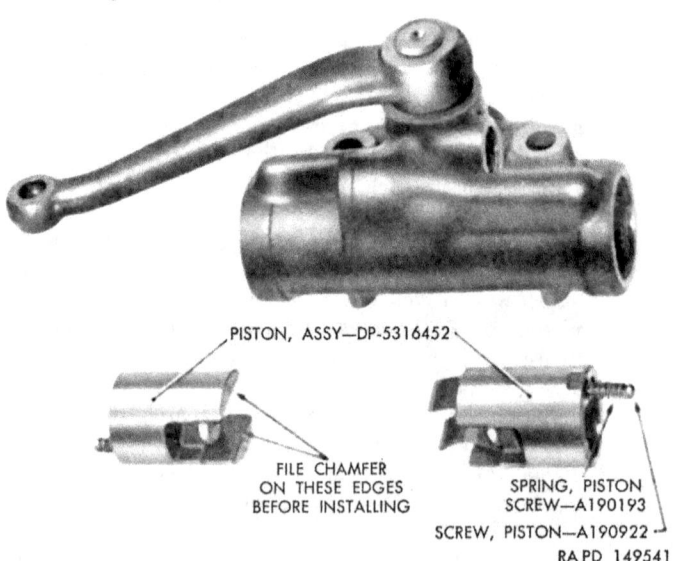

*Figure 211. Piston assembly removed.*

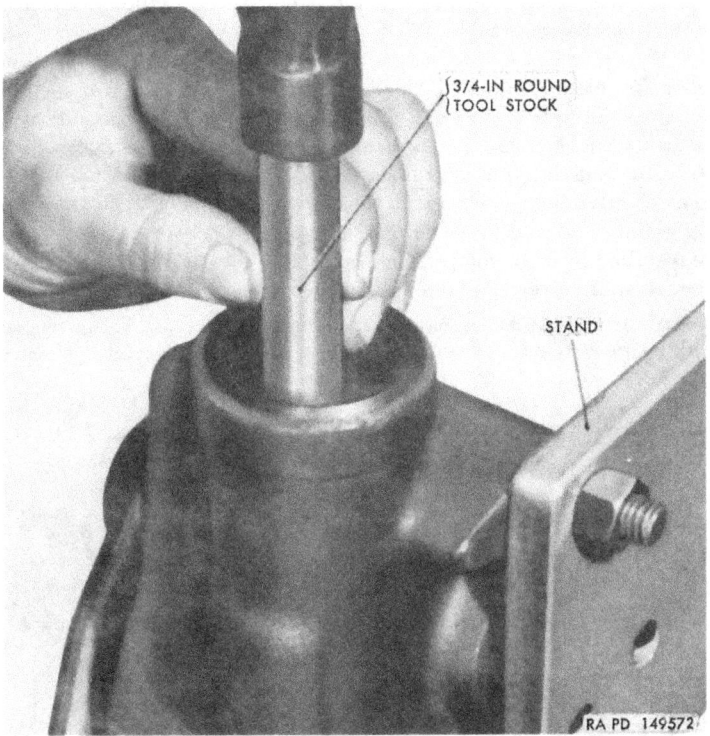

*Figure 212. Installing intake valve spring retainer with shock absorber mounted in stand 41-S-4977-5.*

    *c. Installation of Intake Valves.*
       (1) Mount shock absorber body on shock absorber rebuilding stand 41-S-4977-5 in vertical position. Move shock absorber arm to position piston at top.
       (2) Drop intake valve (K) into piston with raised side up, and place intake valve spring (L) on top of valve with small end of spring next to valve. Place intake valve spring retainer (M) in piston; then using a piece of ¾-inch round tool stock (fig. 212) as a driver, drive retainer down until it seats in groove in piston. Make sure retainer is fully seated in groove and is not driven down beyond the groove.
       (3) Turn shock absorber body over and position other end of piston at top; then install the other intake valve, spring, and retainer as in (2) above.
       (4) Install end plug (A) with new end plug gasket (B) in each end of shock absorber body and tighten firmly, using a short-

length of 1 1/16-inch hex tool stock as an adapter for a box end wrench.

*d. Installation of Rebound and Compression Valves.*—Compression valve assembly, marked "G-2" (fig. 213), must be installed in opening opposite the arm end of the shock absorber, and the rebound valve assembly, marked "2 N," in other end. Place each valve assembly in body, making sure guide at inner end of valve enters passage at bottom of valve bore. Secure in place with rebound and compression valve nuts (N) and valve nut gaskets (Q). Nut marked "G 2" must be installed over the compression valve, and nut marked "G 2 N" over the rebound valve. Tighten nuts firmly.

*Note.* If rebound and compression valve nuts are replaced, stamp new nut with proper marking.

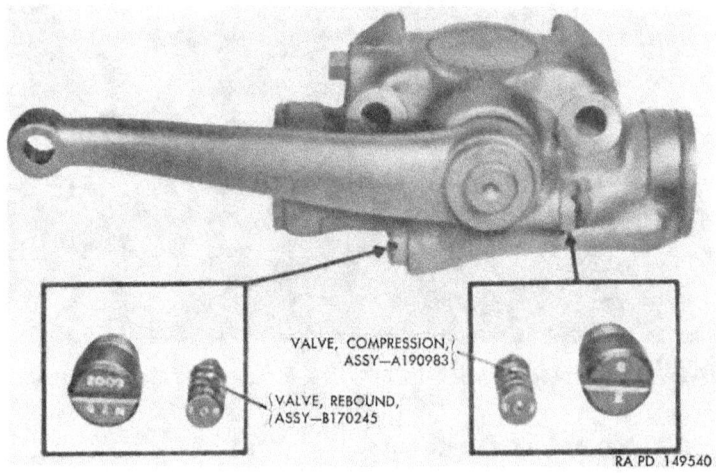

*Figure 213. Shock absorber valve identification.*

*e. Filling Shock Absorber.*—Mount shock absorber assembly on shock absorber rebuilding stand 41-S-4977-5 in operating position. Fill with petroleum base hydraulic oil (OHA), pumping shock absorber arm up and down while adding oil to expel air and to pump oil into internal passages. After all air is expelled and oil remains at level of filler plug hole, install filler plug (G), using a new filler plug gasket (H), and tighten firmly. Remove shock absorber assembly from rebuilding stand.

# CHAPTER 15

# REAR SPRING SUSPENSION

## Section I. DESCRIPTION AND DATA

### 278. Description

*a. General.*—Rear spring suspension system components comprise the main spring assembly, main spring seat and bearings, secondary spring assembly, torque rod assemblies, and attaching parts. Rear spring suspension components are illustrated in figure 214.

*b. Main and Secondary Springs.*—Main and secondary spring assemblies are identical. Each spring assembly consists of ten spring leaves, secured together by a center bolt and four rebound clips. Main spring is mounted on spring seat which in turn is mounted on a shaft on opposed tapered roller bearings. Secondary spring is mounted rigidly to bracket on frame side rail. Slipper ends of main spring are inserted in brackets which are integral with axle housings; secondary spring ends contact top of brackets on axle housing under heavy load conditions.

*c. Main Spring Seat.*—Main spring seat is mounted on tapered roller bearings on spring seat shaft. Tapered spring seat shaft is installed in spring seat and torque rod bracket and secured with a plain washer and safety nut. Outer end of shaft is threaded for bearing adjusting nuts, and grooved for tongue on adjusting nut lock. Spring seat inner oil seal, installed on spring seat shaft sleeve, wipes on inside of seal flange which is pressed into inner end of seat.

*d. Torque Rods.*—Six torque rod assemblies, two upper and four lower, transmit driving and braking forces of the two rear axles to the frame. Both ends of all rear torque rods are equipped with tapered end pins the same as used at axle end of front torque rods (fig. 207). End pin and bearing must be replaced as a complete assembly. Brake and vent line shield is welded to top of each upper torque rod.

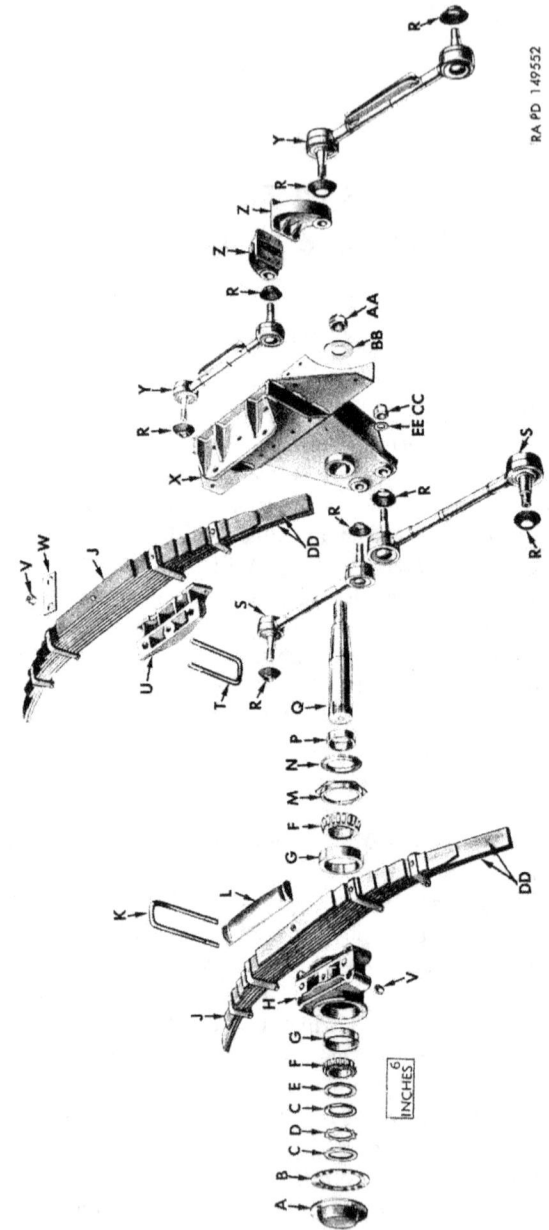

*Figure 214. Rear spring suspension components.*

## 279. Data

Main and secondary spring assemblies:
| | |
|---|---|
| Overall length | 55½ in. |
| Width | 2½ in. |
| Number of leaves | 10 |
| Thickness of leaves: | |
| 4 @ | 0.447 in. |
| 6 @ | 0.401 in. |
| Total thickness | 4.194 in. |

## Section II. REBUILD OF REAR SPRING SUSPENSION COMPONENTS

## 280. General

This section contains inspection and repair procedures of rear spring suspension components which are beyond the scope of the using organization and are not contained in TM 9-819A.

---

*Figure 214.*—Continued.

A—CAP, DUST—7411375
B—GASKET, PAPER—7411265
C—NUT, ADJUSTING—7411379
D—LOCK, ADJUSTING NUT—7411378
E—WASHER, ADJUSTING NUT—7411422
F—CONE, TAPERED ROLLER BEARING—712868
G—CUP, TAPERED ROLLER BEARING—712869
H—SEAT, MAIN SPRING, ASSY—7411381
J—SPRING, MAIN AND SECONDARY, ASSY—7350470
K—BOLT, "U", MAIN SPRING—7350471
L—SPACER, "U" BOLT—7350468
M—FLANGE, OIL SEAL—YT-2278050
N—SEAL, OIL, ASSY—7411380
P—SLEEVE—7411420
Q—SHAFT, MAIN SPRING SEAT—7411383
R—SEAL, DUST—7411374
S—ROD, TORQUE, LOWER, ASSY—2277032
T—BOLT, "U", SECONDARY SPRING—7350472
U—SEAT, SECONDARY SPRING—7350467
V—NUT, ¾-16—7350473
W—SPACER—7350469
X—BRACKET, MOUNTING—YT-2276465
Y—ROD, TORQUE, UPPER, ASSY—7411372
Z—BRACKET, UPPER TORQUE ROD—7411368
AA—NUT, SAFETY, 1½-12—454335
BB—WASHER—7411382
CC—NUT, SAFETY, 1-14—454334
DD—LEAF, No. 1 AND 2—7350466
EE—WASHER—7412849

## 281. Inspection of Rear Spring Suspension Components

*Note.* Key letters noted in parentheses are in figure 214.

*a. Spring Assemblies.*—Examine main and secondary spring assemblies (J) for broken leaves, broken rebound clips, or broken center bolt. If No. 1 and 2 leaves (DD) or center bolt are broken, they may be replaced (par. 282). If rebound clips or spring leaves other than No. 1 and 2 are broken, the complete spring assembly must be replaced.

*b. Main Spring Seat Shaft.*—Examine main spring seat shaft (Q) for damaged threads or distortion. If any damage is evident, replace shaft (par. 283).

*c. Torque Rods.*—Examine upper and lower torque rod assemblies (Y and S) for distortion, damaged threads on end pins, and for loose or damaged pin and bearing assemblies. If end pin and bearing assemblies are damaged or loose, replace (par. 272c and d).

## 282. Repair of Spring Assemblies

*a. Disassembly of Spring.*—Remove nuts from four rebound clip bolts, then remove bolts and spacers. Clamp spring leaves firmly together, using one "C" clamp on each side of center bolt or using an arbor press. Remove nut from center bolt and remove bolt. Release "C" clamp or arbor press slowly to avoid personal injury.

*b. Assembly of Spring.*

(1) Clean all dirt and corrosion from spring leaves, using a wire brush if necessary, then wash in dry-cleaning solvent or volatile mineral spirits. Coat each spring leaf with a thin film of soft graphited grease (GG); grease must cover entire contact area.

(2) Stack spring leaves in correct order with center bolt holes alined, then compress spring leaves using "C" clamps or arbor press. Install center bolt and nut with nut at top and tighten firmly. Install rebound clip spacers and $3/8$-16 x $3 3/8$ bolts and secure with $3/8$-16 nuts. Tighten nuts firmly, pulling ends of rebound clips against ends of spacers. Remove "C" clamps or remove spring assembly from arbor press.

## 283. Replacement of Spring Seat Shaft

*a. Remove Shaft.*—Remove spring assemblies and main spring seat as directed in paragraph 39. Remove safety nut (AA, fig. 214) and plain washer (BB, fig. 214) securing spring seat shaft in spring seat and torque rod bracket. Install a standard $1 1/2$–12 nut on inner end of spring seat shaft to prevent damaging threads; then drive on inner end of shaft with heavy hammer to loosen shaft. Remove nut from shaft and withdraw shaft from bracket.

*b. Install Shaft.*—Make sure tapered end of spring seat shaft and hole in spring seat and torque rod bracket are clean and dry. Insert tapered end of shaft into tapered hole, with groove in outer end of shaft at top. Install plain washer and 1½–12 safety nut on inner end of shaft and tighten to 650 to 700 pound-feet. Install spring seat and springs (par. 49).

# CHAPTER 16

# FRAME AND ASSOCIATED PARTS

## Section I. DESCRIPTION AND DATA

### 284. Description

*a. Frame.*—The frame assembly consists of right and left side members, cross members, gussets, braces, and other miscellaneous brackets, all of which are pressed steel and are riveted in place to form an assembly (fig. 215). Parts which may require frequent removal are held by bolts and self-locking nuts. In some instances several pieces are welded together to form an assembly.

*b. Bumpers.*—Bumper at front of vehicle is pressed steel channel formed into a bumper and bolted to frame with gusset plates. Two rear bumpers, one at each rear corner of frame, are formed from pressed steel and are bolted to frame side member and rear cross member.

*c. Towing Shackles.*—Four towing shackles, one at each corner of frame, are attached to brackets riveted to cross member at rear or welded to bumper gussets at front.

*d. Pintle.*—Pintle assembly is installed at center of rear cross member. Pintle shaft extends through cross member and brackets bolted to each side of cross member.

### 285. Data

Wheelbase:
  M135, M211, M217, and M222 _____ 156 in.
  M215 and M221 _____ 144 in.
Frame overall length:
  M135, M211, M217, and M222 _____ 255 in.
  M215 and M221 _____ 230 in.
Width—front end _____ $34\frac{1}{8}$ in.
Width—rear end _____ $34\frac{1}{8}$ in.
Number of cross members _____ 5

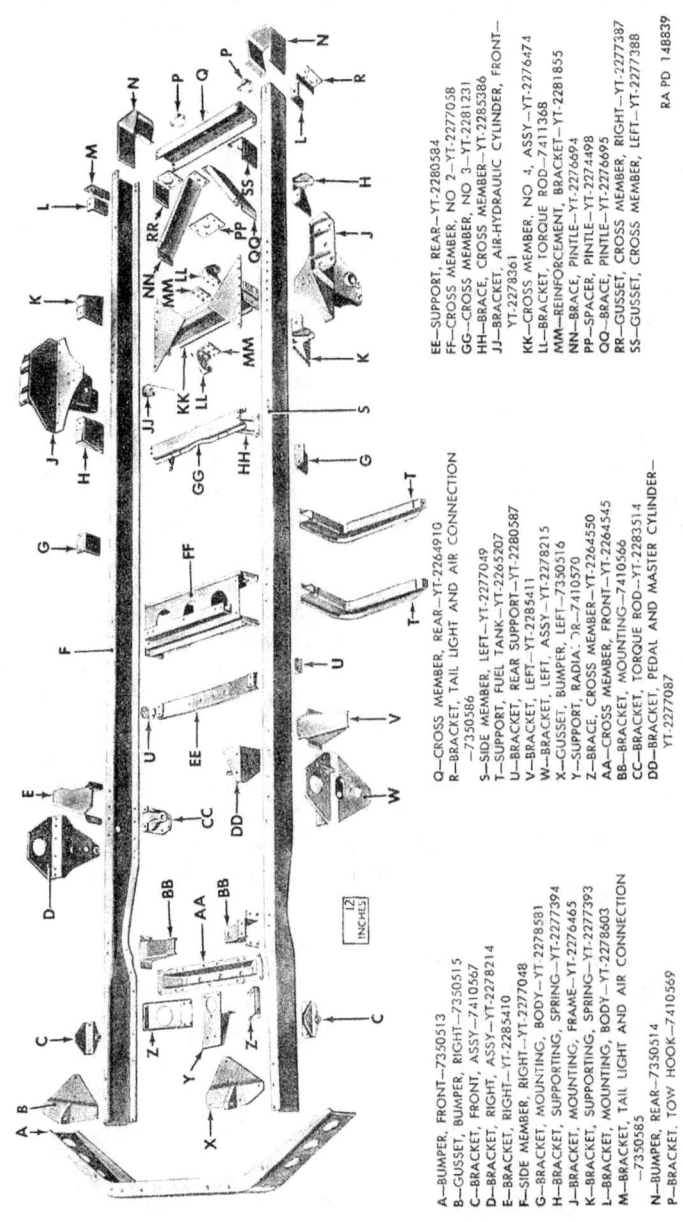

A—BUMPER, FRONT—7350513
B—GUSSET, BUMPER, RIGHT—7350515
C—BRACKET, FRONT, ASSY—7410567
D—BRACKET, RIGHT, ASSY—YT-2278214
E—BRACKET, RIGHT—YT-2285410
F—SIDE MEMBER, RIGHT—YT-2277048
G—BRACKET, MOUNTING, BODY—YT-2278581
H—BRACKET, SUPPORTING, SPRING—YT-2277394
J—BRACKET, MOUNTING, FRAME—YT-2276465
K—BRACKET, SUPPORTING, SPRING—YT-2277393
L—BRACKET, MOUNTING, BODY—YT-2278603
M—BRACKET, TAIL LIGHT AND AIR CONNECTION —7350585
N—BUMPER, REAR—7350514
P—BRACKET, TOW HOOK—7410569
Q—CROSS MEMBER, REAR—YT-2264910
R—BRACKET, TAIL LIGHT AND AIR CONNECTION —7350586
S—SIDE MEMBER, LEFT—YT-2277049
T—SUPPORT, FUEL TANK—YT-2265207
U—BRACKET, REAR SUPPORT—YT-2280587
V—BRACKET, LEFT—YT-2285411
W—BRACKET, LEFT, ASSY—YT-2278215
X—GUSSET, BUMPER, LEFT—7350516
Y—SUPPORT, RADIA. `R—7410570
Z—BRACE, CROSS MEMBER, FRONT—YT-2264550
AA—CROSS MEMBER, MOUNTING—7410566
BB—BRACKET, MOUNTING—7410566
CC—BRACKET, TORQUE ROD—YT-2283514
DD—BRACKET, PEDAL AND MASTER CYLINDER— YT-2277087
EE—SUPPORT, REAR—YT-2280584
FF—CROSS MEMBER, NO 2—YT-2277058
GG—CROSS MEMBER, NO 3—YT-2281231
HH—BRACE, CROSS MEMBER—YT-2285386
JJ—BRACKET, AIR-HYDRAULIC CYLINDER, FRONT— YT-2278361
KK—CROSS MEMBER NO 4, ASSY—YT-2276474
LL—BRACKET, TORQUE ROD—7411368
MM—REINFORCEMENT, BRACKET—YT-2281855
NN—BRACE, PINTLE—YT-2276694
PP—SPACER, PINTLE—YT-2274498
QQ—BRACE, PINTLE—YT-2276695
RR—GUSSET, CROSS MEMBER, RIGHT—YT-2277387
SS—GUSSET, CROSS MEMBER, LEFT—YT-2277388

*Figure 215. Disassembled view of frame and associated parts.*

## Section II. REBUILD OF FRAME AND ASSOCIATED PARTS

### 286. General

Procedure for checking frame alinement and repair or straightening will depend upon the equipment available and skill of available personnel. In some instances suitable checking and straightening equipment may be available, otherwise procedures following must be used.

### 287. Checking Frame Alinement

*a. Layout Procedure.*—The most convenient method of checking frame alinement, where precision equipment is not available, is by marking on the floor all points from which measurements are to be taken. This can be done by tacking or cementing paper to the floor under each point of measurement. Use a "plumb bob" when marking each point. Points of measurement are called out on figure 216. Corresponding points on opposite side of vehicle are indicated by letter *M*. Accuracy is important if satisfactory results are to be obtained. After each point shown has been marked the vehicle should be moved away from the layout on the floor.

*b. Checking Layout.*

   (1) Check frame width at front and rear ends, using marks on floor (fig. 216). If widths correspond with dimensions given in paragraph 285, draw a center line full length of vehicle halfway between marks indicating width at front and rear. If frame width is not as given in paragraph 285, draw a line through points of intersection of any two pairs of equal diagonals (A, B, C, fig. 216). Center line can also be drawn through intersection of any one pair of diagonal lines of equal length and center point of either end of frame.

     *Note.* If extreme front end of frame is damaged, center of front end of frame can be located from point exactly midway between radiator support bolts.

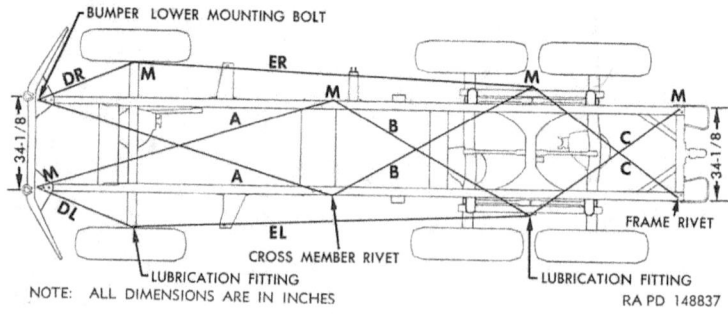

Figure 216. *Method of checking frame alinement.*

(2) With center line properly laid out, measure distance from it to opposite points marked over entire length of chassis. If frame is in proper alinement, measurements should not vary more than one-eighth of an inch at any point.

(3) To locate point at which frame is sprung, measure diagonals marked A, B, and C (fig. 216). If diagonals in each pair are within one-eighth of an inch, that part of frame included between points of measurements may be considered in alinement, and these diagonal lines should intersect within one-eighth of an inch of center line. Variations of more than one-eighth of an inch indicate misalinement.

*c. Checking Front Axle Alinement* (fig. 216).—When it has been determined that frame is properly alined, front axle alinement with frame can be checked as directed below.

(1) Front axle is square with frame if ER equals EL, and DR equals DL.

(2) Front axle has shifted sideways if ER is less than EL, and DR is less than DL, or vice versa.

(3) Front axle is bent, twisted, or shifted if ER is less than EL, and DR is greater than DL, or vice versa.

## 288. Frame Repair

*a. Straightening of Frame.*—The use of heat is not recommended when straightening frames. Heat weakens structural characteristics of frame members and all straightening should be done cold. Frame members which are bent or buckled sufficiently to show strain or cracks after straightening must be reinforced or replaced.

*b. Reinforcing Frame.*—No established rules can be made on the necessity, length, or kinds of reinforcement to install on frames which are bent or broken. Reinforcements can be made with channel, angle, or flat stock. Because of problems encountered when installing channel reinforcements in frame side members, the use of angle reinforcements is recommended. Whenever possible, the reinforcement should extend full length of side member. This may, in some instances, be impractical because of the position of attaching units and existing cross members; therefore, it is necessary that the mechanic use his best judgment to suit the problems encountered. The reinforcement stock thickness should be the same as that of the member being reinforced, and the material of the reinforcement stock should be of the same tensile strength.

*c. Replacing Cross Members and Brackets.*—All cross members, brackets, or gussets that are damaged or broken off must be replaced. Cut off heads and drive out all rivets from part being replaced. Install new part and use hot rivets to secure part in place.

d. *Riveting.*—Specific rules for the spacing of rivets used in reinforcements cannot be given, as such spacing depends entirely upon the number and size of rivets used in attaching reinforcement, bracket, cross members, etc. to the portion of frame being repaired. The mechanic must use his best judgement as to the number, size, and spacing of the rivets. As a general rule rivets should be 50 to 100 percent as thick as the parts being riveted. Rivets that are to be countersunk should protrude the diameter of the rivet through the plates; rivets to be riveted to a round head should protrude twice the diameter of the rivet.

e. *Welding.*—The electric arc welding method is recommended for all frame welding. Heat generated during welding is localized and burning of material is minimized whenever this method is used. Additional advantages are that finished weld can be ground, filed, and drilled as necessary. Welding electrode should be A. W. S. class E-6012.

    (1) *Preparing frame for welding.*—Whenever inspection indicates the necessity of repairing a cracked frame, certain precautions and preparations must be observed. Namely, that a hole must be drilled at starting point of crack (fig. 217) also, crack must be ground out to obtain good weld. Shape reinforcement to fit interior or exterior of the part being repaired. The reinforcement should be of **SAE** 1010 or **SAE** 1020 steel.

    (2) *Welding instructions.*—When reinforcement has been shaped and placed in position, weld at points shown and as instructed by welding symbols in figure 218. Do not weld on fillets or

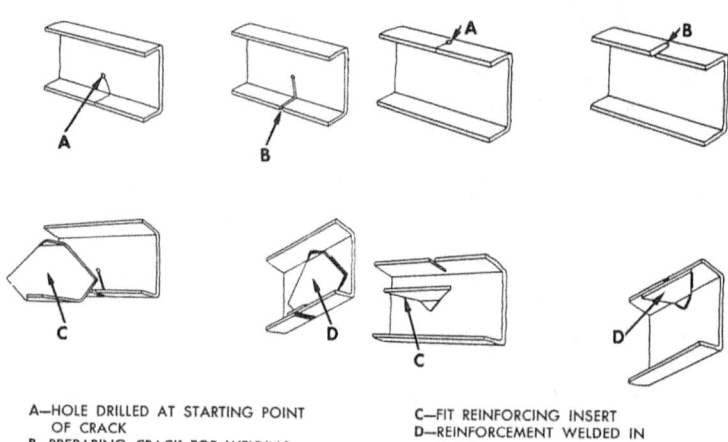

A—HOLE DRILLED AT STARTING POINT OF CRACK
B—PREPARING CRACK FOR WELDING
C—FIT REINFORCING INSERT
D—REINFORCEMENT WELDED IN PLACE

RA PD 149579

*Figure 217. Method of preparing crack for welding.*

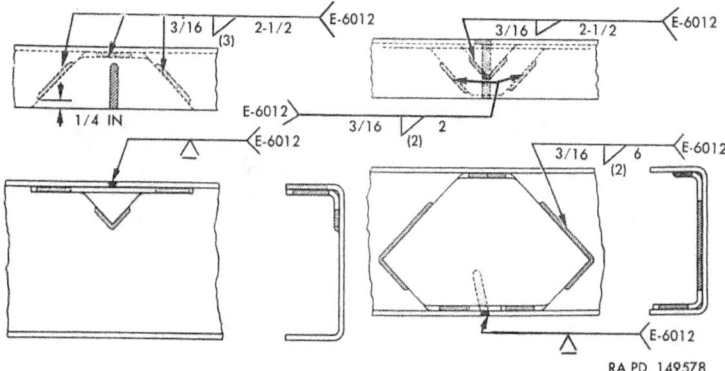

*Figure 218. Method of attaching reinforcement by welding.*

within one quarter of an inch of edges, since welding at these points tends to encourage development of new cracks.

### 289. Rebuild of Pintle

*Note.* Key letters noted in parentheses are in figure 219.

a. *Disassembly.*

(1) *Remove lubrication fittings.*—Remove lubrication fittings (L) from 1¼ x 3⅛ bolt (R) and latch pin (K).

(2) *Removal of lock and latch.*—Remove ⅛ x 1½ cotter pin (S); then remove 1–14 nut (Q). Remove bolt; then lift lock and latch assembly from pintle.

(3) *Removal of pintle latch.*—Use arbor press to remove latch pin (K) attaching pintle latch (G) to pintle lock (J); then lift latch and latch spring (H) from pintle lock.

b. *Cleaning and Inspection.*

(1) *Cleaning.*—Immerse all parts in dry-cleaning solvent or volatile mineral spirits to loosen and remove all grease and dirt. Clean lubricant passages in pintle lock bolt and latch pin.

(2) *Inspection.*

(a) *Pintle, lock, and latch.*—Inspect each of these parts for evidence of cracks, bends, excessive wear, or other damage. Replace defective parts.

(b) *Pintle lock bolt.*—Inspect pintle lock bolt for damage or wear. Replace if defective.

(c) *Latch pin.*—Inspect for damage or excessive wear. Pin must fit tight in lock, also latch must be free on pin. Replace if worn or damaged.

(d) *Latch spring.*—Inspect spring for broken coils and distortion. Replace if damaged.

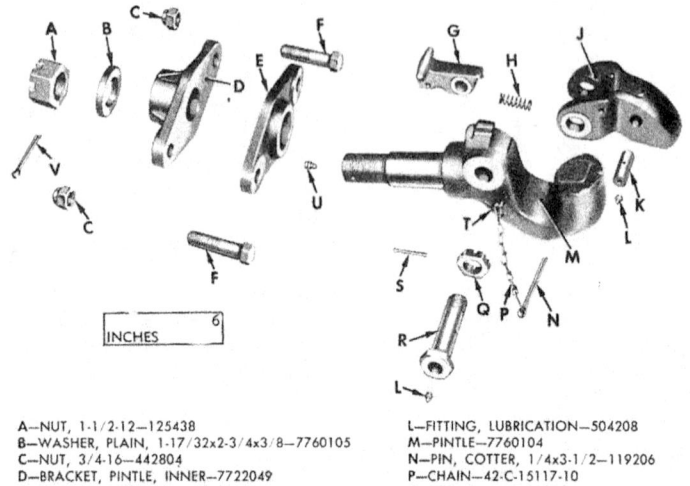

A—NUT, 1-1/2-12—125438
B—WASHER, PLAIN, 1-17/32x2-3/4x3/8—7760105
C—NUT, 3/4-16—442804
D—BRACKET, PINTLE, INNER—7722049
E—BRACKET, PINTLE, OUTER—7722048
F—SCREW, CAP, 3/4-16x3—181836
G—LATCH, PINTLE—7714880
H—SPRING, LATCH—7044253
J—LOCK, PINTLE—7714879
K—PIN, LATCH—7714881
L—FITTING, LUBRICATION—504208
M—PINTLE—7760104
N—PIN, COTTER, 1/4x3-1/2—119206
P—CHAIN—42-C-15117-10
Q—NUT, 1-14—503520
R—BOLT, 1-14x3-1/8—7725844
S—PIN, COTTER, 1/8x1-1/2—103387
T—SCREW, DRIVE, NO 10x1/2—127738
U—FITTING, LUBRICATION—191758
V—PIN, COTTER, 1/4x3—103426

RA PD 148838

*Figure 219. Disassembled view of pintle assembly and attaching parts.*

c. *Assembly.*

(1) *Assembly of lock and latch.*—Install latch spring (H) in pintle lock (J) with spring over boss; then install pintle latch (G) into lock with boss on latch engaging spring. Aline latch pin holes in lock and latch and use arbor press to install latch pin (K) through lock and latch.

(2) *Installation of lock and latch.*—Position assembly of lock and latch in pintle and install 1–14 x 3⅛ bolt (R) attaching assembly to pintle. Install 1–14 nut (Q) on bolt, but do not tighten excessively as this may restrict free movement of lock. Install ⅛ x 1½ cotter pin (S) to secure nut.

(3) *Lubrication.*—Install lubrication fittings (L) in bolt (R) and latch pin (K). Lubricate as directed on official lubrication order (LO 9–819A).

# CHAPTER 17

# ELECTRICAL SYSTEM

## Section I. DESCRIPTION AND DATA

### 290. General

Information contained in this chapter covers only those electrical units which are not covered in other technical manuals. Refer to paragraph 1 for numbers of technical manuals covering electrical units.

### 291. Description

*a. Cable Connections.*—Grouped cable connections are made through multiple plug and receptacle type connectors. Single, and in some cases double, cable connections are made through bayonet type connectors which are held together by interlocking shells (fig. 220). Terminals are crimped onto end of each cable. Rubber grommet is held in place on end of cable by a metal bushing. When cable terminals are inserted into ends of sleeve and the male and female shells are locked together, the shells force the rubber grommets against ends of sleeve, providing a moisture-proof connection.

*b. Horn.*—Horn is air-operated dual diaphragm type, electrically controlled by horn button mounted in center of steering wheel. Solenoid is mounted between horn projectors, with the solenoid air outlet connected directly to the air inlet port in the horn base. Air supply line is connected to inlet port in top of solenoid. When horn circuit through solenoid is completed at the horn button, solenoid acts to admit compressed air into the horn base. Compressed air causes steel diaphragm to vibrate, producing the warning signal.

### 292. Data

Horn:
- Manufacturer_____ Delco-Remy
- Horn model number_____ 1919904
- Solenoid model number_____ 1118165
- Voltage_____ 24
- Type_____ air-operated dual diaphragm

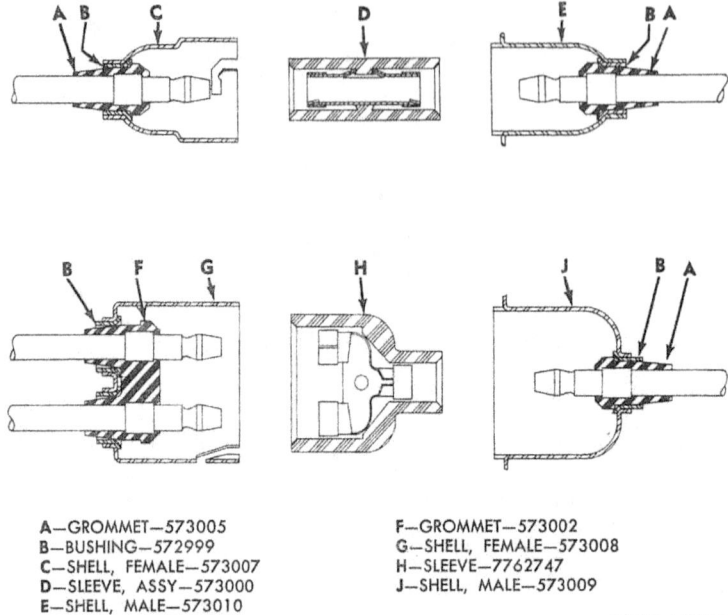

A—GROMMET—573005
B—BUSHING—572999
C—SHELL, FEMALE—573007
D—SLEEVE, ASSY—573000
E—SHELL, MALE—573010
F—GROMMET—573002
G—SHELL, FEMALE—573008
H—SLEEVE—7762747
J—SHELL, MALE—573009

RA PD 149425

*Figure 220. Bayonet type cable connector components.*

## Section II. REPAIR OF CABLE CONNECTIONS

### 293. Bayonet Type Connectors

*a. Inspection.*

(1) Inspect cable terminal for corrosion and clean if necessary. Make sure terminal is securely crimped onto cable and that no wires are broken. If insulation at terminal is cracked or if bare wires are exposed, replace cable or harness assembly.

(2) Examine grommet for evidence of hardening or deterioration. Outer end must form a moisture-proof seal against end of connector sleeve. If any damage is evident, replace grommet (*b* and *c* below).

(3) Inspect shells for distortion and for broken ears on male shells. Replace any shell which is damaged (*b* and *c* below).

*b. Disassembly.*—Slide connector shell back on cable to expose bushing and grommet. Force bushing back off grommet, pull grommet off cable end; then remove bushing and shell.

*c. Assembly.*

(1) To facilitate assembling grommet on cable, apply a thin coat of hydraulic brake fluid to end of cable.

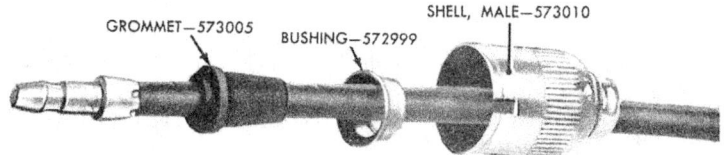

PARTS POSITIONED ON CABLE

BUSHING INSTALLED ON GROMMET

BUSHING AND GROMMET POSITIONED ON END OF CABLE
RA PD 149547

*Figure 221. Installation of cable connector components.*

(2) Place connector shell, bushing, and grommet on end of cable in positions shown in upper view in figure 221.

(3) Slide bushing against shoulder on grommet as shown in center view in figure 221.

(4) Grip grommet and bushing between thumb and finger and pull cable back through grommet until outer end of grommet just covers outer shoulder on cable terminal as shown in lower view in figure 221.

(5) Wipe hydraulic brake fluid off cable terminal after completing assembly.

## 294. Multiple Plug and Receptacle Type Connectors

Refer to paragraph 1 for number of technical manual covering Scintilla multiple plug and receptacle type connectors.

## Section III. REBUILD OF HORN

### 295. Disassembly of Horn

*a.* Disengage cable connector shells from clips on projector bracket.

*b.* Remove two nuts, lock washers, and machine screws attaching projector bracket halves to horn projectors and solenoid. Remove bracket halves.

*c.* Unscrew solenoid assembly from horn base, using wrench on coupling at rear end of solenoid.

*d.* Unscrew projectors from horn base, being careful not to damage projectors with pipe wrench.

### 296. Inspection of Horn

*a. Solenoid.*
  (1) Examine cables for damaged insulation and broken wires. Replace solenoid assembly if either of these conditions is evident. Cable grommets, bushings, and shells can be replaced as described in paragraph 293.
  (2) Solenoid may be tested by connecting cables to a source of direct electrical current and by connecting a compressed air supply line to tapped opening in side of solenoid. Valve must open with 18 volts applied to solenoid, and must close when voltage is removed. Air leakage with valve closed must not exceed one bubble per second with 90 psi air pressure applied. Test for leakage is made with solenoid submerged in water and with unit disconnected from electrical source. If solenoid does not test satisfactorily, replace with new solenoid assembly.

*b. Projectors.*—Inspect projectors for distortion and for damaged threads. Replace if either condition exists. Inside of projectors must be clean and free of obstructions.

*c. Projector Brackets.*—Examine projector bracket halves for distortion, and check for damaged or loose cable connector clips on bracket lower half. Replace with new brackets if damaged.

*d. Horn Base.*—Do not disassemble horn base assembly. If horn is inoperative or operates improperly, and solenoid functions satisfactorily (*a* (2) above), replace horn base assembly.

### 297. Assembly of Horn

*a.* Coat threads on solenoid coupling and on projectors with plastic type gasket cement.

*b.* Thread projectors and solenoid coupling into horn base and tighten securely. Do not distort projectors when tightening.

*c.* Position projector bracket halves on projectors and solenoid, with bracket half having the cable connector clips at the bottom. Secure brackets in place with two No. 10 x 1 machine screws, two No. 10 lock washers, and two No. 10 machine screw nuts. Tighten screws firmly.

*d.* Engage cable connector shells in clips on projector bracket lower half.

# CHAPTER 18

# WINCH AND DRIVE LINE

## Section I. DESCRIPTION AND DATA

### 298. Description

*a. Winch.*—The worm-geared, jaw-clutch, drum winch assembly (fig. 222) is mounted at front of vehicle on support brackets attached to frame side members. Winch is driven by power take-off through two drive shafts. Winch is equipped with a manually-operated clutch control lever and a drum lock poppet knob (fig. 222). The clutch control level is used to engage or disengage the jaw clutch which drives the drum. The drum lock poppet knob is used to lock the drum when winch is not being used. Winch is equipped with two brakes, the drum drag brake and the drive worm automatic brake. The drag brake consists of a flat shoe and lining assembly, spring-loaded to exert a constant drag on end of drum to prevent drum spinning when cable is being pulled off drum. The drive worm automatic brake, which sustains the load when shifting power take-off gears, consists of an external band type brake which acts on a brake disk keyed to end of drive worm.

*b. Drive System.*—The winch drive system (fig. 232) comprises two drive shafts, three universal joints, and a pilot bearing. The rear universal joint consists of a fixed yoke, which is attached to power take-off shaft by a set screw and key, and a slip yoke which is splined to rear drive shaft. Center universal joint consists of two fixed yokes which are attached to front and rear drive shafts by Woodruff keys and straight pins. Front universal joint consists of a fixed yoke, which is attached to winch drive worm by a shear pin, and a slip yoke which is splined to the front drive shaft. The two yokes of each universal joint are assembled together with a universal joint journal, four bushing type bearings, and four snap rings, with a cork seal and a metal seal retainer on each arm of the journal to retain lubricant in bearings. The pilot bearing assembly, consisting of a ball bearing mounted in a bracket attached to left front spring and lower torque rod rear bracket, supports the rear end of the front drive shaft.

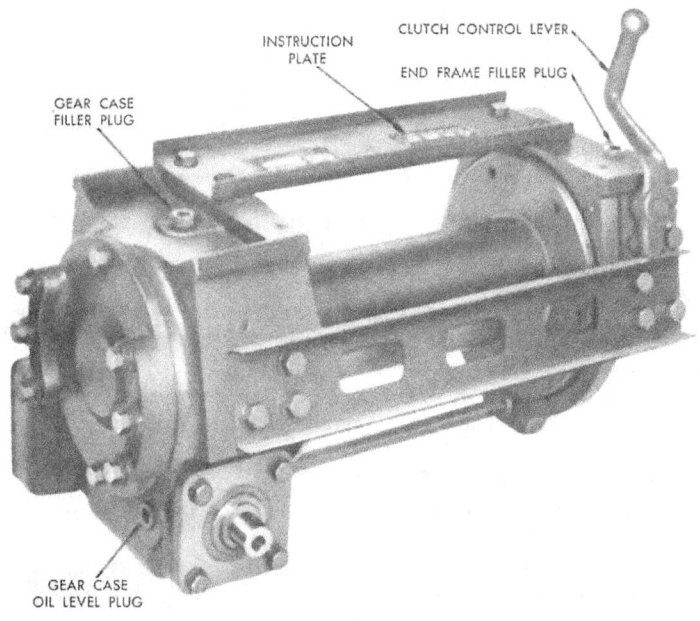
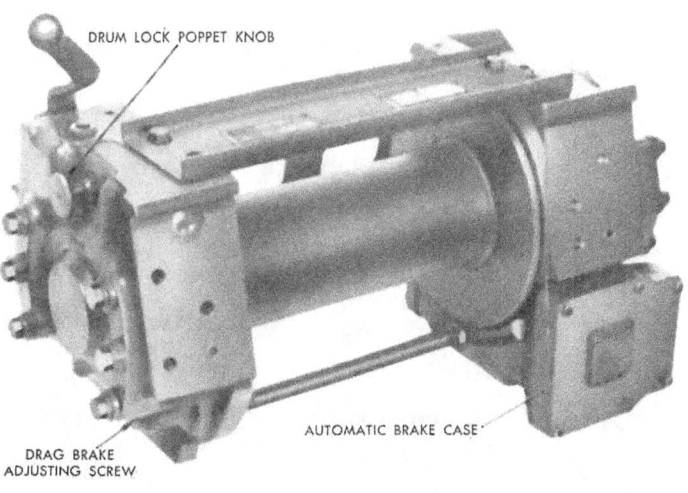

*Figure 222. Front and rear views of winch assembly.*

347

## 299. Data

Winch:
- Manufacturer _____ Gar Wood Industries, Inc
- Model _____ CA514
- Type _____ horizontal drum
- Drive _____ drive shaft from power take-off
- Capacity _____ 10,000 lbs
- Cable _____ 200 ft long, ½-in. diam

## Section II. DISASSEMBLY OF WINCH INTO SUBASSEMBLIES

### 300. Removal of Cable Assembly

*a.* Push clutch control lever (fig. 222) in toward winch drum to disengage drum sliding clutch. Pull out drum lock poppet knob and rotate one-quarter turn to unlocked position.

*b.* Unwind cable from drum; then loosen hex-socket set screw securing cable end in drum. Pull end of cable out of drum.

### 301. Removal of End Frame Assembly

*Note.* Key letters noted in parentheses are in figure 225 unless otherwise indicated.

*a. Drain Lubricant.*—Remove drain plug (TT) from bottom of gear case (SS) and from bottom of end frame (B) and permit all oil to drain.

*b. Remove Tension Channels.*

(1) Remove six ⅝–11 x 1¼ cap screws (J) and ⅝-inch lock washers (H) attaching rear tension channel (W) to gear case and end frame. Remove channel.

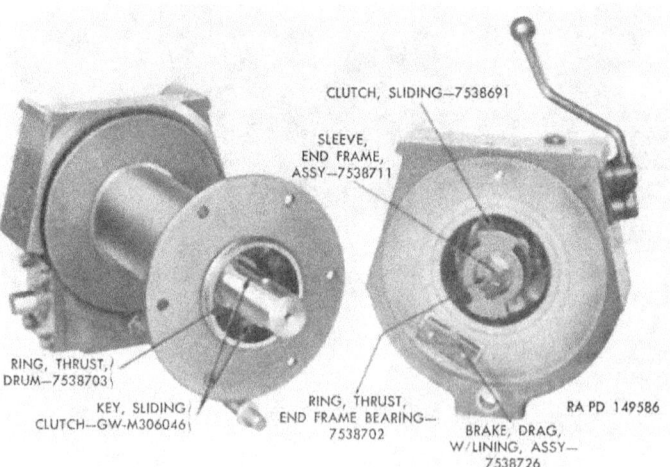

*Figure 223. End frame assembly removed from drum shaft.*

(2) Remove (U) ½-13 x 1¼ cap screws (T) and ½-inch lock washers (U) attaching top tension channel (V) to gear case and end frame. Remove channel.

c. *Remove End Frame Assembly.*—Remove jam nut (AR) from tie rod (AU) at outer side of end frame. Pull end frame assembly off drum shaft. End frame bearing thrust ring (AS) and sliding clutch (AQ) will come off drum shaft with end frame assembly (fig. 223).

### 302. Removal of Drum Assembly

a. *Remove Sliding Clutch Keys.*—Using a sharp chisel against side of sliding clutch key (X) as close to drum shaft as possible, drive key out of shaft as shown in figure 224.

*Note.* Hit chisel sharply to obtain a good bite into side of key.

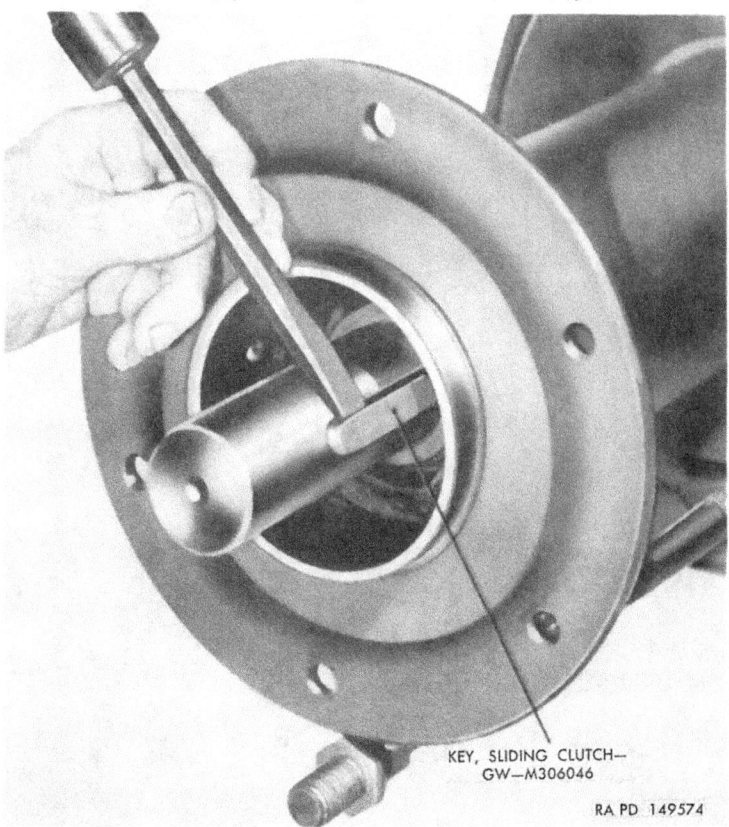

*Figure 224. Removing sliding clutch keys from drum shaft.*

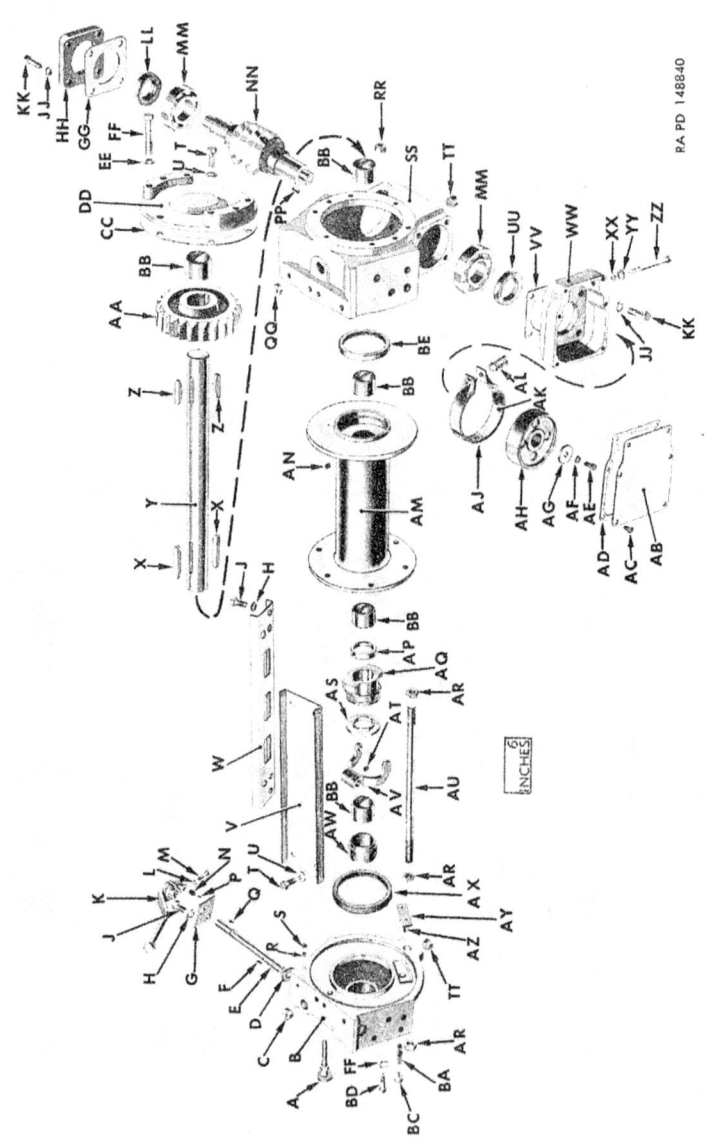

A—KNOB, DRUM LOCK POPPET, ASSY—7538705
B—FRAME, END—7538700
C—PLUG, FILLER, END FRAME—143968
D—SEAL, OIL—7538695
E—SHAFT, SHIFTER YOKE—7538734
F—KEY, WOODRUFF, $5/32$ X $3/4$—103906
G—PLATE, PAWL—7538729
H—WASHER, LOCK, $5/8$-IN.—121574
J—SCREW, CAP, $5/8$–11 X $1 1/4$—180255
K—LEVER, CONTROL CLUTCH—7538708
L—WASHER, LOCK, $5/16$-IN.—120214
M—SCREW, CAP, $5/16$-18 X $1 1/4$—180081
N—SPRING, CLUTCH CONTROL LEVER—104921
P—BALL, CLUTCH CONTROL LEVER—7408727
Q—KEY, WOODRUFF, $5/32$ X $3/4$—103906
R—SCREW, SET, SHIFTER YOKE SHAFT LOCATING—223053
S—PLUG, PIPE, $1/8$-IN.—125947
T—SCREW, CAP, $1/2$–13 X $1 1/4$—180175
U—WASHER, LOCK, $1/2$-IN.—120384
V—CHANNEL, TENSION, TOP—GW-305162
W—CHANNEL, TENSION, REAR—GW-304465
X—KEY, SLIDING CLUTCH—GW-M305046
Y—SHAFT, DRUM—7538692
Z—KEY, DRUM SHAFT—GW-M-306047
AA—GEAR, DRUM SHAFT—7538688
BB—BEARING, BUSHING TYPE—7538696
CC—GASKET, GEAR CASE COVER—7538737
DD—COVER, GEAR CASE—7538698
EE—WASHER, LOCK, $3/8$-IN.—131139
FF—SCREW, SPECIAL, $2 3/8$-IN. LONG—7538710
GG—GASKET, BEARING CAP—7538728
HH—CAP, BEARING—7538732
JJ—WASHER, LOCK, $1/2$-IN.—120384
KK—SCREW, CAP, $1/2$–13 X $1 3/4$—180179
LL—SEAL, OIL—7538704
MM—BEARING, BALL—700338
NN—WORM, DRIVE—7538707
PP—KEY, WOODRUFF, $3/16$ X $3/4$—106751

QQ—PLUG, FILLER, GEAR CASE—143968
RR—PLUG, OIL LEVEL, GEAR CASE—143968
SS—CASE, GEAR—7538699
TT—PLUG, DRAIN—143968
UU—SEAL, OIL—500094
VV—GASKET, AUTOMATIC BRAKE CASE—7538728
WW—CASE, AUTOMATIC BRAKE—7538687
XX—GASKET, "O" RING—7538682
YY—WASHER, PLAIN, $7/16$-IN.—126388
ZZ—SCREW, CAP, ADJUSTING, $3/8$–24 X 4—189331
AB—COVER, AUTOMATIC BRAKE CASE—7538701
AC—SCREW, W/EXT-TEETH LOCK WASHER, $5/16$-18 X $5/8$—187527
AD—GASKET, AUTOMATIC BRAKE CASE COVER—7538736
AE—SCREW, CAP, $5/16$–24 X $3/4$—181595
AF—WASHER, LOCK, $5/16$-IN.—120214
AG—WASHER, PLAIN, $3/8$-IN.—5273972
AH—DISK, AUTOMATIC BRAKE—7538882
AJ—BAND, AUTOMATIC BRAKE, W/LINING, ASSY—7538731
AK—LINING, AUTOMATIC BRAKE BAND—7538730
AL—SPRING, ADJUSTING, AUTOMATIC BRAKE—5277602
AM—DRUM, ASSY—7538690
AN—SCREW, SET, HEX-SOCKET, $1/2$–13 X $3/4$—222589
AP—RING, THRUST, DRUM—7538763
AQ—CLUTCH, SLIDING—7538691
AR—NUT, JAM—124882
AS—RING, THRUST, END FRAME BEARING—7538702
AT—SCREW SET, $3/8$–16 X $1/2$—223089
AU—ROD, TIE—7538735
AV—YOKE, SHIFTER, CLUTCH—7538723
AW—SLEEVE, END FRAME—GW-304419
AX—SEAL, OIL—7538694
AY—LINING, DRAG BRAKE—7538727
AZ—BRAKE, DRAG, W/LINING, ASSY—7538726
BA—SPRING, DRAG BRAKE—7538712
BC—SCREW, ADJUSTING DRAG BRAKE—7538733
BD—SCREW, SPECIAL $1 3/4$-IN. LONG—7538709
BE—SEAL, OIL—7538893

*Figure 225. Winch assembly components.*

Side of key will be damaged, but can be smoothed up with a fine cut mill file. Remove both sliding clutch keys (X) in same manner.

*b. Remove Drum Thrust Ring.*—Slide drum thrust ring (AP) off end of drum shaft.

*c. Remove Drum Assembly.*—Slide drum assembly (AM) off end of drum shaft. If drum does not readily slide off shaft, edges of keyways may have been slightly raised when removing keys. Dress down edges of keyways with a fine-cut mill file; then slide drum off shaft.

## Section III. REBUILD OF END FRAME ASSEMBLY

### 303. Disassembly of End Frame

*Note.* Key letters noted in parentheses are in figure 225.

*a.* Lift drag brake with lining assembly (AZ) and drag brake spring (BA) out of end frame. Remove drag brake adjusting screw (BC) from end frame.

*b.* Move clutch control lever (K) out to engaged position, then disengage sliding clutch (AQ) from clutch shifter yoke (AV) and remove sliding clutch. Lift end frame bearing thrust ring (AS) and end frame sleeve (AW) and bushing type bearing (BB) out of end frame.

*c.* Loosen $\frac{5}{16}$–18 x $1\frac{1}{4}$ cap screw (M) clamping clutch control lever (K) to clutch shifter yoke shaft. Spread lever slightly; then lift lever off shaft, at the same time removing clutch control lever ball (P) and clutch control lever ball spring (N) as shown in figure 230. Remove $\frac{5}{32}$ x $\frac{3}{4}$ Woodruff key (Q) from shaft.

*d.* Using a $\frac{3}{16}$-inch hex set screw wrench, remove socket-head $\frac{1}{8}$-inch pipe plug (S) from end frame to gain access to clutch shifter yoke shaft locating set screw (R). Using a $\frac{5}{32}$-inch hex set screw wrench, remove set screw from end frame.

*e.* Using a $\frac{3}{16}$-inch hex set screw wrench, loosen $\frac{3}{8}$–16 x $\frac{1}{2}$ set screw (AT) securing clutch shifter yoke (AV) on clutch shifter yoke shaft (E). Withdraw shaft from end frame far enough to permit removing clutch shifter yoke (AV) from shaft; then remove $\frac{5}{32}$ x $\frac{3}{4}$ Woodruff key (F) from shaft and remove shaft from end frame.

*f.* Remove $\frac{5}{8}$–11 x $1\frac{1}{4}$ cap screw (J) and $\frac{5}{8}$-inch lock washer (H) attaching pawl plate (G) to end frame and remove plate.

*g.* Remove drum lock poppet knob assembly (A) by unscrewing retaining nut from end frame.

*h.* Do not remove end frame oil seal (AX) or shifter yoke shaft oil seal (D) from end frame unless replacement is necessary as indicated by inspection (par. 304*b*).

## 304. Cleaning, Inspection, and Repair of End Frame Components

*Note.* Key letters noted in parentheses are in figure 225.

a. *Cleaning.*—Wash all parts except drag brake with lining assembly (AZ) in dry-cleaning solvent or volatile spirits. Blow out all tapped holes, clutch shifter yoke shaft bore, and interior of end frame with compressed air. Wipe all small parts dry.

b. *Inspection.*
   (1) *End frame.*—Examine end frame (B) for evidence of cracks or damaged threads in tapped holes. Replace end frame if any damage is evident. Examine end frame oil seal (AX) and shifter yoke shaft oil seal (D) in end frame for evidence of worn or deteriorated seal lip or other damage. Replace seals if damaged (*c* below).
   (2) *End frame sleeve assembly.*—Check bushing type bearing (BB) in end frame sleeve (AW) for wear (par. 355). Replace bearing (*c* below) if worn beyond specified limits.
   (3) *Thrust rings.*—Check drum thrust ring (AP) and end frame bearing thrust ring (AS) for wear by measuring thickness of rings. If not within limits listed in paragraph 355, replace with new parts.
   (4) *Sliding clutch.*—Examine sliding clutch (AQ) for visible wear or damage. Check for burred edges on clutch lugs and file smooth as necessary.
   (5) *Clutch shifter yoke.*—Examine clutch shifter yoke (AV) for evidence of distortion or for wear on lugs which engage sliding drum clutch. Replace with new part if any damage is evident.
   (6) *Drum lock poppet knob.*—Examine drum lock poppet knob assembly (A) for damaged tapered inner end, weakened spring, or damaged threads on retaining nut. Replace with new assembly if damaged in any way.
   (7) *Springs.*—Check clutch control lever ball spring (N) and drag brake spring (BA) for free length and compression (par. 355). Replace with new parts if not within specified limits.
   (8) *Drag brake.*—Examine drag brake with lining assembly (AZ) for damaged or worn lining. If lining is oil soaked or worn down close to rivet heads, replace lining (*c* below).

c. *Repair.*
   (1) *Oil seal replacement.*—To replace either the end frame oil seal (AX) or the clutch shifter yoke shaft oil seal (D), pry old seal out of end frame. Clean all old sealing compound out of oil seal bore in end frame. Coat outside diameter of

new oil seal case with plastic type gasket cement before installing. Either oil seal must be installed with the seal lip pointing inward. Press end frame oil seal (AX) in until seal case is flush with end frame; press shifter yoke shaft oil seal (D) in until it bottoms in end frame.

(2) *End frame sleeve bearing replacement.*—To replace bushing type bearing (BB) in end frame sleeve (AW), press old bearing out of sleeve. Press new bearing in until centered in sleeve. Use extreme care when installing bearing not to damage inside surface, as bearing is designed to provide proper inside diameter when pressed into place and is not finished after installation. Any burs formed on edge of bearing during installation must be scraped off.

(3) *Drag brake lining replacement.*—Drive old rivets out of drag brake with lining assembly (AZ) and remove drag brake lining (AY). Position new lining on brake shoe and secure with two new 9/64 x ½ tubular rivets. Make sure lining is tight against shoe and that rivets are properly upset.

### 305. Assembly of End Frame

*Note.* Key letters noted in parentheses are in figure 225 unless otherwise indicated.

*a.* Insert shifter yoke shaft (E) through oil seal (D) into end frame (B).

*Note.* End of shaft having set screw hole opposite keyway must be inserted in end frame.

Push shaft through far enough to install Woodruff key (F) and tap key into keyway. Figure 226 shows shaft in place with Woodruff key installed. Pull shaft out as far as Woodruff key permits, position clutch shifter yoke (AV), with ⅜-16 x ½ set screw (AT) removed, in end frame with ends of yoke pointing away from end frame.

*b.* Insert shaft through shifter yoke with Woodruff key in shaft entering keyway in shifter yoke. Aline set screw hole in shifter yoke with hole in shaft, install ⅜-16 x ½ set screw (AT) and tighten firmly, using a $\frac{3}{16}$-inch hex set screw wrench.

*c.* Sight down set screw hole in end frame and aline locating groove in shifter yoke shaft (E) with set screw hole in end frame. Figure 226 illustrates set screw hole in end frame and locating groove in shaft. Install shifter yoke shaft locating set screw (R) in end frame and tighten firmly, using a $\frac{5}{32}$-inch hex set screw wrench. Check action of clutch shifter yoke shaft; dog point on locating set screw must engage groove in shaft to locate shaft longitudinally, but must permit free radial movement of shaft. Install ⅛-inch pipe plug (S) in end frame over locating set screw and tighten firmly, using a $\frac{3}{16}$-inch hex set screw wrench.

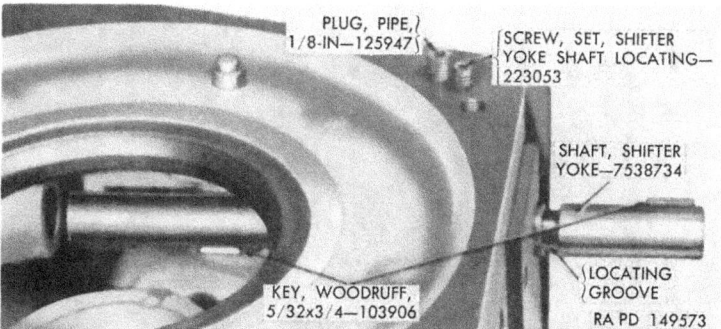

*Figure 226. Clutch shifter yoke shaft and Woodruff keys installed.*

d. Place pawl plate (G) over end of shifter yoke shaft (E) and attach to end frame with one 5/8–11 x 1 1/4 cap screw (J) and 5/8-inch lock washer (H).

e. Insert drum lock poppet knob assembly (A) into end frame, thread retaining nut into end frame, and tighten firmly. Position knob in released position with lugs on knob engaging shallow notches in retaining nut.

f. Tap 5/32 x 3/4 Woodruff key (Q) into keyway in outer end of shifter yoke shaft (E). Place clutch control lever (K) on shaft with Woodruff key in shaft entering keyway in lever. Do not tighten 5/16–18 x 1 1/4 cap screw (M) in clutch control lever at this time. Clutch control lever ball and spring (P and N), and drag brake with lining assembly (AZ), drag brake spring and adjusting screw (BA and BC) will be installed at time winch is assembled from subassemblies.

## Section IV. REBUILD OF GEAR CASE AND AUTOMATIC BRAKE

### 306. Disassembly of Automatic Brake

*Note.* Key letters noted in parenthesis are in figure 225, unless otherwise indicated.

a. Remove six 5/16–18 x 5/8 screws with external-teeth lock washers (AC) attaching automatic brake case cover (AB) to automatic brake case (WW). Remove automatic brake case cover and gasket (AB and AD).

b. Remove 3/8–24 x 4 adjusting cap screw (ZZ), 7/16-inch plain washer (YY), and "O" ring gasket (XX). Remove automatic brake band with lining assembly (AJ) and automatic brake adjusting spring (AL) from automatic brake case (WW).

c. Remove 5/16–24 x 3/4 cap screw (AE), 5/16-inch lock washer (AF), and 3/8-inch plain washer (AG) securing automatic brake disk (AH) on drive worm (NN). Using a suitable jaw type puller in manner

355

*Figure 227.* Using puller to remove automatic brake disk.

shown in figure 227, pull automatic brake disk off drive worm. Remove Woodruff key (PP) from drive worm.

*d.* Remove four ½–13 x 1¾ cap screws (KK) and ½-inch lock washers (JJ) attaching automatic brake case (WW) to gear case (SS). Remove automatic brake case (WW) and automatic brake case gaskets (VV). Note number of gaskets used so the same number may be installed at assembly.

## 307. Disassembly of Gear Case

*Note.* Key letters noted in parenthesis are in figure 225.

*a.* Remove four ½–13 x 1¾ cap screws (KK) and ½-inch lock washers (JJ) attaching bearing cap (HH) to gear case (SS). Remove bearing cap and oil seal assembly (HH and LL) and bearing cap gaskets (GG). Note number of gaskets used so the same number may be used at assembly.

*b.* Remove two ½–13 x 1¼ cap screws (T) and ½-inch lock washers (U) attaching gear case cover (DD) to gear case. If the six 2¾-inch long special screws (FF) were not left off when winch was removed from vehicle and mounting brackets removed, remove special screws and 9/16-inch lock washers (EE). Remove gear case cover (DD) and gear case cover gaskets (CC). Note number of gaskets used so the same number may be used at assembly.

*c.* Using a heavy hammer and hardwood block against one end of drive worm (NN), or using an arbor press, force drive worm out of

356

gear case (SS). One ball bearing (MM) will remain on drive worm and the other ball bearing will remain in gear case. Remove ball bearing (MM) from drive worm and from gear case, using a suitable bearing puller.

*d.* Remove drum shaft gear and drum shaft (AA and Y) from gear case. Do not remove drum shaft gear (AA) from drum shaft (Y) unless replacement of one of the parts is necessary as indicated by inspection (par. 308*b*).

*e.* Loosen jam nut (AR) which is against gear case (SS). Unscrew tie rod (AU) from gear case (SS).

## 308. Cleaning, Inspection, and Repair of Gear Case and Automatic Brake Components

*Note.* Key letters noted in parentheses are in figure 225 unless otherwise indicated.

*a. Cleaning.*—Wash all parts except automatic brake band and lining assembly (AJ) in dry-cleaning solvent or volatile mineral spirits. Blow ball bearings and gear case dry with compressed air. Wipe all other parts dry.

*b. Inspection.*
  (1) *Gear case.* Examine gear case (SS) for cracks and for damaged threads in tapped holes. Either of these conditions require replacing with new part. Check bushing type bearing (BB) in gear case for damage and for wear (par. 355). If bearing is damaged or is worn beyond specified limits, replace bearing (*c* below).
  (2) *Gear case cover.*—Examine bushing type bearing (BB) in gear case cover (DD) for damage and for wear (par. 355). If bearing is damaged or is worn beyond specified limits, replace bearing (*c* below).
  (3) *Drum shaft and gear.*—Inspect drum shaft (Y) for distortion and for wear at bearing contact surfaces (par. 355). If damaged in any way or if worn beyond specified limits, replace shaft (*c* below). Examine drum shaft gear (AA) for scored or broken teeth. If any damage is evident, replace gear (*c* below).
  (4) *Drive worm.*—Examine drive worm (NN) for scored or otherwise damaged threads. Also check for damaged threads in automatic brake end and for elongated shear pin hole in drive end. If any damage is evident, replace with new part.
  (5) *Ball bearings.*—Examine ball bearings (MM) for evidence of damaged balls or races. Rotate bearings to detect roughness. If any damage is evident, replace with new bearings. If bearings are satisfactory, coat with oil to prevent rusting prior to installation.

(6) *Automatic brake band and lining assembly.*—Inspect automatic brake band with lining assembly (AJ) for damaged or worn lining. If lining is damaged or worn down close to rivet heads, replace lining (*c* below). Check for damaged threads in nut which is welded to upper end of band. If threads are damaged, replace band and lining assembly.

(7) *Automatic brake disk.*—Examine automatic brake disk (AH) for scored or rough braking surface. If any damage is evident, replace with new part.

(8) *Automatic brake case and oil seal.*—Inspect automatic brake case (WW) for damaged threads in tapped holes. Any damage necessitates replacement with new part. Examine oil seal (UU) in brake case for damaged or deteriorated sealing lip. If any damage is evident, replace oil seal (*c* below).

(9) *Bearing cap and oil seal.*—Examine oil seal (LL) in bearing cap (HH) for damaged or deteriorated sealing lip. If any damage is evident, replace oil seal (*c* below).

(10) *Automatic brake adjusting spring.*—Check automatic brake adjusting spring (AL) for free length and compression (par. 355). If not within specified limits, replace with new part.

(11) *Tie rod.*—Examine tie rod (AU) for damaged threads or distortion. If only slightly bent, tie rod can be straightened. If badly bent or if threads are damaged, replace tie rod.

(12) *Gaskets.*—Any gaskets which were damaged during disassembly must be discarded and new gaskets obtained for assembly. Make sure number of gaskets used at each point is the same as the number removed at time of disassembly.

*c. Repair.*

(1) *Gear case bearing replacement.*—Press old bushing type bearing (BB) out of gear case (SS), using care not to damage bearing bore in case. Carefully press new bearing into place, using care not to damage inside surface of bearing. Proper inside diameter of bearing is provided when bearing is pressed into place and no finishing is necessary. Scrape off any burs which may have formed on edge of bearing during installation.

(2) *Gear case cover bearing replacement.*—Using a punch or chisel at one of the notches in hub of gear case cover (DD), collapse bushing type bearing (BB) at one side and remove bearing from cover. Use extreme care not to damage bearing bore in cover. Carefully press new bearing into cover until edge of bearing is flush with edge of bearing bore.

Scrape off any burs which may have formed on edge of bearing during installation.

(3) *Drum shaft or drum shaft gear replacement.*—To replace either the drum shaft (Y) or the drum shaft gear (AA), press shaft out of gear with an arbor press. Make sure drum shaft keys (Z) are fully seated and secure in shaft. Position gear on shaft with side of gear having notches in hub down and with keyways in gear alined with keys in shaft. Press gear onto shaft, using an arbor press, to dimension shown in figure 228.

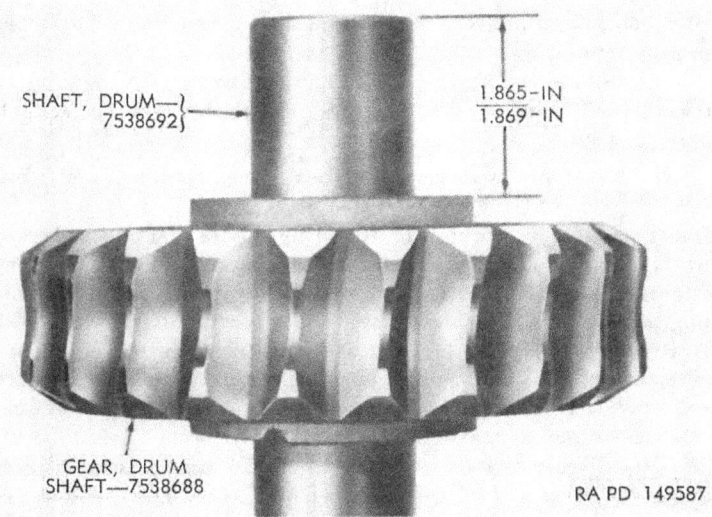

*Figure 228. Drum shaft gear installed on drum shaft.*

(4) *Automatic brake band lining replacement.*—Punch rivets out of automatic brake band with lining assembly (AJ) and remove automatic brake band lining (AK) from band. Clean all dirt and corrosion from band. Position new lining inside of brake band. Lining should not extend beyond center of end rivet holes in band more than one-half inch. Spread band to slightly larger than normal diameter while installing brake lining to assure good band-to-lining contact. Using brake band as a template, and using a conventional brake relining machine, drill nine $9/64$-inch diameter holes in lining, countersink holes to a depth of one-eighth of an inch with a $3/8$-inch drill; then install and upset nine $9/64$ x $9/32$ tubular rivets.

(5) *Automatic brake case or bearing cap oil seal replacement.*—Press old oil seal (UU or LL) out of automatic brake case (WW) or bearing cap (HH). Clean all traces of sealing compound out of seal bore. Coat outside of new oil seal case with plastic type gasket cement. Press oil seal into brake case or bearing cap, with lip of seal pointing toward gear case side.

### 309. Assembly of Gear Case

*Note.* Key letters noted in parentheses are in figure 225.

*a.* Insert drum shaft (Y) through bearing in gear case (SS) from outer end and push through until drum shaft gear (AA) is positioned in gear case.

*b.* Install one ball bearing (MM) on drive worm (NN) and install the other ball bearing in gear case (SS), using arbor press and suitable bearing drivers.

*Note.* End of drive worm having the shear pin hole must be at side of gear case adjacent to gear case oil level plug (RR).

Insert end of drive worm through bearing bore in gear case and screw worm over drum shaft gear until end of drive worm enters bearing in gear case and bearing on drive worm is positioned at bearing bore in gear case. Using an arbor press and a bearing driver which will exert force on ball bearing inner and outer races, press bearing into gear case, at the same time pressing drive worm into bearing which was previously installed in gear case. Make sure inner race of both bearings are seated against shoulders on drive worm.

*c.* Install gear case cover (DD), using the same number of gear case cover gaskets (CC) that were removed (par. 307*b*), and secure with two ½–13 x 1¼ cap screws (T) and ½-inch lock washers (U), applying plastic type gasket cement to cap screw threads before installing. Tighten cap screws firmly. If the six 2¾-inch long special screws (FF) were removed at time of disassembly, coat threads with plastic type gasket cement and install with ⁹⁄₁₆-inch lock washer (EE) on each cap screw.

*d.* Position bearing cap gasket (GG) on gear case, using same number of gaskets that were removed (par. 307*a*). Install bearing cap and oil seal assembly (HH and LL) on drive worm, using a piece of thin shim stock wrapped around drive worm to guide oil seal lip over shoulder on drive worm. Position bearing cap against gear case. Coat threads of four ½–13 x 1¾ cap screws (KK) with plastic type gasket cement; then attach bearing cap to gear case with the four cap screws and ½-inch lock washers (JJ). Tighten cap screws firmly.

*e.* Install one jam nut (AR) on each end of tie rod (AU) and run nuts down to end of threads. Coat threads at end of tie rod having the shortest threads with plastic type gasket cement; then thread this end of tie rod into tapped hole in bottom of gear case. Turn tie rod into gear case several turns; then tighten jam nut against gear case.

### 310. Assembly of Automatic Brake

*Note.* Key letters noted in parentheses are in figure 225.

*a.* Place automatic brake case gaskets (VV) on gear case, using the same number of gaskets that were removed (par. 306*d*). Coat threads of four ½–13 x 1¾ cap screws (KK) with plastic type gasket cement. Install automatic brake case and oil seal (WW and UU) on drive worm, wrapping thin shim stock around drive worm to guide oil seal lip over shoulder on drive worm. Attach automatic brake case to gear case with the four ½–13 x 1¾ cap screws (KK) and ½-inch lock washers (JJ).

*b.* Check action of drive worm by turning with hand. Drive worm must turn freely with no binding. If any binding is felt, add one automatic brake case gasket (VV) between automatic brake case and gear case and again check action of drive worm. Add additional gaskets if required to relieve binding. Tap on both ends of drive worm each time gasket is added before checking action.

*c.* Tap ³⁄₁₆ x ¾ Woodruff key (PP) into keyway in drive worm (NN); then install automatic brake disk (AH) on drive worm, with keyway in brake disk engaging key in drive worm. Drive brake disk onto drive worm until it bottoms against shoulder on drive worm and secure in place with ⅜-inch plain washer( AG), ⁵⁄₁₆-inch lock washer (AF), and ⁵⁄₁₆–24 x ¾ cap screw (AE). Tighten cap screw firmly.

*d.* Install automatic brake band with lining assembly (AJ) over brake disk with end having nut at top, inserting automatic brake adjusting spring (AL) between lower end of brake band and brake case as brake band is slid into place. Place ⁷⁄₁₆-inch plain washer (YY) and "O" ring gasket (XX) over ⅜–24 x 4 adjusting cap screw (ZZ), insert cap screw up through brake case, adjusting spring, and lower end of brake band and thread into nut which is welded to upper end of brake band. Using a long drift in shear pin hole in drive worm to turn drive worm, tighten adjusting screw until worm cannot be easily turned in reverse direction. Final adjustment must be made under load after winch is installed on vehicle.

*e.* Position automatic brake case cover gasket (AD) and automatic brake case cover (AB) on automatic brake case (WW) and attach with six ⁵⁄₁₆–18 x ⅝ screws with external-teeth lock washers. Tighten screws firmly.

## Section V. REBUILD OF DRUM ASSEMBLY

### 311. Cleaning and Inspection

*Note.* Key letters noted in parentheses are in figure 225.

*a.* Thoroughly clean all grease, dirt, and corrosion from exterior and interior of drum assembly (AM).

*b.* Examine drum for cracks or distortion and replace with new drum assembly if these conditions are evident.

*c.* Examine oil seal (BE) in gear case end of drum for damage or deterioration. Replace oil seal (par. 312*a*) if either of these conditions is evident.

*d.* Check bushing type bearings (BB) in drum for damage and for wear (par. 355). If worn beyond specified limits, replace bearings (par. 312*b*).

### 312. Repair of Drum

*Note.* Key letters noted in parentheses are in figure 225.

*a. Oil Seal Replacement.*—Pry oil seal (BE) out of drum. Thoroughly clean seal seat in drum. Press new seal into drum, with seal lip pointing inward, until side of seal is flush with end of drum.

*b. Bearing Replacement.*—Press bushing type bearing (BB) out of each end of drum, using suitable bearing driver, and being careful not to damage bearing bores in drum. Press new bearings into place with suitable bearing driver and arbor press, using extreme care not to damage inside diameter of bearings. Proper inside diameter of bearings is provided when bearings are pressed into place and no finishing is necessary. Scrape off any burs which may have formed on edges of bearings during installation.

## Section VI. ASSEMBLY OF WINCH FROM SUBASSEMBLIES

### 313. Installation of Drum Assembly

*Note.* Key letters noted in parentheses are in figure 225.

*a.* Place drum assembly (AM) over drum shaft (Y) with end of drum having oil seal next to gear case, carefully guiding shaft through bearings in drum. Push drum all the way onto shaft until oil seal is in place on hub of gear case.

*b.* Place drum thrust ring (AP) over end of drum shaft with notches facing outward and alined with keyways in shaft.

*c.* Turn drum to position the clutch lugs inside of drum directly over the keyways in drum shaft. Install one sliding clutch key (X) in keyway in drum shaft, using a thick chisel as a wedge between clutch lug in drum and key to force inner end of key down into keyway. Drive outer end of key down, using a brass rod and hammer.

File off any burs which may have formed on edges of key with a fine cut mill file. Blow out all filings with compressed air. Install sliding clutch key in opposite side of drum shaft in the same manner, then check action of sliding clutch (AQ) over shaft and keys. Sliding clutch must slide freely. If any binding is evident, dress off edges of keys with a fine cut mill file until free action is obtained.

### 314. Installation of End Frame Assembly

*Note.* Key letters noted in parentheses are in figure 225.

*a.* Position gear case, drum shaft, and drum assembly with tie rod at top. Place sliding clutch (AQ) and end frame bearing thrust ring (AS) on end of drum shaft in position shown in figure 229.

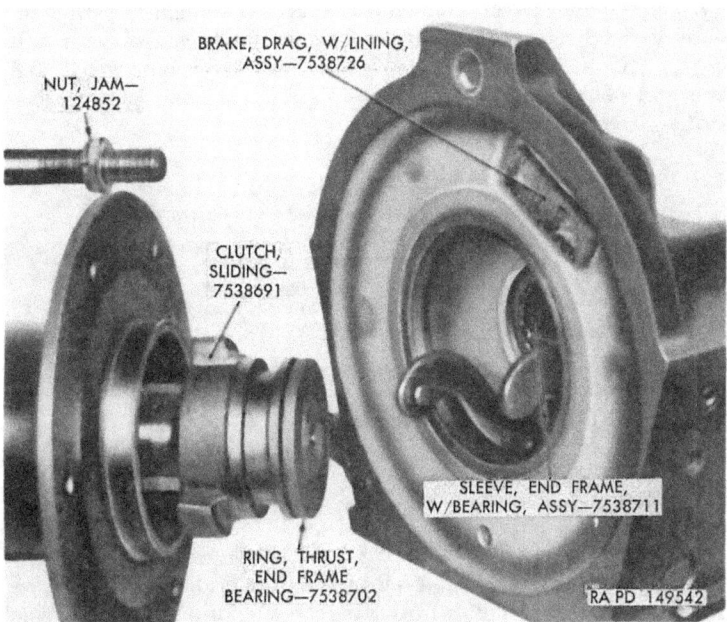

*Figure 229. Installing end frame and components on drum shaft.*

*b.* Place end frame sleeve (AW) and bushing type bearing (BB) in end frame with groove in sleeve over dowel in end frame. Position drag brake and lining assembly in depression in end frame (fig. 229).

*c.* Swing clutch control lever (K) to move ends of clutch shifter yoke as far as possible out of inner side of end frame as shown in figure 229. With parts positioned as shown, lift end frame assembly and engage lugs on clutch shifter yoke in groove in sliding clutch; then push end frame onto drum shaft, at the same time guiding end of tie rod into hole in end frame. Make sure notches in end frame bearing thrust ring (AS) aline with sliding clutch keys (X) to permit end frame to go all the way on.

*d.* Turn the assembly over, then place top tension channel (V) on end frame and gear case and attach with four ½–13 x 1¼ cap screws (T) and ½-inch lock washers (U). Do not tighten cap screws until rear tension channel is installed (*e* below).

*e.* Place rear tension channel (W) on end frame and gear case and attach each end with three ⅝–11 x 1¼ cap screws (J) and ⅝-inch lock washers (H). Tighten cap screws firmly; then tighten top tension channel attaching cap screws. Turn jam nut (AR) on tie rod out against end frame; then install other jam nut on outer end of tie rod. Tighten nuts against both sides of end frame.

*f.* Pull clutch control lever (K) off shaft far enough to permit installing clutch control lever ball (P) and clutch control lever ball spring (N) as shown in figure 230. Push lever onto shaft and secure in place with ⁵⁄₁₆–18 x 1¼ cap screw (M) and ⁵⁄₁₆-inch lock washer (L).

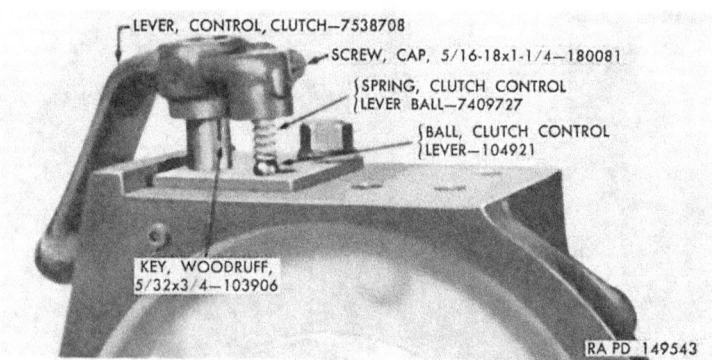

*Figure 230. Installation of clutch control lever, spring, and ball.*

*g.* Install drag brake spring (BA) and drag brake adjusting screw (BC) in end frame. Tighten adjusting screw against adjusting spring until a slight drag is felt as drum is turned by hand. Final adjustment must be made after winch is installed on vehicle as directed in TM 9–819A.

*h.* Install drain plugs (TT) into bottom of end frame and gear case and tighten securely. Add lubricant to end frame and gear case as directed on Lubrication Order 9–819A.

### 315. Installation of Cable Assembly

*a.* Winch cable must be installed after winch assembly is installed on vehicle. Coat cable with used engine oil before winding on drum.

b. Insert end of cable through hole in drum and secure in place by tightening ½–13 x ⅝ hex-socket screw (AN, fig. 225) firmly against cable.

  c. Connect other end of cable to a load, preferably another vehicle, to keep cable taut while winding on drum. Leave all controls in neutral and brake released on vehicle being used as a load.

  d. Make sure drum lock poppet knob at winch is pulled out and turned to disengaged position.

  e. Pull clutch control lever outward away from drum to engaged position.

  f. Place transfer lever in "DOWN-NEUTRAL" position and apply parking brake. Place transmission control lever in "N" (neutral) position and start engine.

  g. Lower the power take-off lever to "DOWN-FORWARD" position. With engine idling, move transmission control lever to "F–2 LOW RANGE" position.

  h. Use hand throttle to accelerate engine to not over one-third throttle to wind cable. Cable should be guided to make sure coils on each layer of cable are tight together.

  i. After cable is completely wound on drum, place transmission control lever in "N" (neutral) position. Pull out drum lock poppet knob and turn one-quarter turn; then release knob to lock drum. Insert cable chain through one front towing shackle and place chain hook in other front towing shackle.

## Section VII. REBUILD OF WINCH DRIVE SYSTEM

### 316. General

Winch drive system is removed from the vehicle as a complete assembly and disassembled into subassemblies as directed in TM 9–819A. The procedures contained in this section cover disassembly, cleaning, inspection, and assembly of the universal joint assemblies, and inspection of the pilot bearing assembly and drive shafts. The component parts of all three universal joint assemblies are identical except for the yokes, and procedures contained herein apply to either of the three universal joint assemblies. Disassembled view of winch drive system components is shown in figure 232.

### 317. Disassembly of Universal Joint

  *Note.* Key letters noted in parentheses are in figure 232 unless otherwise indicated.

  a. Grip one yoke of universal joint in a vise and remove four snap rings (A) securing bushing type bearings (B) in yokes.

  b. Remove 90-degree lubrication fitting (F) from journal (E).

c. Support one yoke in vise jaws; then strike other yoke with hammer as shown in left view, figure 231 to remove one bushing type bearing. Turn yoke over and strike on opposite side to remove opposite bearing. Remove yoke from which bearings were removed from journal.

d. Install soft metal jaw protectors in vise; then support ends of journal on vise jaws as shown in right view, figure 231. Strike yoke with hammer as shown to remove one bushing type bearing; then turn yoke over and strike opposite side to remove opposite bearing.

e. Remove journal (E) from yoke; then remove cork seals (C) and seal retainers (D) from journal.

*Figure 231. Disassembly of winch drive system universal joint.*

## 318. Cleaning, Inspection, and Repair of Drive Line Components

*Note.* Key letters noted in parentheses are in figure 232.

a. *Cleaning.*—Cork seals (C), seal retainers (D), shear pin (H), and $5/16$ x 2 straight pin (X) should be discarded and new parts obtained for assembly. Wash all other parts, including pilot bearing assembly (S) and front and rear drive shafts (N and V) in dry-cleaning solvent or volatile mineral spirits. Make sure lubricant passages through journals (E) are open, and that all old lubricant is removed from inside of bushing type bearings (B).

b. *Inspection.*

(1) *Journals and bearings.*—Check inside diameter of bushing type bearings (B) and outside diameter of arms on journal (E) for wear (par. 356). If worn beyond specified limits, replace with new parts.

(2) *Yokes.*—Examine shear pin hole through hub of front fixed yoke (G) for elongation. If hole is worn, new hole may be drilled as directed in c below. Examine set screw hole in rear fixed yoke (U) for damaged threads. Replace yoke if threads are damaged. Check fit of slip yoke (J) on front

and rear drive shafts (N and V). Slip yokes must slide freely on shafts, but backlash between yoke and shaft splines must not be excessive (par. 356). If backlash is excessive, check grooves in slip yoke and teeth on drive-shaft for wear (par. 356) and replace parts as required.

(3) *Drive shafts.*—Check front and rear drive shafts (N and V) for run-out. If run-out is excessive (par. 356) replace shaft. Fit of shaft splines in slip yoke is checked in (2) above.

(4) *Pilot bearing.*—Hold pilot bearing case and rotate inner race by hand to check for roughness or looseness in bearing. Check set screw holes in bearing inner race for damaged threads. If rough action, looseness, or damaged threads are evident, replace pilot bearing assembly.

*c. Repair.*—If shear pin hole in front fixed yoke (G) is elongated ((2) above), a new hole can be drilled at right angle to the old hole. Hole must be drilled exactly on centerline and square with bore of yoke. Using a drill press, drill and ream a 0.312 to 0.314-inch diameter hole, with centerline of hole one-half inch in from end of yoke hub. Old holes should be plugged to prevent installing shear pin in worn holes.

### 319. Assembly of Universal Joint

*Note.* Key letters noted in parentheses are in figure 232.

*a.* The following procedures cover assembly of either of the three universal joint assemblies. Refer to figure 232 for yoke to be used in universal joint assembly at winch, pilot bearing, or power take-off.

*b.* Install new seal retainer (D) and new cork seal (C) on each arm of journal (E). Push retainers down against shoulders on journal and make sure cork seals are fully seated in retainers. Install 90° lubrication fitting (F) in journal, with end of fitting pointing between two arms of journal for accessibility.

*c.* Grip hub of one yoke in vise and insert journal into yoke. Start one bushing type bearing (B) into each side of yoke, guiding ends of journal into bearings. Drive bearings in below snap ring grooves; then install snap rings (A), making sure they are fully seated.

*d.* Position other yoke in vise, insert free ends of journal in yoke; then install bushing type bearings (B) and snap rings (A) as in *c* above.

*e.* Assembly of winch drive line and installation in the vehicle is described in TM 9–819A.

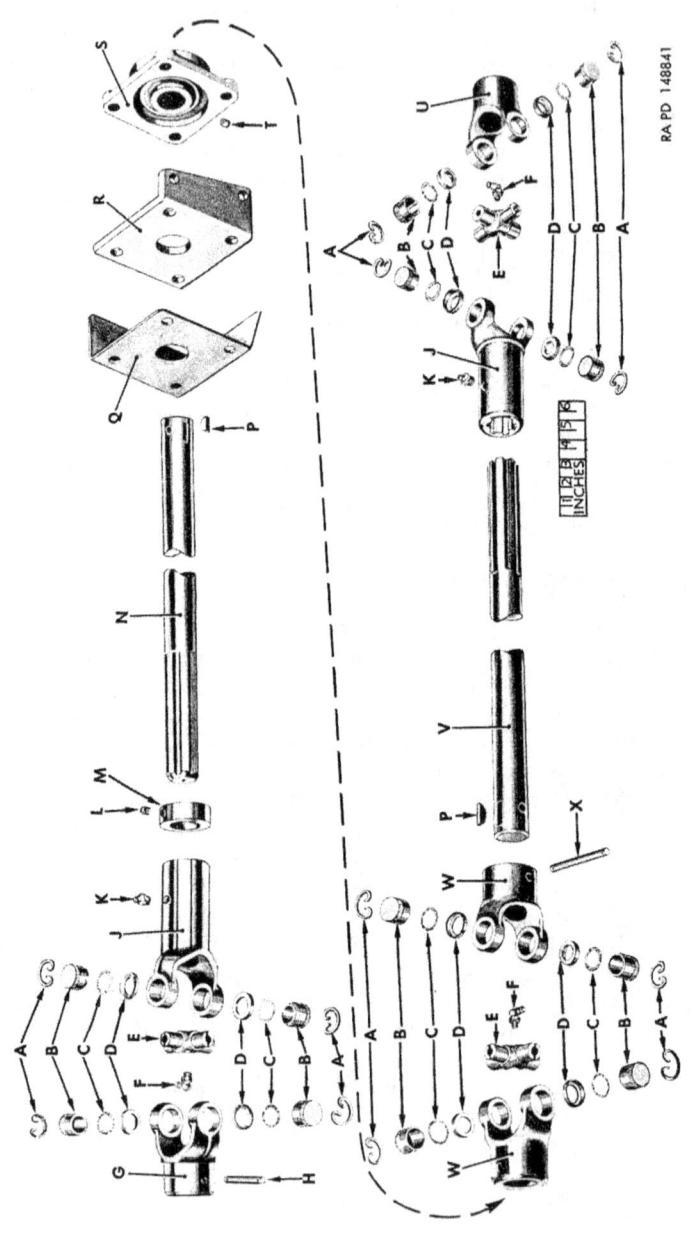
RA PD 148841

A—RING, SNAP—DT-BB-9754
B—BEARING, BUSHING TYPE—BM-L-12S-2
C—SEAL, CORK—7365355
D—RETAINER, SEAL—GW-6-Y-1924
E—JOURNAL—YT-2185106
F—FITTING, LUBRICATION, 90-DEG—504207
G—YOKE, FIXED—BM-L-12-SYR-20-2
H—PIN, SHEAR—7412755
J—YOKE, SLIP—BM-L-12-SYS-20
K—FITTING, LUBRICATION, STRAIGHT—504208
L—SCREW, SET, HEX-SOCKET—426874
M—STOP, FRONT DRIVE SHAFT—7411760
N—SHAFT, DRIVE, FRONT—7412758
P—KEY, WOODRUFF, ¼ X 1—113782
Q—SHIELD, PILOT BEARING—YT-228374
R—BRACKET, PILOT BEARING—7412753
S—BEARING, PILOT, ASSY—7412751
T—SCREW, SET, HEX-SOCKET—7412752
U—YOKE, FIXED—BM-L-12-SYR-20-1
V—SHAFT, DRIVE, REAR—7412759
W—YOKE, FIXED—BM-L-12-SYR-20-13
X—PIN, STRAIGHT, 5/16 X 2—142535

*Figure 232. Winch drive system components.*

# CHAPTER 19

# CAB AND ASSOCIATED PARTS

## Section I. DESCRIPTION AND DATA

### 320. Description

*a. General.*—Cab consists of an open-top structure enclosing driver's compartment. Cab is all-steel construction, except glass and paulins, and is assembled from several subassemblies into a complete unit. Subassemblies include cowl and dash unit, sides, rear panel, floor panel, doors, windshield, and paulins. Design of cab facilitates replacement of subassemblies, using only standard common tools. Weather protection and driver comfort is provided by windshield, doors, cowl ventilators, roof paulin, and rear curtain. Equipment also includes two windshield wipers, one inside and two outside rearview mirrors. Components of cab are illustrated in figures 233, 234, 235, and 238.

*b. Panels.*—Basic cab is composed of floor pan and sills, back panel, cowl, and side panels. The above panel assemblies are bolted together to form a major assembly, and other cab items such as doors, windshield, seats, etc., are assembled to this basic unit. Since each panel assembly is bolted in place, they can be replaced separately whenever necessary.

*c. Doors.*—Cab door consists of a frame assembly to which is assembled frame and glass, regulator, door lock and remote control, check arm, and handles. Door is designed so that it can be adjusted horizontally or vertically whenever necessary; also, each of the components can be readily replaced.

*d. Windshield.*—Windshield consists of two inner frame and glass assemblies hinged at top to outer frame assembly. Outer frame is hinged to cab cowl by means of bolts and hinge brackets. Each inner frame assembly can be adjusted independently of the other; also, complete windshield assembly can be hinged forward.

*e. Windshield Wipers.*—Two windshield wipers, mounted at top of each inner frame, are air-operated type, and are controlled by a single valve mounted at left of instrument panel. Valve is pressure-regulating type; therefore, wiper speed is not affected by fluctuations in system air pressure.

*f. Roof Paulin and Rear Curtain.*—Roof paulin is one-piece type which serves as cab roof. Paulin is secured to windshield header panel by a bead which slides into a retaining channel. Roof paulin is secured by a lashing rope to roof and side panels, also to hooks on back panel. Rear curtain is one-piece type with rear window. Curtain is attached to top bow by means of screws and washers. Top bow legs fit into sockets at rear corner of cab and are held by ring nut at each side. Curtain lashing rope threads through loops in side of curtain and engages hooks on rear panel.

*g. Seats.*—Cab is equipped with two seats, one driver and one companion. Driver's seat is adjustable in fore and aft direction, while companion seat is held in fixed position. Both seats are easily replaced, and no tools are required to remove or install driver's seat. Companion seat is supported by risers attached to floor pan and back panel. Seat back is hinged to lay forward on seat, also seat can be tilted up against seat back. Latch, attached to back panel, is used to hold seat back and seat in upright position.

### 321. Data

*a. Windshield Wiper Motor.*

| | |
|---|---|
| Make | Trico |
| Type | air |
| Model | 22193–ZQ |

*b. Windshield Wiper Blade.*

| | |
|---|---|
| Number | L–778–42ZQ |
| length | 12 in. |

### Section II. GENERAL REBUILD OF CAB

### 322. General Repair

*a. General.*—Cab is of all metal construction with exception of seat cushions, roof paulin, back curtain, and door glass. Repair is limited to straightening, dinging, patching, stitching, riveting, and welding operations, in addition to replacement of parts and subassemblies. Operations as outlined in this section are based on the assumption that cab has been removed from the chassis. However, many replacement operations can be performed when cab is on chassis, such as roof paulin, roof panels, rear curtain, roof bow, seats, and seat cushions. When replacing individual components of cab assembly, parts adjacent to the component being replaced should not be moved out of position further than necessary to permit replacement of the component to assure proper alinement after installation.

b. *Repair of Cab Panels.*
   (1) *Patching and welding.*—Prepare the hole or break for patching by trimming off all curled edges with a cutting torch or by any other suitable method. Do not cut away more metal than is necessary to make a smooth flat surface upon which to place the patch.
   (2) *Patch preparation.*—Cut a patch of the same thickness as the part to which it is to be applied. Patch must be of sufficient size to overlap at least one inch all around the hole or break.
   (3) *Patch application.*—Position the patch over the hole or break, making sure that patch and weld will not affect function of part to be repaired, nor interfere with other parts of vehicle. Tack weld patch to panel at several evenly spaced points. On the opposite side of panel, run a continuous fillet weld all around the patch or the hole, depending on which side patch is placed. Use shielded arc for welding; if electric welding equipment is not available, gas welding may be used.
   (4) *Paint application.*—After welding is completed, chip away all slag resulting from welding and then apply paint. Refer to TM 9–2851 for painting and methods of painting.
c. *Repair of Cab Paulin and Rear Curtain.*
   (1) *General.*—When patching tears or holes in cab paulin or rear curtain, cut a patch from No. 8 duck or from an old paulin. Patch must be large enough to extend well beyond the edges of the tear or hole. Edges of patch and hole should be turned under and several rows of stitches should be applied around the edges of the patch and edges of hole. All stitching must be made with a minimum of four stitches per inch. Reinforcements of extra thicknesses of duck must be used at points subject to tension. Metal grommets must be installed at all points where lashing ropes pass through roof paulin or rear curtain.
   (2) *Waterproofing.*—When new material is used for patching cab paulin or rear curtain, or if the paulin or curtain have lost their water-repellent qualities, the material should be treated with a suitable waterproof, weatherproof, and mildew-resistant compound. Follow instructions on the compound container for applying.

## 323. Replacement of Cab Paulin, Roof Panels, Rear Curtain, and Bow

*Note.* Key letters noted in parentheses are in figure 233 unless otherwise indicated.

*a. General.*—The cab paulin, roof panels, rear curtain, and roof bow can be replaced as stated in the following subparagraphs. The sequence of procedures for replacing these components are listed in logical sequence.

*b. Removal of Cab Paulin.*—Untie cab paulin side and rear ropes (L and M) from bow retainer ring nuts (M, fig. 234) at each side of cab. Pull ropes from rear curtain loop straps (N) in sides of rear curtain assembly (K), then disengage ropes from hooks on upper right side roof panel (B) and upper left side roof panel (EE). Unhook cab paulin rear rope (M) from lashing hooks on back of back panel (Q, fig. 234). Lift cab paulin (A) over cab roof bow assembly (PP) and windshield onto engine hood. From either side, pull bead of canvas paulin from channel at upper front side of windshield outer frame assembly (GG) and then remove paulin.

*c. Installation of Cab Paulin.*—Slide bead of cabin paulin (A) into channel on upper front side of windshield outer frame assembly (GG). Draw paulin over windshield and roof bow into position. Engage cab paulin rear rope (M) in lashing hooks on back of back panel (Q, fig. 234). Pull rear lashing rope taut; then tie rope ends to bow retainer ring nuts (M, fig. 234). Engage each cab paulin side rope (L) in hooks on roof panels, thread ropes through rear curtain loop straps (N) in ends of rear curtain assembly (K), and engage ropes in hooks on side panels. Pull each rope taut; then tie to bow retainer ring nuts (M, fig. 234).

*d. Removal of Roof Panels.*—With cab paulin (A) removed (*b* above) or rolled forward, push upper right and left side roof panels (B and EE) straight upward to disengage from windshield and roof bow anchors and from studs on right and left rear side roof panels (G and QQ). Lift each roof rear side panel straight upward to disengage studs from right and left side panels (J and Y, fig. 234).

*e. Installation of Roof Panels.*—Insert studs of right and left rear side roof panels (G and QQ) into right and left side panels (J and Y, fig. 234). Engage roof upper right and left side panels (B and EE) to anchors on windshield outer frame assembly (GG) and cab roof bow assembly (PP) and over studs of roof side panels.

*f. Removal of Rear Curtain and Roof Bow.*—With cab paulin (A) removed (*b* above) or rolled forward, and upper right and left side roof panels (B and EE) and right and left side roof panels (G and QQ) removed (*d* above), untie cab paulin rear rope (M) from each bow retainer ring nut (M, fig. 234). Disengage rear curtain rope (J) from hooks at inside of back panel (Q, fig. 234). Loosen both bow retainer ring nuts (M, fig. 234) and remove bow retaining screw from each leg of roof bow; then pull cab roof bow assembly (PP), with rear curtain assembly (K) straight upward out of roof bow retainers

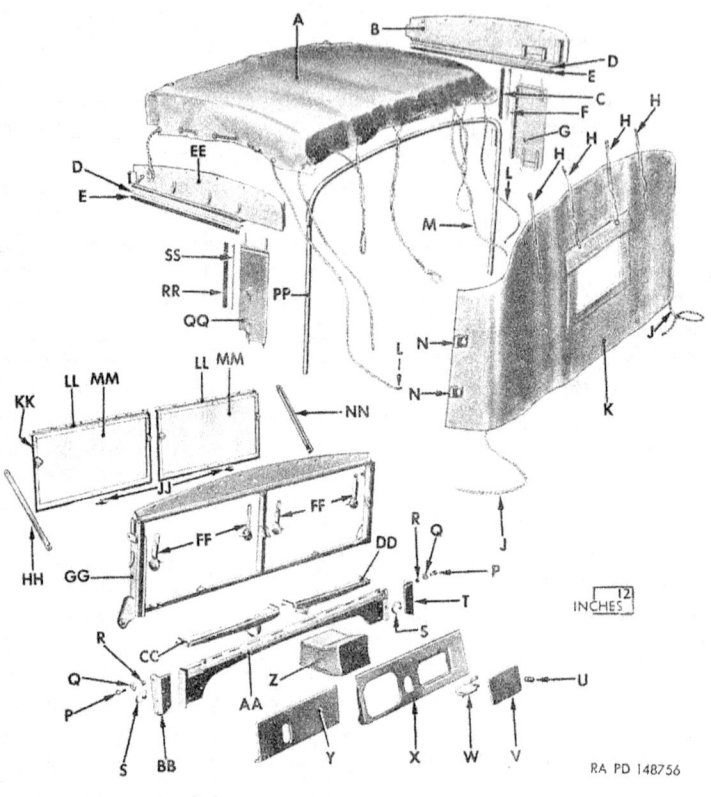

A—PAULIN, CAB—7410694
B—PANEL, ROOF, UPPER RIGHT SIDE—YT-2279109
C—SEAL, REAR, RIGHT WINDOW—7410726
D—RETAINER, DOOR GLASS UPPER SEAL—YT-2280711
E—SEAL, UPPER, DOOR GLASS—7410725
F—RETAINER, RIGHT WINDOW REAR SEAL—YT-2280707
G—PANEL, ROOF, RIGHT REAR SIDE—YT-2279160
H—STRAP, REAR CURTAIN—7410706
J—ROPE, REAR CURTAIN—YT-2280313
K—CURTAIN, REAR, ASSY—7410693
L—ROPE, SIDE, CAB PAULIN—YT-2280309
M—ROPE, REAR, CAB PAULIN—YT-2280310
N—STRAP, LOOP, REAR CURTAIN—7410709
P—BOLT, SPECIAL, 5/8-18 X 1 7/8—7410701
Q—WASHER, SPACING—YT-2266216
R—NUT, 5/8-18—442803
S—NUT, RING, CAB LIFTING—7410647
T—HINGE, RIGHT LOWER, OUTER FRAME, ASSY—7410699

*Figure 233. Instrument boards, windshield, and cab covering components.*

(L, fig. 234). Remove 11 screws which retain back curtain to roof bow and remove curtain from bow.

*g. Installation of Rear Curtain and Roof Bow.*—Position upper edge of rear curtain assembly (K) over top of cab roof bow assembly (PP) and secure with eleven screws. Insert legs of roof bow down through roof bow retainers (L, fig. 234) and install stop screw in each leg of bow below bow retainers, then raise bow until stop screws contact retainers. Tighten bow retainer ring nuts (M, fig. 234). Install roof panels (*e* above) and cab paulin (*c* above).

## 324. Replacement of Seats and Cushions

*Note.* Key letters noted in parentheses are in figure 235.

a. *Driver's Seat.*
 (1) *Removal and disassembly.*—Loosen wing nut on driver's seat lock bolt assembly (R) at rear of driver's seat frame (W); then swing lock bolt forward and downward. Raise and tilt seat forward and while retaining seat in this position, disengage driver's seat support (S) from driver's seat support bracket (T) on cab floor pan; then remove driver's seat from cab. Remove nuts from four studs which attach driver's seat support (S) to driver's seat left and right adjuster assemblies

*Figure 233.*—Continued.

U—KNOB, COMPARTMENT DOOR—CV-3674709
V—DOOR, COMPARTMENT, ASSY—YT-2280772
W—HINGE, COMPARTMENT DOOR, ASSY—YT-2259208
X—PANEL, INSTRUMENT, RIGHT, ASSY—YT-2277795
Y—PANEL, INSTRUMENT, LEFT, ASSY—YT-2277802
Z—COMPARTMENT, INSTRUMENT BOARD, ASSY—YT-2266252
AA—BAR, CROSS, WINDSHIELD—YT-2267239
BB—HINGE, LEFT LOWER, OUTER FRAME, ASSY—7410700
CC—DUCT, DEFROSTER, LEFT—YT-2280776
DD—DUCT, DEFROSTER, RIGHT—YT-2280775
EE—PANEL, ROOF, UPPER LEFT SIDE—YT-2279110
FF—ARM, ADJUSTING, WINDSHIELD—7373327
GG—FRAME, OUTER, WINDSHIELD, ASSY—7410695
HH—ARM, SUPPORT, WINDSHIELD, LEFT—YT-2281283
JJ—HANDLE, WINDSHIELD INNER FRAME, ASSY—7373330
KK—WEATHERSTRIP, WINDSHIELD, INNER FRAME—7410667
LL—FRAME, INNER, WINDSHIELD, ASSY—7410696
MM—GLASS, WINDSHIELD—YT 2280793
NN—ARM, SUPPORT, WINDSHIELD, RIGHT—YT-2281282
PP—BOW, CAB ROOF, ASSY—7410664
QQ—PANEL, ROOF, LEFT REAR SIDE—YT-2279161
RR—SEAL, REAR, LEFT WINDOW—7410727
SS—RETAINER, LEFT WINDOW REAR SEAL—YT-2280706

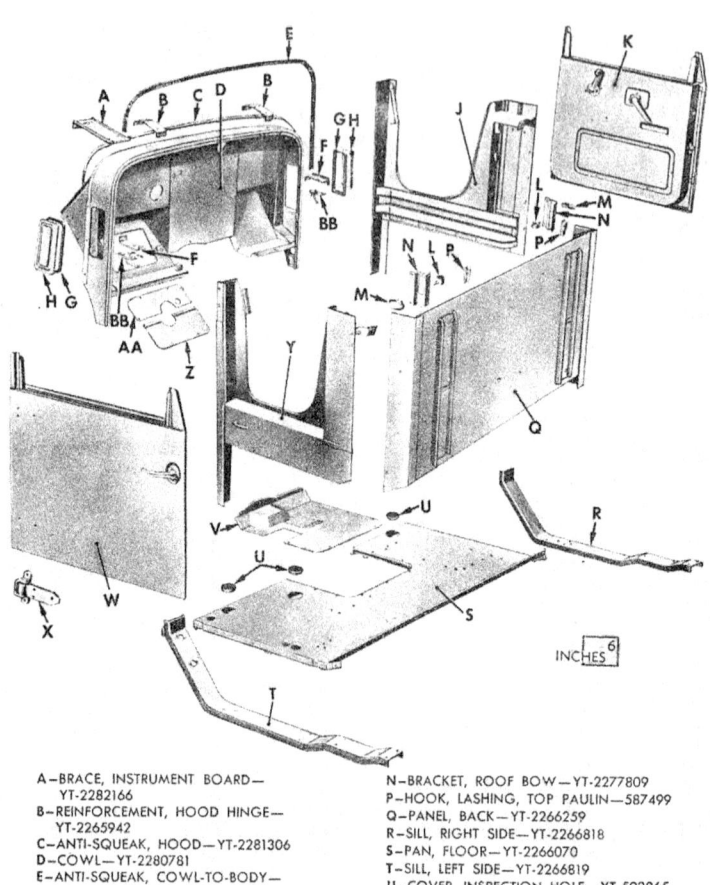

A—BRACE, INSTRUMENT BOARD—YT-2282166
B—REINFORCEMENT, HOOD HINGE—YT-2265942
C—ANTI-SQUEAK, HOOD—YT-2281306
D—COWL—YT-2280781
E—ANTI-SQUEAK, COWL-TO-BODY—YT-2281260
F—BRACKET, VENTILATOR LOCK—YT-2265692
G—SEAL, VENTILATOR, LID—7410703
H—LID, VENTILATOR, COWL—YT-2265693
J—PANEL, RIGHT SIDE—YT-2266099
K—DOOR, RIGHT, ASSY—YT-2280742
L—RETAINER, ROOF BOW—YT-2181035
M—NUT, RING, BOW RETAINER—7410647
N—BRACKET, ROOF BOW—YT-2277809
P—HOOK, LASHING, TOP PAULIN—587499
Q—PANEL, BACK—YT-2266259
R—SILL, RIGHT SIDE—YT-2266818
S—PAN, FLOOR—YT-2266070
T—SILL, LEFT SIDE—YT-2266819
U—COVER, INSPECTION HOLE—YT-592865
V—PAN, FLOOR, FRONT—YT-2281263
W—DOOR, LEFT, ASSY—YT-2280743
X—HINGE, LEFT, DOOR, ASSY—YT-2284512
Y—PANEL, LEFT SIDE—YT-2266100
Z—PLATE, LOWER PEDAL—YT-2276826
AA—PLATE, UPPER PEDAL—YT-2279150
BB—LOCK, VENTILATOR, ASSY—YT-2280768

RA PD 148757

*Figure 234. Cab cowl, doors, floor, and panel components.*

(U and V). Remove nuts from four studs which attach adjusters to driver's seat frame (W) and remove adjusters from seat frame. If adjusters have been bent or if they fail to function properly, replace with new assemblies.

(2) *Assembly and installation.*—Attach driver's seat left and right adjuster assemblies (U and V) to driver's seat frame

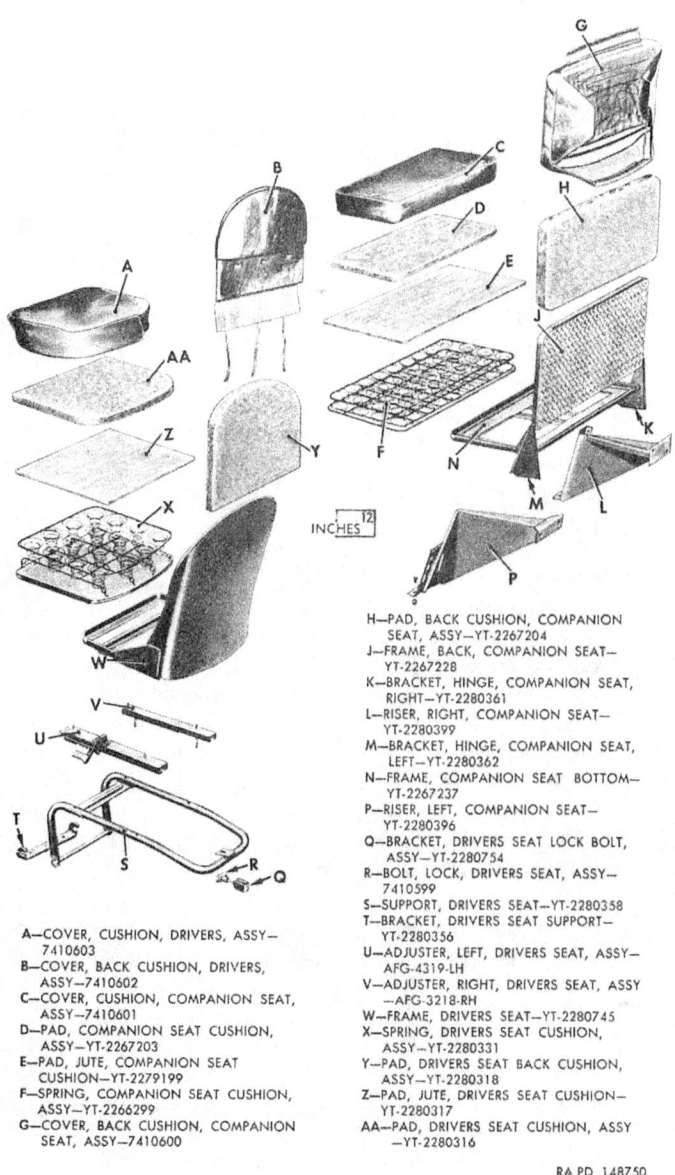

A—COVER, CUSHION, DRIVERS, ASSY—7410603
B—COVER, BACK CUSHION, DRIVERS, ASSY—7410602
C—COVER, CUSHION, COMPANION SEAT, ASSY—7410601
D—PAD, COMPANION SEAT CUSHION, ASSY—YT-2267203
E—PAD, JUTE, COMPANION SEAT CUSHION—YT-2279199
F—SPRING, COMPANION SEAT CUSHION, ASSY—YT-2266299
G—COVER, BACK CUSHION, COMPANION SEAT, ASSY—7410600
H—PAD, BACK CUSHION, COMPANION SEAT, ASSY—YT-2267204
J—FRAME, BACK, COMPANION SEAT—YT-2267228
K—BRACKET, HINGE, COMPANION SEAT, RIGHT—YT-2280361
L—RISER, RIGHT, COMPANION SEAT—YT-2280399
M—BRACKET, HINGE, COMPANION SEAT, LEFT—YT-2280362
N—FRAME, COMPANION SEAT BOTTOM—YT-2267237
P—RISER, LEFT, COMPANION SEAT—YT-2280396
Q—BRACKET, DRIVERS SEAT LOCK BOLT, ASSY—YT-2280754
R—BOLT, LOCK, DRIVERS SEAT, ASSY—7410599
S—SUPPORT, DRIVERS SEAT—YT-2280358
T—BRACKET, DRIVERS SEAT SUPPORT—YT-2280356
U—ADJUSTER, LEFT, DRIVERS SEAT, ASSY—AFG-4319-LH
V—ADJUSTER, RIGHT, DRIVERS SEAT, ASSY—AFG-3218-RH
W—FRAME, DRIVERS SEAT—YT-2280745
X—SPRING, DRIVERS SEAT CUSHION, ASSY—YT-2280331
Y—PAD, DRIVERS SEAT BACK CUSHION, ASSY—YT-2280318
Z—PAD, JUTE, DRIVERS SEAT CUSHION—YT-2280317
AA—PAD, DRIVERS SEAT CUSHION, ASSY—YT-2280316

RA PD 148750

*Figure 235. Cab seat components.*

(W) with four 5/16-24 nuts. Attach driver's seat support (S) to seat adjusters with four 5/16-24 nuts. Position driver's seat in cab, with driver's seat support (S) engaged in driver's seat support bracket (T). At rear of seat, swing driver's seat lock bolt assembly (R) upward into engagement with bracket on driver's seat support (S). Tighten wing nut firmly.

  b. *Driver's Seat Cushions.*
   (1) *Removal and disassembly.*—Raise seat cushion from driver's seat frame (W) to remove. Untie cover hold-down strap at back of driver's seat frame; then pull strap from grommet in seat back frame. Lift driver's back cushion cover assembly (B) and driver's rest back cushion pad assembly (Y) straight up from driver's seat frame (W) to remove. Untie straps at back of seat back cover; then pull driver's seat back cushion pad assembly (Y) from cover. Untie lacing cord at bottom of seat cushion; then disengage cord from hooks on bottom of driver's seat cushion spring assembly (X). Remove driver's cushion cover assembly (A) from springs and padding.
   (2) *Assembly and installation.*—Position driver's cushion cover assembly (A) on driver's seat cushion pad assembly (AA) and driver's seat cushion jute pad (Z) and driver's seat cushion spring assembly (X). Engage lacing cord in hooks on bottom of spring assembly, pull cord taut, and tie. Insert driver's seat back cushion pad assembly (Y) in driver's back cushion cover assembly (B), fold flap up, and tie outer lacing straps to loops. Position seat back cover and pad on driver's seat frame (W) with lower lacing strap inserted through grommet in seat frame. Tie straps together at back of seat frame to hold seat back cover and pad securely in place. Position seat cushion in seat frame.
  c. *Replacement of Companion Seat and Risers.*
   (1) *Removal.*—Remove six cap screws and nuts which attach companion seat right and left hinge brackets (K and M) to companion seat right and left risers (L and P). Lift seat from risers and remove seat from cab. Remove four cap screws and nuts which attach seat left riser to floor pan and back panel, and remove five cap screws and nuts which attach seat right riser to floor pan, side panel, and back panel; then remove risers from cab.
   (2) *Installation.*—Install companion seat right riser (L) to floor pan, side panel, and back panel, using five cap screws and nuts. Install companion seat left riser (P) to floor pan and back panel, using four cap screws and nuts. Position companion seat on seat risers. Install three cap screws and nuts which

attach each companion seat right and left hinge bracket (K and M') to risers.
  d. *Companion Seat Cushions.*
    (1) *Removal and disassembly.*—Pull bottom of seat back cushion forward and upward. Slide seat back cushion to side to disengage bead on cushion from metal channel on companion seat back frame (J). Raise companion seat bottom frame (N) to vertical position. Disengage seat spring latches by tapping with hammer. Lower companion seat bottom frame (N) to horizontal position. Lift seat cushion from seat bottom frame. Untie lacing cord at bottom of seat cushion, disengage cord from hooks on companion seat cushion spring assembly (F) and lift companion seat cushion cover assembly (C) from companion seat cushion pad assembly (D) companion seat cushion jute pad (E), and spring assembly. Untie lacing cord at back of seat back cushion, pull cord from loops in cover, then lift companion seat back cushion pad assembly (H) from companion seat back cushion cover assembly (G).
    (2) *Assembly and installation.*—Position companion seat back cushion pad assembly (H) in companion seat back cushion cover assembly (G), thread cord through loops, pull cord taut, and tie. Position companion seat cushion cover assembly (C) over companion seat cushion pad assembly (D), companion seat cushion jute pad (E), and companion seat cushion spring assembly (F), and engage loops of cord in hooks on spring bottom. Pull cord taut and tie. Position seat cushion and companion seat bottom frame (N), raise seat cushion and frame to vertical position and engage latches by tapping with hammer; then lower seat to horizontal position. Holding seat back cushion upside down, engage bead on seat back cushion in metal channel on companion seat back frame (J). Slide bead fully into channel; then lower cushion into place.

## 325. Replacement of Back Panel

  a. *General.*—The back panel (Q, fig. 234) can be removed from the cab regardless of whether cab is on chassis or removed from chassis. However, in either instance, the cab paulin, rear curtain, roof top and side panels, roof bow, companion seat and driver's seat, parking brake and bracket, and transfer and power take-off control levers and bracket must be removed from cab before removing the back panel.
  b. *Removal of Back Panel.*
    (1) Remove the cab paulin (A, fig. 233), rear curtain assembly (K, fig. 233), roof upper right and left side panels (B and EE,

fig. 233), right and left rear side roof panels (G and QQ, fig. 233), cab roof bow assembly (PP, fig. 233) as directed in paragraph 323. Remove driver's seat and support, and companion seat and brackets as directed in paragraph 324. Remove parking brake lever and bracket and transfer and power take-off control levers with bracket from cab.

(2) Remove all cap screws which attach back panel (Q, fig. 234) to right and left side panels (J and Y, fig. 234), floor pan (S, fig. 234), companion seat right and left risers (L and P, fig. 235), roof bow brackets (N, fig. 234), bow retainer ring nut (M, fig. 234), roof bow retainer (L, fig. 234), and cab right and left side sills (R and T, fig. 234). If operation is performed when cab is on chassis, remove four cap screws which attach cab to rear mounting springs and remove cab rear center mounting cap screw; then remove cab back panel by forcing panel back from side panels.

c. *Back Panel Installation.*
(1) Position back panel (Q, fig. 234) to back of cab, making sure ends of side panels enter between flange on back panel and guide straps welded to inner side of back panel.
(2) Attach back panel to right and left side panels (J and Y, fig. 234), using 23 cap screws and nuts.
(3) Attach back panel to floor pan (S, fig. 234) using 14 cap screws and nuts.
(4) Attach back panel to right and left side sills (R and T, fig. 234), using four cap screws and nuts.
(5) Attach back panel to companion seat right and left risers (L and P, fig. 235), using four cap screws and nuts.
(6) Attach back panel to roof bow brackets (N, fig. 234), using four cap screws and nuts.
(7) Attach parking brake lever bracket to back panel, using four cap screws and nuts.
(8) Attach transfer and power take-off control lever bracket to back panel, using three cap screws and nuts.
(9) Install driver's and companion seats (par. 324), and install roof bow, rear curtain, roof side and top panels, and cab paulin (par. 323).

## 326. Replacement of Side Panels

*Note.* Key letters noted in parentheses are in figure 234 unless otherwise indicated.

*a. General.*—In order to replace right or left side panel (J or Y) on cab, it is first necessary to remove the following cab components: cab paulin, roof panels, and rear curtain and cab roof bow (par. 323); driver's seat (par. 324); doors (par. 331); tool box, running boards, tail pipes, and heat shield; and complete windshield assembly with lower hinges (par. 334).

*b. Removal of Side Panels.*—Remove cap screws and nuts which attach back panel (Q) to floor pan (S), companion seat right and left risers (L and P, fig. 235), right and left side sills (R and T), transfer control lever bracket, and to side panel which is being replaced. Remove cap screws, nuts, and washers which attach side panel to floor pan (S), cowl (D), instrument panel, and windshield cross bar (AA, fig. 233). Remove side panel by forcing back panel (Q) back and

*Figure 236. Removing cab side panel.*

away from side panel; then pull out bottom of side panel until lower reinforcement on inside of side panel clears front corner of floor pan (S); then with rubber mallet drive cab side panel (fig. 236) down and away from windshield cross bar. Remove side panel.

*c. Installation of Side Panels.*—Position side panel to side of cab and insert upper front corner of panel under end flange of windshield cross bar (AA, fig. 233); then with small pry bar to guide side panel at windshield cross bar, and with a rubber mallet, drive side panel up and into position against floor pan and cowl. Install back panel (Q) to side panel (par. 325c (2)). Install cap screws, nuts, and washers which attach side panel to floor pan (S), cowl (D), instrument panel, and windshield cross bar (AA, fig. 233). Install door (par. 331); install complete windshield assembly with lower hinges (par. 334); install driver's seat (par. 324); install roof bow with rear curtain, roof panels and cab paulin (par. 323); and install running boards, tail pipes, and heat shield on cab.

### 327. Replacement of Floor Pan and Cab Sills

*Note.* Key letters noted in parentheses are in figure 234.

*a. General.*—Floor consists of floor pan (S) and right and left side sills (R and T) bolted together. Floor pan, of sheet steel, is reinforced by steel channels which are spot welded to under side of floor pan. Seats and risers must be removed from cab (par. 324) before removing floor pan. Also remove transfer shift control levers and bracket by removing cap screws and nuts which attach bracket to back panel, and disconnect electrical wiring harness which passes through floor. Position cab on its back to permit removal of floor pan and sills. Brace or support cab in such a manner that cab will not become out of alinement while making replacement.

*b. Removal of Cab Floor Pan and Sills.*—Remove cap screws and nuts which attach each cab sill to front of cowl (D). Remove two cap screws and nuts which attach each cab sill to back panel (Q). Remove all cap screws, nuts, and washers which attach floor pan (S) to back panel (Q), to right and left side panels (J and Y), and to cowl (D); then remove floor pan and sills.

*c. Disassemble Floor Pan.*—Remove six cap screws and nuts which attach each right and left side sill (R and T) to floor pan (S).

*d. Assemble Floor Pan.*—Install six cap screws and nuts which attach each right and left side sill (R and T) to floor pan (S).

*e. Installation of Floor Pan and Sills.*—Position floor pan to cab and install cap screws and nuts which attach right and left side sills (R and T) to cowl (D) and back panel (Q). Install cap screws, nuts, and washers which attach floor pan (S) to cowl (D), to right and left side panels (J and Y), and to back panel (Q). Tip cab to upright position; then install wiring harness through floor pan. Install seats (par. 324); then install transfer shift control levers and bracket to floor pan (S) and back panel (Q) with cap screws and nuts.

### 328. Replacement of Cowl

*Note.* Key letters noted in parentheses are in figure 234 unless otherwise indicated.

*a. General.*—The cowl assembly, which includes two side ventilator assemblies, can be replaced by first removing generator-regulator, air cleaner, instrument panels with instruments, and necessary air lines and vent lines from cab, providing they had not been removed previously.

*b. Removal of Cowl Assembly.*—Remove cap screws and nuts which attach cowl (D) to floor pan (S), right and left side sills (R and T), right and left side panels (J and Y), and windshield cross bar (AA, fig. 233) ; then remove cowl assembly and cowl-to-body antisqueak (E).

*c. Installation of Cowl Assembly.*—Position cowl (D) and cowl-to-body antisqueak (E) to front of cab. Install 13 cap screws and nuts which attach cowl to windshield cross bar (AA, fig. 233). Install 10 cap screws and nuts which attach cowl to right and left side panels (J and Y), and install seven cap screws and nuts which attach cowl to floor pan (S). Install right and left instrument panel assemblies (X and Y, fig. 233) with instrument to cab right and left side panels (J and Y) with five cap screws and nuts, and to windshield cross bar (AA, fig. 233) with seven cap screws and nuts. Install air cleaner, generator-regulator, air lines, and vent lines on cowl.

*d. Cowl Ventilators.*
  (1) *General.*—Cowl ventilators, one on each side of cowl, are controlled by handle at each ventilator. The ventilator lid, lid bracket, and hinge are of sheet metal and are welded together in a unit assembly. Each ventilator lid assembly is attached to cowl with four screws, nuts, and washers. The control link assembly connecting ventilator lid bracket and lock lever is attached with two nuts and washers. Ventilator lid can be set in three open positions. A rubber seal is cemented around inner side of lid with synthetic rubber cement to seal ventilator.
  (2) *Ventilator removal.*—Remove nut from each end of ventilator control link and disengage link from bracket on lid and lock lever. Remove four screws, nuts, and washers which attach ventilator lid hinge to cab cowl, and then remove cowl ventilator lid (H). Remove two cap screws which attach ventilator lock assembly (BB) to ventilator lock bracket (F) ; then remove lock assembly. Remove three cap screws and nuts which attach ventilator lock bracket (F) to cowl and remove lock bracket from cowl.
  (3) *Disassemble ventilator lock.*—Remove nut and washer which retains ventilator lock control lever to lock handle; then remove lever and handle.
  (4) *Assemble ventilator lock.*—Insert ventilator lock handle through lock bracket, lock plate, and lock lever; then install $3/8$-inch internal-teeth lock-washer and $3/8$–24 nut on control handle. Tighten nut firmly.

(5) *Ventilator installation.*

*Note.* Right ventilator lock assembly (BB) is installed under ventilator lock bracket (F) which in turn is installed with flanged edge facing upward. The left ventilator lock assembly and bracket is installed in opposite manner.

Install each ventilator lock bracket (F) to cowl with three cap screws and nuts. Attach ventilator lock assembly (BB) to bracket, using two cap screws. Install ventilator lid and hinge to cowl using four screws, nuts, and washers. Engage ends of control link in bracket on lid and in lever on lock, and secure link with two nuts and external-teeth lock washers. Tighten nuts after making necessary adjustments mentioned in (6) below.

(6) *Ventilator adjustment.*—The upward and downward adjustment of ventilator lock assemblies is obtained by shifting ventilator lock bracket (F) up or down after loosening three cap screws which attach bracket to cowl. The desired tightness or fit of cowl ventilator lid (H) and ventilator lid rubber seal to cowl (D) is obtained by loosening two cap screws which attach ventilator lock assembly (BB) to ventilator lock bracket (F) and shifting lock assembly inward or outward. After making these adjustments, tighten cap screws and nuts firmly.

## 329. Sheet Metal

Replacement information covering sheet metal components of vehicle, such as hood, hood-to-cowl extension panels, fenders, fender skirts, running boards, and brush guard is contained in TM 9–819A.

## Section III. REBUILD OF CAB DOORS

## 330. General

Following text covers removal, installation, and adjustment of cab door assembly; also replacement of components such as frame and glass, regulator, lock and remote control, etc.

## 331. Replacement of Cab Door

*Note.* **Key** letters noted in parentheses are in figure 238 unless otherwise indicated.

*a. Removal.*—Remove arm pivot pin (J). Remove three $5/16$–24 x $3/4$ cap screws (H) with external-teeth lock washer (H) attaching two hinges to each door; then remove door assembly.

*Note.* Door assembly can also be removed by driving out pin between two halves of hinge or by removing two screws with external-teeth lock washer attaching hinge to cab.

*b. Installation.*—Position door assembly and attach each hinge to door with three 5/16–24 x 3/4 screws with external-teeth lock washers. Install arm pivot pin (J).

*c. Adjustment.*—Provision for horizontal and vertical adjustment is provided and should be made whenever door is installed, or at any other time as required.

    (1) *Horizontal adjustment.*—Loosen three screws which attach each hinge to door; then shift door forward or rearward, as required. When correct position is obtained, tighten screws securely.

    (2) *Vertical adjustment.*—Loosen two screws which attach each hinge to cab, then shift door up or down as required. When correct position is obtained, tighten screws securely.

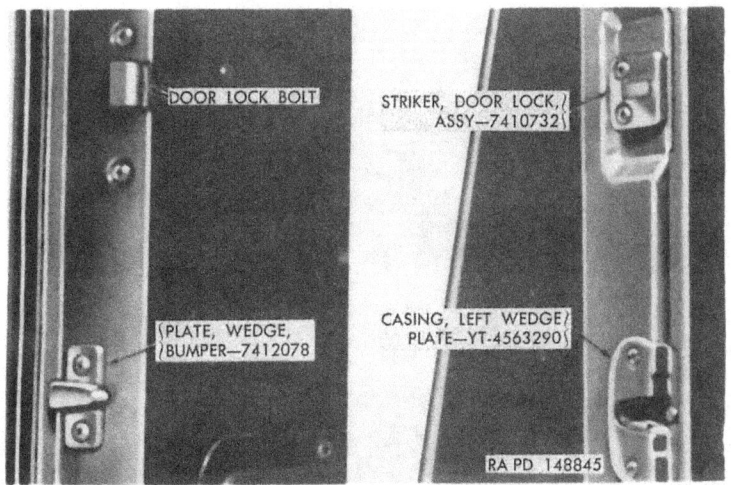

*Figure 237. Cab door lock striker and wedge plate adjustments.*

    (3) *Door lock striker adjustment.*—Loosen two screws attaching door lock striker assembly (fig. 237) to cab; then move striker in or out as required. Tighten screws firmly; then close door to determine if looseness, rattling, or improper latching have been corrected. Repeat adjustment procedure, if necessary.

    (4) *Door wedge plate adjustment.*—To properly position bumper wedge plate (fig. 237) loosen two screws attaching plate to door. Close door which will move plate up or down and properly locate plate in wedge plate casing. Open door carefully so as not to move wedge plate; then tighten two screws securely. Close and open door several times, noting

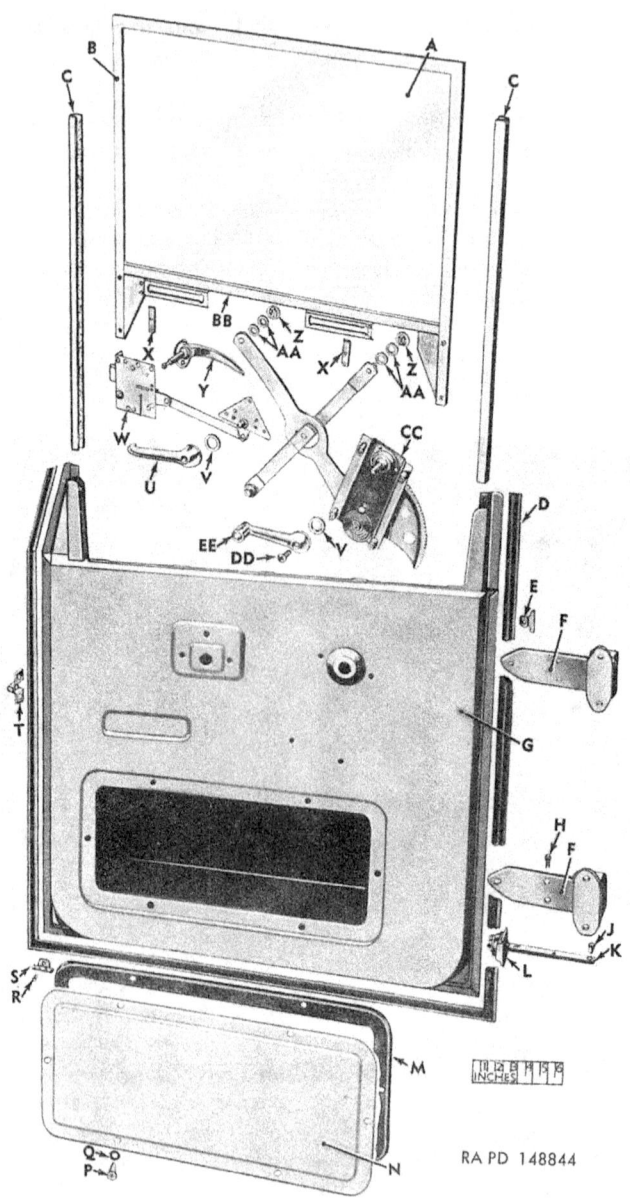

*Figure 238. Cab door components.*

if wedge plate is properly centered. Repeat adjustment procedure, if necessary.

## 332. Replacement of Cab Door Components

*Note.* Key letters noted in parentheses are in figure 238.

*a. General.*—Following operations can be performed either with door installed or removed, whichever is the more practical.

*b. Removal of Door Frame and Glass.*
   (1) *Remove door panel plate.*—Remove six screws and lock washers attaching door panel plate (N) to inside of door, then remove panel plate and panel plate weatherstrip (M).

---

*Figure 238.*—Continued.

A—GLASS, WINDOW—YT-2266258
B—{ FRAME, LEFT, W/GLASS—7410711
    FRAME, RIGHT, W/GLASS—7410710
C—CHANNEL, RUN, GLASS—7410633
D—WEATHERSTRIP, DOOR—7410734
E—CLIP, WEATHERSTRIP—7373283
F—{ HINGE, LEFT DOOR, ASSY—YT-2284512
    HINGE, RIGHT DOOR, ASSY—YT-2284511
G—{ PANEL, LEFT DOOR,—YT-2280729
    PANEL, RIGHT DOOR—YT-2280728
H—SCREW, CAP, W/EXT-TEETH LOCK WASHER, 5/16-24 X 3/4—425640
J—PIN, PIVOT, ARM—7410720
K—ARM, CHECK, ASSY—7410629
L—RETAINER, CHECK ARM—7410712
M—WEATHERSTRIP, PANEL PLATE—YT-2280719
N—PLATE, DOOR PANEL—YT-2266094
P—SCREW, TAPPING, CROSS-RECESS, NO 14-10 X 3/4—162077
Q—WASHER, LOCK, EXT-TEETH, 1/4-IN.—121753
R—SCREW, TAPPING, CROSS-RECESS, NO 6-18 X 3/4—161790
S—CLIP, WEATHERSTRIP—7410634
T—PLATE, WEDGE, BUMPER—7412078
U—HANDLE, REMOTE CONTROL—7410713
V—SPACER, FIBRE—YT-2280643
W—{ CONTROL, REMOTE, LEFT LOCK, ASSY—7410636
    CONTROL, REMOTE, RIGHT LOCK, ASSY—7410635
X—STOP, GLASS FRAME—YT-2280376
Y—HANDLE, ASSY—7410715
Z—FASTENER, REGULATOR ARM—7410638
AA—WASHER, LEATHER—7410719
BB—{ CHANNEL, LEFT SASH, ASSY—7410632
     CHANNEL, RIGHT SASH, ASSY—7410631
CC—{ REGULATOR, LEFT WINDOW, ASSY—7410723
     REGULATOR, RIGHT WINDOW, ASSY—7410722
DD—SCREW, CROSS-RECESS, 1/4-20 X 1/2—160515
EE—HANDLE, REGULATOR, ASSY—7410714

(2) *Remove door frame stops.*—Lower the frame assembly and remove two screws with external-teeth lock washers attaching two glass frame stops (X) to frame.

(3) *Disconnect regulator.*—Press regulator arm fastener (Z) from stud on both regulator arms; then remove leather washers (AA) and pull regulator arm studs from slot in frame lower channel.

(4) *Remove frame and glass.*—Lift frame and glass assembly straight up to complete removal.

c. *Installation of Door Frame and Glass.*

(1) *Position door frame and glass.*—Position door frame and glass assembly into channels with flat side of frame lower channel toward inside of door. Push frame and glass assembly down until lower channel slots are opposite regulator arm studs.

(2) *Connect regulator arms.*—Install one new leather washer (AA) on each regulator arm stud; then position arm studs through slots in frame lower channel. Install another new leather washer (AA) on each of the regulator arm studs. Secure arms to frame by installing a regulator arm fastener (Z) to each of the arm studs.

(3) *Install frame stops.*—Install two glass frame stops (X) to frame lower channel, using two No. 8–32 x $\frac{1}{2}$ screws with external-teeth lock washers at each stop.

(4) *Install door panel plate.*—Position door panel plate (N) and panel plate weatherstrip (M) over opening in door panel and secure with six No. 14–10 x $\frac{3}{4}$ cross-recess tapping screws (P) and $\frac{1}{4}$-inch external-teeth lock washers (Q).

d. *Replacement of Window Glass.*

(1) *Remove door frame and glass.*—Remove door frame and glass assembly as directed in *b* above.

(2) *Remove glass.*—Remove four screws and nuts attaching frame lower channel to frame side channels. Remove lower channel. Remove glass and filler from frame channel. Be sure that channel is cleaned of all glass and filler.

(3) *Install door glass.*—Install a piece of filler 63-inches long on top and two sides of glass. Install glass and filler into frame, using rubber mallet, if necessary, to seat glass properly. Install a piece of filler 25 inches long at bottom of glass. Install frame lower channel and secure with four No. 6–32 x $\frac{5}{16}$ screws and No. 8–32 sleeve nuts.

(4) *Install door frame and glass.*—Install door frame and glass assembly as directed in *c* above.

e. *Replacement of Glass Run Channel.*
  (1) *Remove door frame and glass.*—Remove door frame and glass assembly as directed in *b* above.
  (2) *Remove run channel.*—Use chisel to cut rivet near top of channel, then remove glass run channel (C) from guide in door.
  (3) *Install run channel.*—Coat channel in door and outside of run channel with rubber cement. Install run channel in guide channel and secure with rivet near top of channel.
  (4) *Install door frame and glass.*—Install door frame and glass assembly as directed in *c* above.
f. *Removal of Window Regulator.*
  (1) *Remove door panel plate.*—Remove door panel plate as directed in *b*(1) above.
  (2) *Disconnect regulator.*—Disconnect regulator from door frame as directed in *b*(3) above; then lift frame and glass assembly to top so that regulator assembly can be removed.
  (3) *Remove regulator.*—Remove $\frac{1}{4}$–20 x $\frac{1}{2}$ cross-recess screw (DD), then pull regulator handle assembly (EE) and fibre spacer (V) from regulator. Remove four screws and external-teeth lock washers attaching regulator to door inside panel. Remove right or left window regulator assembly (CC) through opening in door inside panel.
g. *Installation of Window Regulator.*
  (1) *Position regulator.*—Position right or left window regulator assembly (CC) on inside of door panel with handle stem through opening. Install four No. 12–32 x $\frac{1}{2}$ cross-recess screws with external-teeth lock washers. Tighten screws. Install fibre spacer (V), window regulator handle assembly (EE) and secure with $\frac{1}{4}$–20 x $\frac{1}{2}$ cross-recess screw (DD) and $\frac{1}{4}$-inch lock washer.
  (2) *Connect regulator arms.*—Install one new leather washer (AA) on each regulator arm stud; then position studs through slots in frame lower channel. Install another new leather washer (AA) on each of the regulator arm studs. Secure arms to frame by installing a regulator arm fastener (Z) to each of the arm studs.
  (3) *Install door panel plate.*—Install door panel plate as directed in *c*(4) above.
h. *Removal of Lock and Remote Control.*
  (1) *Remove door panel plate.*—Remove door panel plate as directed in *b*(1) above.
  (2) *Remove door handles.*—Remove screw and lock washer attaching remote control handle (U) at inside of door; then remove handle and fibre spacer (V). Remove two screws and

lock washers attaching handle assembly (Y) to outside of door, then pull handle from lock.

(3) *Remove lock and remote control.*—Crank window up until against stops, then lower approximately two inches. Remove three screws and lock washers at door lock, also remove three screws attaching right or left lock remote control assembly (W) to inner panel. Remove lock remote control assembly through opening in door inner panel.

*i. Installation of Lock and Remote Control.*

(1) *Install lock and remote control.*—Position lock remote control assembly on inside of door panel with lock through opening in edge of door and remote control handle stud through opening in door inside panel. Install six No. 12–32 x ½ cross-recess screws with external-teeth lock washers attaching assembly to door and tighten screws securely.

(2) *Install door handles.*—Install handle assembly (Y) at outside of door and secure with two No. 10–32 x ½ cross-recess screws and No. 10 lock washers. At inside of door, install fibre spacer (V) and remote control handle (U). Secure handle with one ¼–20 x ⅞ cross-recess screw and ¼-inch lock washer.

(3) *Install door panel plate.*—Install door panel plate as directed in *c*(4) above.

*j. Removal of Check Arm and Retainer.*

(1) *Remove door panel plate.*—Remove door panel plate as directed in *b*(1) above.

(2) *Remove check arm and retainer.*—Remove arm pivot pin (J) attaching arm to bracket. Remove two screws and lock washers attaching check arm retainer (L) to door at inside of door panel.

*k. Installation of Check Arm and Retainer.*

(1) *Installation.*—Insert check arm assembly (K) through opening and position check arm retainer (L) against bracket. Secure retainer with two No. 10–24 x ½ cross-recess screws and No. 10 lock washers. Install arm pivot pin (J) attaching arm to bracket.

(2) *Installation of door panel plate.*—Install door panel plate as directed in *c*(4) above.

*l. Weatherstrip Replacement.*

(1) *Removal.*—Remove screws; then remove six weatherstrip clips (E and S). Use knife or other similar tool to remove weatherstrip. Thoroughly clean weatherstrip contact surface on door.

(2) *Installation.*—Apply rubber cement to metal surface of cab door, also to contact surface of door weatherstrip (D) and

permit to dry a few minutes. Carefully install door weatherstrip (D) to cab door and be sure that it is properly located. Secure with six weatherstrip clips (E and S) using No. 6–18 x ¾ cross-recess tapping screws.

  *m. Handle Replacement.*
    (1) *Removal.*—At remote control handle (U) and regulator handle assembly (EE), remove screw and lock washer retaining each handle; then pull handle to remove; also remove fibre spacers (V). At handle assembly (Y), remove two screws and lock washers; then pull handle to remove from lock.
    (2) *Installation.*—Install fibre spacer (V); then position remote control handle (U) or regulator handle assembly (EE) to their respective units. Secure remote control handle with one ¼–20 x ⅞ cross-recess screw and ¼-inch lock washer, or regulator handle with one ¼–20 x ½ cross-recess screw and ¼-inch lock washer. Position handle assembly (Y) into lock; then secure with two No. 10–32 x ½ cross-recess screws and No. 10 external-teeth lock washers.

## Section IV. REBUILD OF WINDSHIELD ASSEMBLY

### 333. General

Windshield consists of two inner frame and glass assemblies hinged to an outer frame. Each inner frame assembly is independent of the other and can be adjusted or replaced without disturbing the opposite one. Entire assembly can be positioned as desired (TM 9–819A) or replaced as instructed in paragraph 334*b*.

### 334. Replacement of Windshield Components

*Note.* Key letters noted in parentheses are in figure 233 unless otherwise indicated.

  *a. General.*—Following paragraphs provide repair and replacement instructions of windshield assemblies or components. In most instances these instructions are applicable when windshield is installed or removed from vehicle.

  *b. Replacement of Windshield (Complete).*
    (1) *Removal.*—Remove cab top paulin as directed in paragraph 323. Remove windshield wiper blades (*e*(1) below) and wiper motors (*e*(2) below). Remove wiper motor air hose by removing six screws and clips. Remove ⅝–18 nut (R) and washer from two ⅝–18 x 1⅞ special bolts (P); then remove two bolts and washers. Lift complete windshield assembly from lower hinge on cab.
    (2) *Installation.*—Locate complete windshield assembly on cab with frame hinges on outside of lower hinges. Insert special

bolts through hinges from the outside, using 1½-inch outside diameter plain washer between hinge halves, and install plain washer and nut. Do not tighten nut excessively, since this would prevent proper windshield raising or lowering. Refer to TM 9-819A for method of adjusting, particularly if new assembly is being installed. Install wiper motors ($e(3)$ below) and windshield wiper blades ($e(4)$ below). Install wiper motor air hose and secure with six clips.

c. *Replacement of Windshield Frame and Glass.*
  (1) *Removal.*—Remove windshield wiper blade ($e(1)$ below) and wiper motor ($e(2)$ below). Unlock windshield inner frame handle assembly (JJ) and open inner frame sufficiently to remove two screws and two spring type washers attaching windshield adjusting arms (FF) to each side of inner frame. Support frame and glass assembly while seven cap screws with lock washers attaching inner frame hinge to windshield outer frame assembly (GG) are removed; then remove inner frame and glass assembly.
  (2) *Installation.*—Install weatherstrip in groove on outside of inner frame hinge. Position frame and glass in outer frame and install seven No. 10-32 x ½ screws with external-teeth lock washers attaching inner frame hinge to outer frame. Screws should only be installed fingertight until frame and glass is centered in opening; then tighten screws securely. Open frame sufficiently to connect windshield adjusting arms (FF) to each side of frame, using special washer between arm and frame; then secure each arm with one ¼-28 x ½ screw. Close frame and lock with handle. Install wiper motor ($e(3)$ below) and windshield wiper blade ($e(4)$ below).

d. *Replacement of Windshield Glass.*
  (1) *Removal.*—Remove individual windshield inner frame assembly (LL) and windshield glass (MM) as an assembly (*b* above). Remove upper hinge from frame by sliding hinge endwise. Remove nut, lock washer, and screw, also two screws and lock washers at upper corner of frame. Remove frame upper channel. Remove windshield glass (MM) and seal from frame channel; then clean channel thoroughly.
  (2) *Installation.*—Install a piece of seal 98 inches in length around edge of glass, then trim end to proper length. Install windshield glass (MM) and seal into windshield inner frame (LL), using rubber mallet if necessary to seat glass properly. Position frame upper channel, being sure that hinge edge is toward outside of frame. Install one No. 10-32 x ⅞ screw, No. 10 internal-teeth lock washer, and No. 10-32 crown nut, and two No. 8-32 x 7/16 screws and No. 8 internal-teeth lock

washers at upper corners of frame. Install upper hinge to frame by sliding hinge over formed upper channel. Install frame and glass assembly as directed in $c(2)$ above.

*e. Replacement of Windshield Wiper Blade and Motor.*

(1) *Removal of wiper blade.*—Pull wiper arm away from windshield, then swing lower end of blade outward and upward until free of wiper arm.

(2) *Removal of wiper motor.*—Remove wiper blade as directed in (1) above. Remove nut attaching wiper arm to wiper motor shaft; then pull arm from shaft. Disconnect air hose from wiper motor. Remove two screws attaching wiper motor to windshield frame; then remove wiper motor assembly.

(3) *Installation of wiper motor.*—Position wiper motor to rear of windshield frame with shaft through opening in frame. Install two No. 10–32 x ⅞ cross-recess screws with external-teeth lock washers and tighten screws securely. Install wiper arm on motor shaft, making sure arm is correctly positioned for wiping arc; then install and tighten wiper arm retaining nut. Connect air hose to wiper motor.

(4) *Installation of wiper blade.*—Pull wiper arm away from windshield, then position blade parallel with arm and with blade hook pointing toward wiper motor, engage blade hook with arm. Swing upper end of blade outward and downward into position.

*f. Rebuild of Windshield Wiper Motor.*—Refer to TM 9–1819B for rebuild of windshield wiper motors.

*g. Replacement of Windshield Weatherstrip and Seal.*

(1) Weatherstrips used between upper hinge of inner frames and outer frame are ¼-inch round sponge rubber 35 inches in length. Partially open windshield and lock at adjusting arms. Loosen seven screws attaching inner frame upper hinge to outer frame, which will permit inner frame to be lowered sufficiently to replace seal. Tighten seven screws securely and close windshield.

(2) Windshield inner frame weatherstrip (KK) is one piece specially formed to fit in groove at sides and bottom of frame. To replace, partially open windshield and lock at adjusting arms. Pull weatherstrip from frame, then use wire brush to thoroughly clean groove in frame. Position weatherstrip around frame with corners properly located; then force weatherstrip into frame groove using narrow blunt tool.

(3) Seals of rubber weatherstrip, 11½-inches long, are used at vertical edges of windshield outer frame to provide seal

between windshield and cab door. To replace, open cab door: then pull or lift seal from retainer. Clean retainer thoroughly; then slide new seal into place in retainer.
(4) Seal of sponge rubber is used across bottom of outer frame to provide seal between windshield frame and cowl. To replace, tilt complete windshield forward into lowered position (TM 9-819A) to expose seal. Remove seal and clean windshield lower frame of all rubber and sealing compound. Apply rubber cement to seal and frame then permit to dry until tacky. Install seal in frame. Close windshield which will hold seal firmly in place until cement is dry.

# CHAPTER 20

# CARGO BODY

## Section I. DESCRIPTION AND DATA

### 335. Description

*a. General.*—The following description applies to both of the cargo bodies used on 2½-ton 6x6 cargo trucks M135 and M211, with differences between models noted in the text. Construction and components of both bodies are illustrated in figures 239 and 240.

*b. Body.*—Cargo bodies are of all steel welded construction. Metal tail gate, with steps bolted to edge, is hinged to rear end of body at bottom. Paulin and end curtain lashing hooks are welded or bolted to body sides, front end, and tail gate. Stake pockets are provided in body side and front end channels for installation of side and front end racks. On cargo truck M135, the body bed is raised to provide clearance for the tires. On cargo truck M211, which is equipped with smaller dual rear tires, the bed of the body is flat and unobstructed.

*c. Side and Front End Racks.*—Side and front end racks consist of wood slats bolted to upright metal stake pockets; lower ends of stake pockets engage pockets in body side and front end channels to support racks. The four lower slats of the side racks are bolted to hinge straps which are hinged to the upright stake pockets, forming troop seats when in lowered position. When locked in raised position, seats form an integral part of the side racks. When in lowered position, seats are supported by legs which are hinged to seat straps.

*d. Paulin, End Curtains, and Bows.*—Top paulin and end curtains are supported by five bow assemblies, each consisting of a top bow, two side stakes, and two metal corners. Top bow and side stakes are secured in metal corners by cross-recess head screws. Lower ends of side stakes are inserted into the side rack stake pockets. On cargo truck M211, all bow assemblies are interchangeable. On cargo truck M135, the two end bows are longer than the three center bows due to the raised wheelhouses, and are not interchangeable. None of the bow assemblies are interchangeable between models. End curtains are reinforced and equipped with metal grommets for lashing ropes.

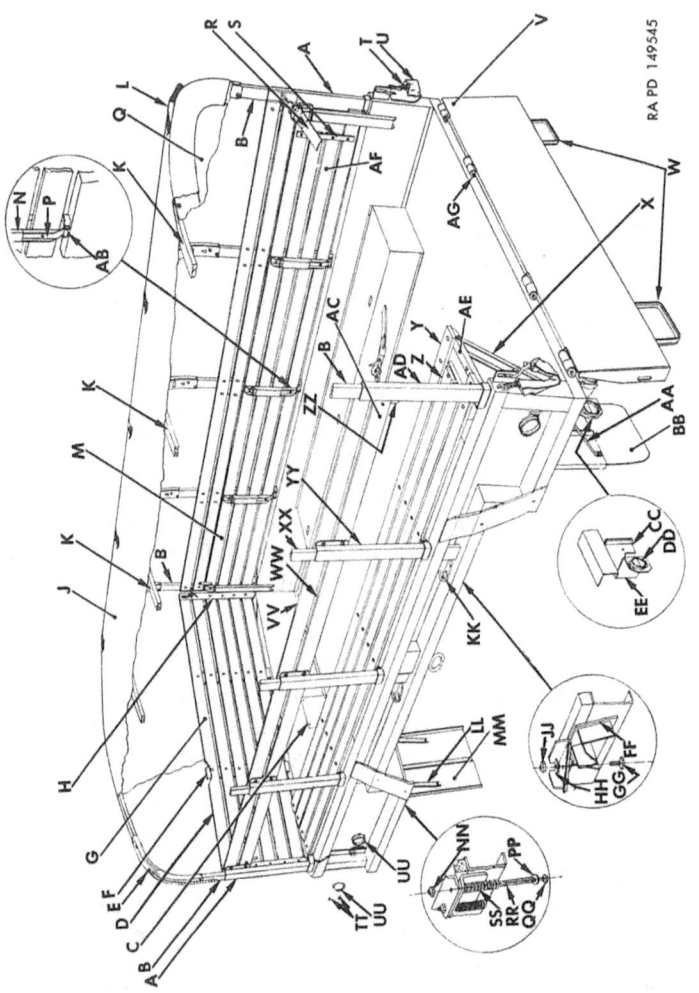

A—POCKET, SIDE RACK STAKE, LEFT FRONT AND RIGHT REAR—7397966
B—STAKE, END. BOW—7372781
C—COVER, FRONT PANEL—7397999
D—SLAT, FRONT END RACK—7397993
E—CORNER, BOW—540404
F—POST, FRONT END RACK—7370126
G—RACK, FRONT END, ASSY—7372787
H—{CHANNEL, FRONT END RACK, RIGHT, ASSY—7397991
   {CHANNEL, FRONT END RACK, LEFT, ASSY—7397990
J—PAULIN, TOP—7411556
K—BOW, TOP—7372780
L—STRAP, ROLL UP—545558
M—RACK, RIGHT SIDE, ASSY—7529475
N—LEG, SEAT, SHORT—7372791
P—HINGE, STRAP, SEAT—7061094
Q—CURTAIN, END—7411557
R—STRAP, SAFETY, ADJUSTABLE—545546
S—ANGLE, SEAT—7370149
T—CHAIN, TAIL GATE, ASSY—7411555
U—COVER, TAIL GATE CHAIN—7061088
V—GATE, TAIL, ASSY—7370211
W—STEP, TAIL GATE—7370210
X—LEG, SEAT, LONG—7372790
Y—SLAT, SEAT, OUTER—7061092
Z—SLAT, SEAT, INNER—7370148
AA—BRACE, REAR SPLASH SHIELD—7397979
BB—SHIELD, SPLASH, REAR, ASSY—7397987
CC—BRACKET, TAIL LIGHT—7372776
DD—REFLECTOR, REAR (RUBY)—506101
EE—BRACKET, REAR REFLECTOR—7372775
FF—BRACKET, SECONDARY SPRING—YT-2277394
GG—SCREW, CAP, ⅜-16 x 2¼—181833
HH—WASHER, PLAIN—131017
JJ—NUT, ¾-16—442804
KK—HOOK, LASHING, PAULIN—7064246-P
LL—BRACE, FRONT SPLASH SHIELD—7411552
MM—{SHIELD, SPLASH, LEFT FRONT—7372797
    {SHIELD, SPLASH, RIGHT FRONT—7370201
NN—NUT, ⅝-18—442803
PP—WASHER, PLAIN—130999
QQ—BOLT, ⅜-18 x 9—188956
RR—SPRING, HOLD DOWN, INNER—7372792
SS—SPRING, HOLD DOWN, OUTER—7372793
TT—SCREW, W/EXT-TEETH LOCK WASHER, ¼-20 x ½—7412828
UU—REFLECTOR, FRONT (AMBER)—506102
VV—SLAT, SIDE RACK, UPPER—7397994
WW—SLAT, SIDE RACK, LOWER—7397995
XX—STAKE, CENTER, BOW—7372784
YY—POCKET, SIDE RACK STAKE, CENTER—7397998
ZZ—CLAMP, SEAT—7372788
AB—PIN, SEAT STRAP HINGE—7370134
AC—RACK, LEFT SIDE, ASSY—7529275
AD—POCKET, SIDE RACK STAKE, RIGHT FRONT AND LEFT REAR—7397997
AE—SEAT, TROOP, LEFT SIDE, ASSY—7061090
AF—SEAT, TROOP, RIGHT SIDE, ASSY—7061091
AG—SHAFT, HINGE, TAIL GATE—7373205

*Figure 239. M135 cargo body components.*

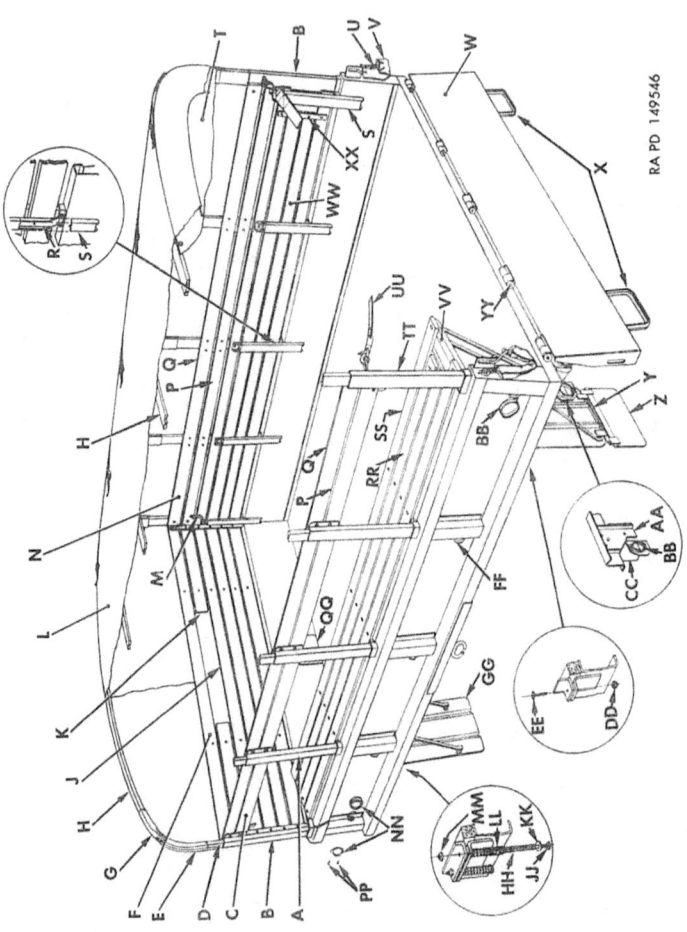

A—POCKET, SIDE RACK STAKE, CENTER—7370384
B—POCKET, SIDE RACK STAKE, LEFT FRONT AND RIGHT REAR—7397996
C—RACK, LEFT SIDE, ASSY—7370381
D—STAKE, BOW—7372781
E—STRAP, ROLL UP—545558
F—RACK, FRONT END ASSY—7370377
G—CORNER, BOW—540404
H—BOW, TOP—7370394
J—SLAT, FRONT RACK, LONG—7370378
K—SLAT, FRONT RACK, SHORT—7370419
L—PAULIN, TOP—8328220
M—CLAMP, SEAT—7372788
N—RACK, RIGHT SIDE, ASSY—7370380
P—SLAT, SIDE RACK, LOWER—7397995
Q—SLAT, SIDE RACK, UPPER—7397994
R—HINGE, STRAP, SEAT—7061094
S—LEG, SEAT—7372790
T—CURTAIN, END—8345032
U—CHAIN, TAIL GATE, ASSY—7411555
V—COVER, TAIL GATE CHAIN—7061088
W—GATE, TAIL, ASSY—7370841
X—STEP, TAIL GATE—7370210
Y—ROD, HINGE, REAR SPLASH SHIELD—7370372
Z—SHIELD, REAR SPLASH, LOWER, ASSY—7370371
AA—BRACKET, TAIL LIGHT—7370395
BB—REFLECTOR, REAR (RUBY)—506101
CC—BRACKET, REAR REFLECTOR—7372775
DD—NUT, 3/4–16—442804
EE—SCREW, CAP, 3/4–16 x 2¼—181833
FF—HOOK, LASHING, PAULIN—7064246-P
GG—{SHIELD, SPLASH, LEFT FRONT—7370426
     SHIELD, SPLASH, RIGHT FRONT—7370389
HH—SPRING, HOLD DOWN, INNER—7372792
JJ—BOLT, 5/8–18 x 9—188956
KK—WASHER, PLAIN—130999
LL—SPRING, HOLD DOWN, OUTER—7372793
MM—NUT, 3/4–18—442803
NN—REFLECTOR, FRONT (AMBER)—506102
PP—SCREW, W/EXT-TEETH LOCK WASHER, ¼–20 x ½—7412828
QQ—COVER, FRONT PANEL—7397999
RR—SLAT, SEAT, INNER—7370148
SS—SLAT, SEAT, OUTER—7061092
TT—POCKET, SIDE RACK STAKE, RIGHT FRONT AND LEFT REAR—7397997
UU—STRAP, SAFETY, ADJUSTABLE—545546
VV—SEAT, TROOP, LEFT SIDE, ASSY—7370383
WW—SEAT, TROOP, RIGHT SIDE, ASSY—7376383
XX—ANGLE, SEAT—7370149
YY—SHAFT, HINGE, TAIL GATE—7373205

*Figure 240. M211 cargo body components.*

Top paulin has reinforced metal grommets at sides and hemmed tunnel at each end for lashing ropes, with leather reinforcements used at each corner. End curtains and top paulin are not interchangeable between models.

## 336. Data

*a. Cargo Truck M135.*

| | |
|---|---|
| Inside body width | 80 in. |
| Width—center to center of stake pockets | 85$\frac{5}{32}$ in. |
| Inside body length | 147 in. |
| Inside body height | 14 in. |
| Width between wheelhouses | 48½ in. |
| Height between wheelhouses | 8¼ in. |
| Height—floor to paulin bow at center | 60 in. |

*b. Cargo Truck M211.*

| | |
|---|---|
| Inside body width | 88 in. |
| Width—center to center of stake pockets | 91$\frac{3}{32}$ in. |
| Inside body length | 147 in. |
| Inside body height | 14 in. |
| Height—floor to paulin bow at center | 60 in. |

## Section II. REBUILD OF CARGO BODY

## 337. General

The following instructions do not contain detailed step-by-step procedures on body repair. Since the nature and extent of damage to the body will vary, no definite repair procedure can be established. Successful repair of body will depend to a great extent upon the use of proper welding equipment and material, and upon the ability of the welder. As a general rule, except for major repairs removal of the body from the chassis is not necessary. Since the following instructions are only general in nature, they will apply to the cargo body used on both the M135 and the M211.

## 338. Straightening

Steel parts can be repaired by straightening, brazing, or welding; however, badly damaged parts should be replaced. Heat should not be used when straightening parts of the body. Heat weakens the structural characteristics of metal; therefore, all straightening should be done with parts cold. Any part buckled or bent sufficiently to show strains or cracks after straightening should be replaced or reinforced.

## 339. Patching and Welding

*a. Preparation.*—Prepare the hole or break for patching by trimming off all curled edges with a cutting torch.

*b. Equipment.*—Electric welding equipment should be used exclusively in the repair of the body. Use shielded arc method; the heat of the weld is localized and burning of the material is minimized with this method. In the event electric welding equipment is not available, gas welding or brazing may be used.

*c. Welding Materials.*—Welding rods of electrode classification No. E-6012 A. W. S. (American Welding Society) should be used. When patching the body, the patch should be of the same thickness as the panel to which the patch is being applied.

*d. Safety Precautions.*—Welding or cutting should not be undertaken in areas where fire is forbidden, nor should work of this nature be performed near inflammable materials unless proper precautions are taken to prevent fire. During operation in an inadequately ventilated place, the fumes, suffocating gases, and toxic gases generated in the welding process, or the reduction of oxygen in the air, may overcome the operator. For this reason, welding should not be attempted in such places unless adequate forced ventilation is provided.

*e. Preparing Patch.*—Cut a patch of the same material and thickness as the panel to which the patch is to be applied. The patch must be of sufficient size to overlap at least two inches all around the hole. Form patch to fit contour of mating part.

*f. Applying Patch.*—Position patch over hole and tack weld patch to one side of panel or part at several evenly spaced points. On the other side of the panel or part, run a continuous fillet weld all around the edge of the hole. When welding body panels, patch can be applied to the inside or outside of the panel. The edges of the fracture inside the body should be hammered smooth to fit against the patch and then welded solidly to patch. All slag must be chipped off and the area primed and painted in accordance with TM9-2851.

### 340. Repair of Wood Parts

*a. General.*—Wood parts can be repaired by gluing and splicing with wood or metal cleats. Use only waterproof glue. When splicing cleats are used, make sure cleats do not affect function of repaired part or cause interference with adjacent or mating parts. Parts which are subjected to considerable strain should not be repaired, but should be replaced with new parts. If fabricated wood parts are not available, new parts can be made using any suitable kiln-dried hard wood. Cut to size and mark location of holes, using the old part as template. If old part is too badly damaged to use as a template, fabricate a new part (*b* below).

*b. Fabricating New Parts.*
  (1) Cut wood to as near correct size and shape as possible by comparison with or by measurement of the same part on another vehicle.
  (2) Position new part, making sure other parts are correctly alined, and temporarily fasten in place with "C" clamps.
  (3) Using mating parts as templates, mark location of holes or drill through holes in mating parts. Bolt new part in place, using same size bolts used for attaching same part on other vehicles.

*c. Painting Wood Parts.*—New wood parts must be sealed, primed, and painted in accordance with applicable instructions in TM 9–2851.

## 341. Repair of Top Paulin and End Curtains

*a. General.*—When patching tears or holes in top paulin or end curtains, cut a patch from No. 8 duck or from an old paulin. Patch must be large enough to extend well beyond the edges of the tear or hole. Edges of patch and hole should be turned under, and several rows of stitches should be applied around the edges of the patch and edges of hole. All stitching must be made with a minimum of four stitches per inch. Reinforcements of leather of of extra thicknesses of duck must be used at points subjected to tension. Leather reinforcements at each corner must be riveted as well as stitched. Metal grommets must be installed at all points where lashing ropes pass through paulin or end curtains. Roll-up strap buckles are riveted to leather pads and the leather pads are stitched to the paulin.

*b. Waterproofing.*—When new material is used for patching top paulin or end curtains, or if the paulin or curtains have lost their water-repellent qualities, the material should be treated with a suitable waterproof, weatherproof, and mildew-resistant compound. Follow instructions on the compound container for applying.

# CHAPTER 21

# REPAIR AND REBUILD STANDARDS

## 342. General

The repair and rebuild standards included herein give the minimum, maximum, and key clearances of new parts. In the "sizes and fits of new parts" column, the letter "L" indicates a loose fit clearance and the letter "T" indicates a tight fit (interference).

*Note.* All dimensions are in inches unless otherwise indicated.

## 343. Transfer Assembly

*a. Input Shaft Components.*

| Fig. No. | Ref. letter | Point of measurement | Sizes and fits of new parts |
|---|---|---|---|
| 241 | A | Outside diameter of input shaft gear bearing. | 2.1035 to 2.1040 |
| 241 | B | Inside diameter of input shaft gear bore. | 2.106 to 2.107 |
| 241 | C | Width of fork groove in sliding gear | 0.337 to 0.342 |

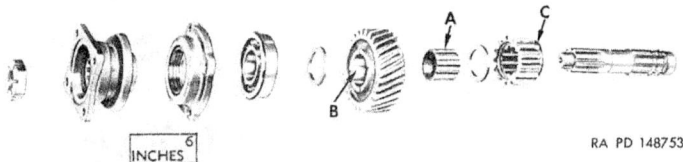

*Figure 241. Repair and rebuild standard points of measurement for transfer input shaft components*

*b. Front Axle Output Shaft Components.*

| Fig. No. | Ref. letter | Point of measurement | Sizes and fits of new parts |
|---|---|---|---|
|  |  | Front axle output shaft gear: |  |
| 242 | A | Inside diameter of bore | 1.8550 to 1.8560 |
| 242 | B | Width of fork groove | 0.384 to 0.389 |
| 242 | C | Outside diameter of front axle output shaft at gear. | 1.8520 to 1.8530 |
| 242 | A–C | Fit of gear on shaft | 0.002L to 0.004L |

403

Figure 242. Repair and rebuild standard points of measurement for transfer front axle output shaft components.

c. *Shifting Mechanism.*

| Fig. No. | Ref. letter | Point of measurement | Sizes and fits of new parts |
|---|---|---|---|
| 243 | A | Poppet ball spring: | |
|  |  | Free length | 2 (max.) |
|  |  | Working load when compressed to 1⅜ | 30 to 34 lb. |
| 243 | B | Shifter shaft front spring: | |
|  |  | Free length | 4 13/16 |
|  |  | Working load when compressed to 1 29/32 | 20 to 24 lb. |
| 243 | C | Shifter shaft rear spring: | |
|  |  | Free length | 3 1/16 |
|  |  | Working load when compressed to 1 5/16 | 10 to 14 lb |
|  |  | Front axle shifter shaft: | |
| 243 | D | Outside diameter at front axle shifter fork | 0.8710 to 0.8725 |
| 243 | E | Outside diameter at pilot (rear) end | 0.7445 to 0.7465 |
| 243 | F | Sliding gear shifter shaft OD of shaft | 0.7465 to 0.7475 |
|  |  | Front axle shifter fork: | |
| 243 | G | Inside diameter of bore | 0.8745 to 0.8755 |
| 243 | H | Width of fork pad | 0.339 to 0.349 |
| H, fig. 243 |  | Fit of shifter fork in front | |
| B, fig. 242 |  | Axle output shaft gear groove | 0.035L to 0.050L |
|  |  | Sliding gear shifter fork: | |
| 243 | J | Width of fork pad | 0.323 to 0.328 |
| J, fig. 243 |  | Fit of shifter fork in rear | |
| C, fig. 241 |  | Sliding gear groove | 0.009L to 0.019L |

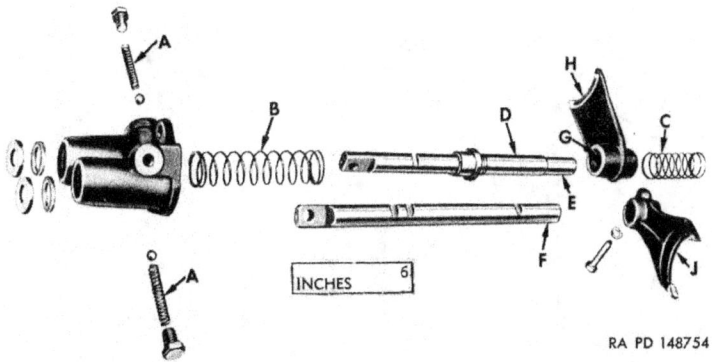

*Figure 243.* Repair and rebuild standard points of measurement for transfer shifting mechanism.

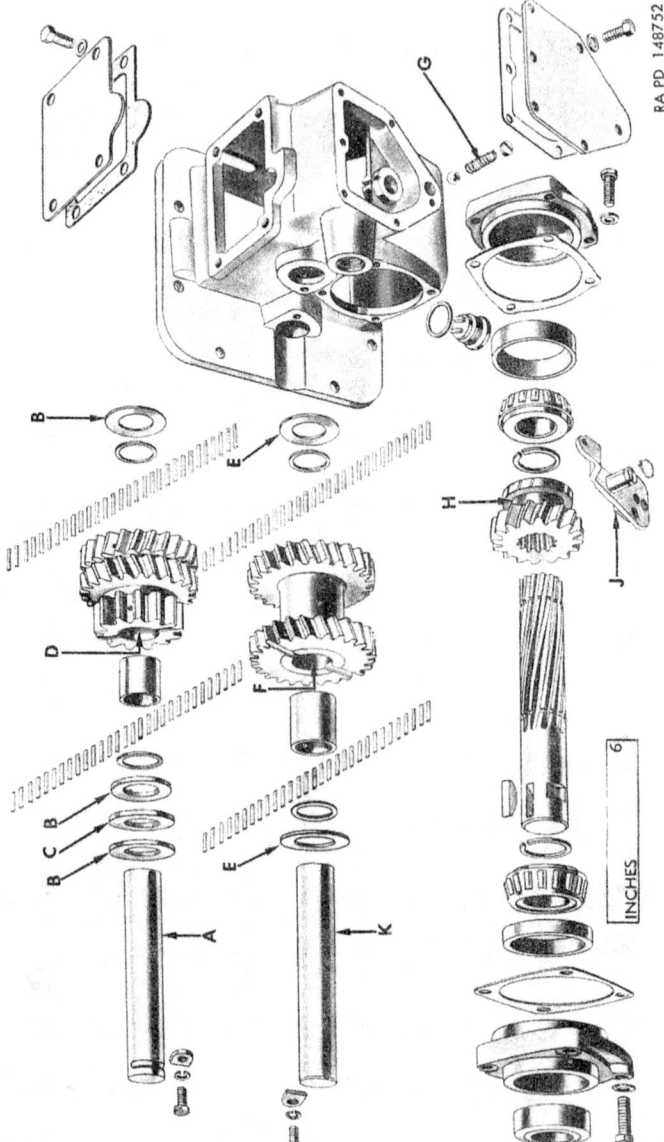

*Figure 244. Repair and rebuild standard points of measurement for power take-off components.*

406

### 344. Power Take-Off Assembly

| Fig. No. | Ref. letter | Point of measurement | Sizes and fits of new parts |
|---|---|---|---|
| 244 | D | Inside diameter of idler gear bore | 1.3620 to 1.3625 |
| 244 | F | Inside diameter of reverse gear bore | 1.3620 to 1.3625 |
| 244 | A | Outside diameter of idler shaft | 1.1105 to 1.1110 |
| 244 | K | Outside diameter of reverse gear shaft | 1.1105 to 1.1110 |
| 244 | B | Thickness of idler gear thrust washers | 0.120 to 0.125 |
| 244 | E | Thickness of reverse gear thrust washers | 0.120 to 0.125 |
| 244 | C | Thickness of idler gear spacer | 0.176 ot 0.181 |
| 244 | H | Width of groove in sliding gear | 0.505 ot 0.515 |
| 244 | J | Width of shoe on shifter plate assembly | 0.500 |
| 244 | J–H | Fit of shoe in sliding gear groove | 0.005 to 0.015 |
| 244 | G | Free length of shifter detent ball spring | 1 1/16 |

### 345. Power Take-Off Accessory Drive Assembly

| Fig. No. | Ref. letter | Point of measurement | Sizes and fits of new parts |
|---|---|---|---|
| 245 & 246 | A | Outside diameter of shifter shaft | 0.685 to 0.686 |
| 245 & 246 | B | Inside diameter of shifter shaft bore in housing | 0.688 to 0.690 |
| 245 & 246 | B–A | Fit of shifter shaft in housing bore | 0.002L to 0.005L |
| 245 & 246 | C | Free length of shifter poppet ball spring | 1 1/16 |
| 245 & 246 | D | Outside diameter of drive shaft at front bearing | 0.9995 to 1.000 |
| 245 & 246 | E | Thickness of thrust washer | 0.122 to 0.124 |
| 245 & 246 | F | Width of fork groove in sliding gear | 0.505 to 0.515 |
| 245 & 246 | G | Width of shifter fork pads | 0.495 to 0.505 |
| 245 & 246 | G–F | Fit of fork pads in sliding gear groove | 0.020L (max) |

### 346. Front Axle Steering Knuckle, Support, and Universal Joint Assembly

| Fig. No. | Ref. letter | Point of measurement | Sizes and fits of new parts |
|---|---|---|---|
| 247 | A | Outside diameter of oil seal sleeve | 3.245 to 3.255 |
| 247 | B | Indise diameter of bushing type bearing | 1.786 to 1.788 |
| 247 | C | Thickness of thrust washers | 0.155 to 0.157 |
| 247 | D | Trunnion bearing shim thickness | 0.002, 0.005, 0.010, 0.020 |
| 247 | E | Outer shaft bearing surface | 1.770 to 1.772 |
| 247 | F | Inner shaft oil seal surface | 1.8115 to 1.8125 |
| 247 | G | Diameter of outer balls—select fit | 1.372, 1.373, 1.374, 1.375, 1.376, 1.377, 1.378. |
| 247 | H | Diameter of center ball | 1.2495 to 1.2505 |

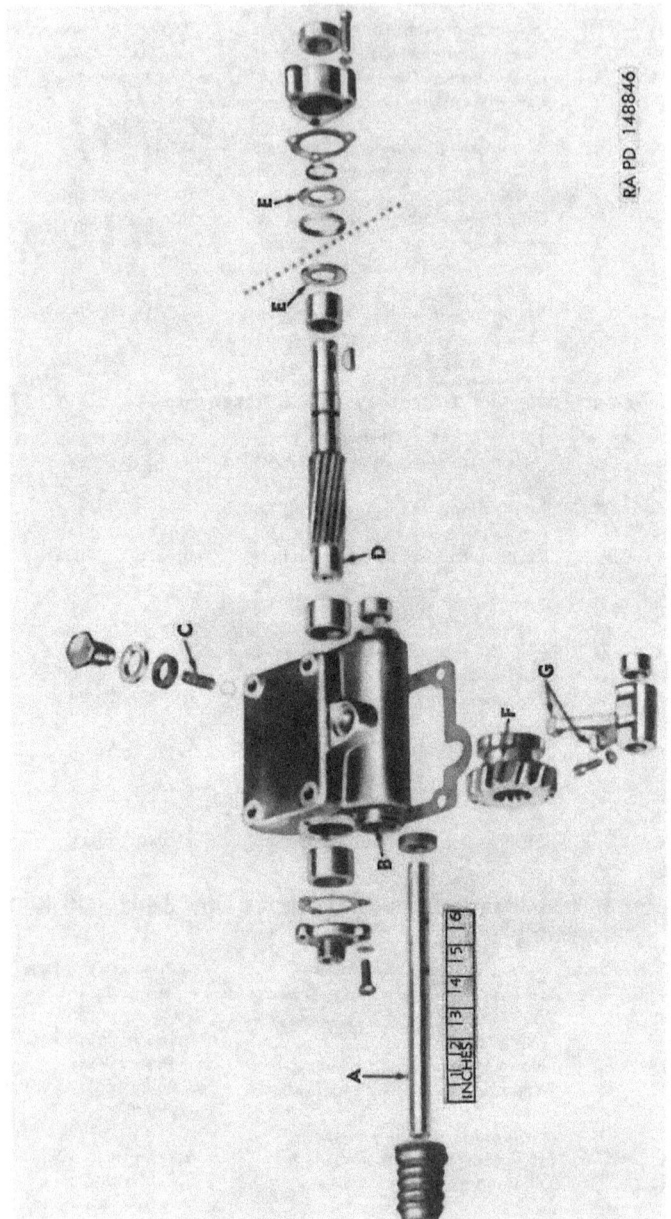

*Figure 245. Repair and rebuild standard points of measurement for accessory drive used on tank trucks.*

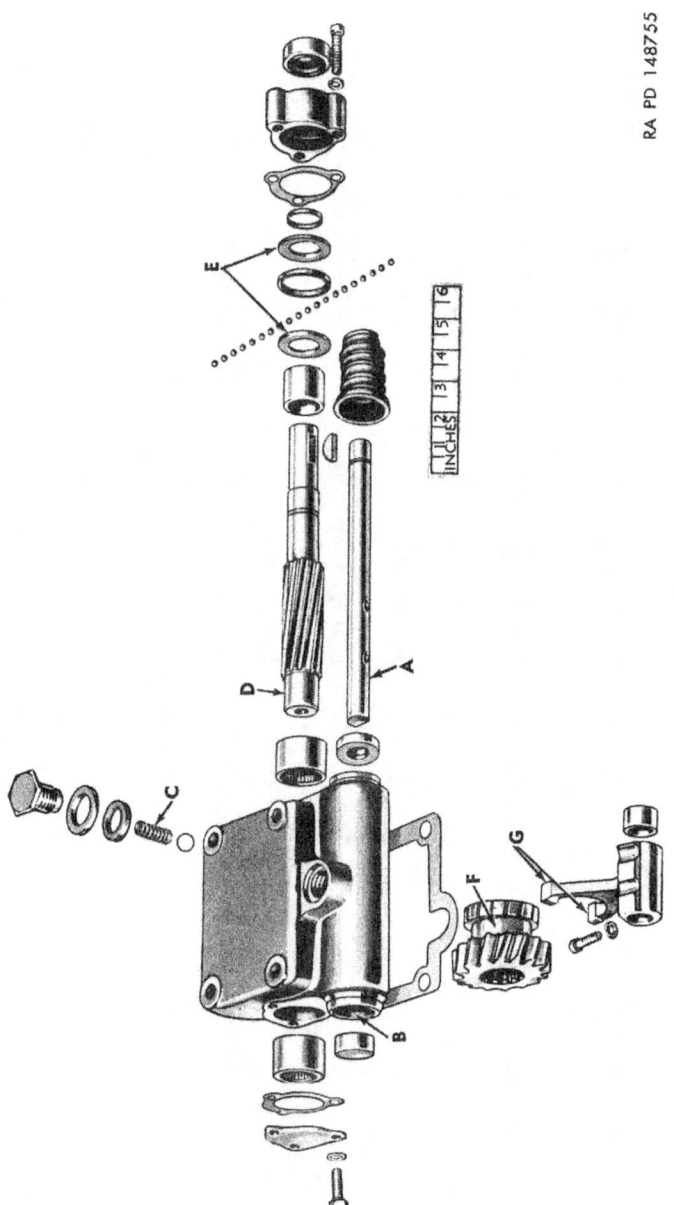

Figure 246. Repair and rebuild standard points of measurement for accessory drive used on dump truck.

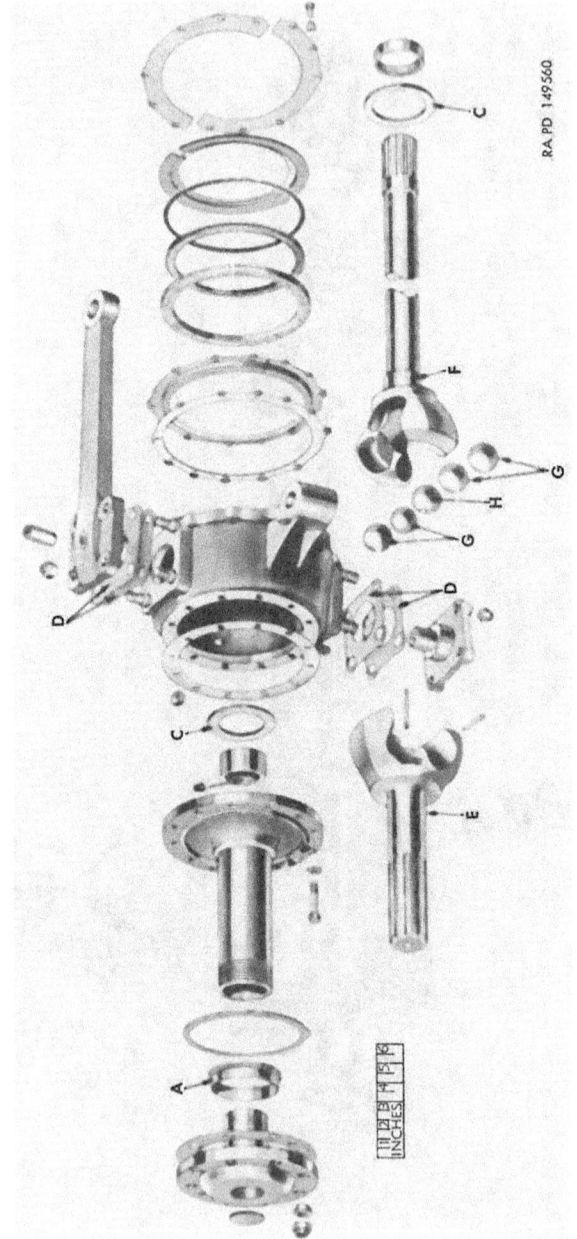

*Figure 247. Repair and rebuild standard points of measurement for front axle steering knuckle, support, and universal joint assembly.*

## 347. Axle Differential Assembly

| Fig. No. | Ref. letter | Point of measurement | Sizes and fits of new parts |
|---|---|---|---|
| 248 | J–K | Drive gear to drive pinion backlash | 0.005 to 0.008 |
| 248 | A | Diameter of side gear pilot in case and case cover. | 2.193 to 2.195 |
| 248 | B | Thickness of side gear thrust washers. | 0.058 to 0.062 |
| 248 | C | Diameter of side gear pilot | 2.189 to 2.191 |
| 248 | D | Diameter of spider pinion bore | 0.880 to 0.881 |
| 248 | F–D | Fit of spider arm in spider pinion | 0.005L to 0.007L |
| 248 | E | Thickness of spider pinion thrust washer. | 0.058 to 0.062 |
| 248 | F | Diameter of spider arms | 0.874 to 0.875 |
| 248 | G | Thickness of thrust pad | 0.1845 to 0.1885 |
| 248 | H | Outside diameter of flange sleeve | 2.599 to 2.601 |

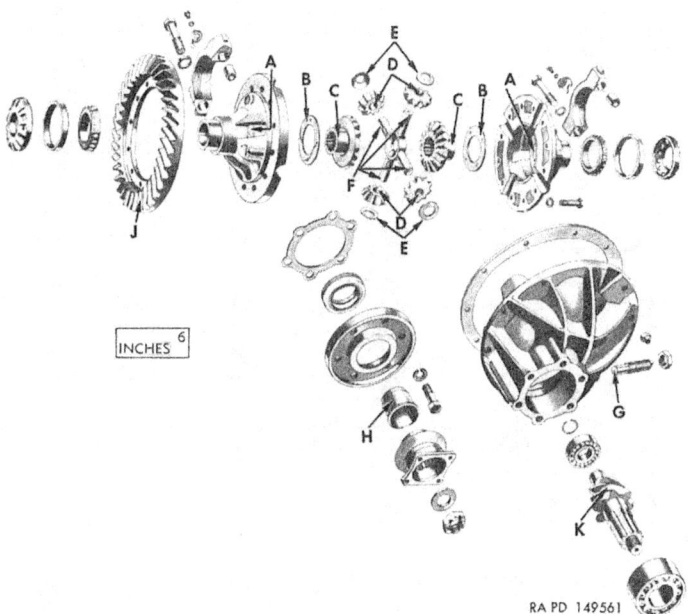

*Figure 248. Repair and rebuild standard points of measurement for axle differential assembly.*

411

### 348. Service Brakes

| Fig. No. | Ref. letter | Point of measurement | Sizes and fits of new parts |
|---|---|---|---|
| 249 | A | Backing plate flange—runout not to exceed (total dial indicator reading) | 0.010 |
| 249 | B | Diameter of anchor pin | 0.866 to 0.868 |
| 249 | C | Diameter of anchor pin bore in anchor block | 0.877 to 0.879 |
| 249 | B-C | Fit of anchor pin in anchor block | 0.009L to 0.013L |
| 249 | D | Brake shoe return spring: | |
|  |  | Free length | $5\frac{7}{16}$ |
|  |  | Working load when compressed to $5\frac{15}{16}$ | 50 to 60 lb |

### 349. Master Cylinder

| Fig. No. | Ref. letter | Point of measurement | Sizes and fits of new parts |
|---|---|---|---|
| 250 | B | Piston return spring: | |
|  |  | Approximate free length | $4\frac{11}{16}$ |
|  |  | Working load when compressed to $2\frac{5}{8}$ in. | $11\frac{1}{4}$ to $12\frac{3}{4}$ lb |
| 250 | C | Inside diameter of cylinder bore | 1.750 to 1.753 |
|  |  | Maximum diameter after honing | 1.757 |
| 250 | A-C | Fit of piston in cylinder bore | 0.005L (max) |

### 350. Wheel Cylinders

| Fig. No. | Ref. letter | Point of measurement | Sizes and fits of new parts |
|---|---|---|---|
| 251 | B | Piston return spring: | |
|  |  | Approximate free length | $2\frac{7}{8}$ |
|  |  | Working load when compressed to $1\frac{3}{8}$ | $2\frac{1}{4}$ to $3\frac{1}{4}$ lb |
| 251 | C | Inside diameter of cylinder body bore | 1.250 to 1.253 |
|  |  | Maximum diameter after honing | 1.257 |
| 251 | A-C | Fit of piston in cylinder body bore | 0.005L (max) |

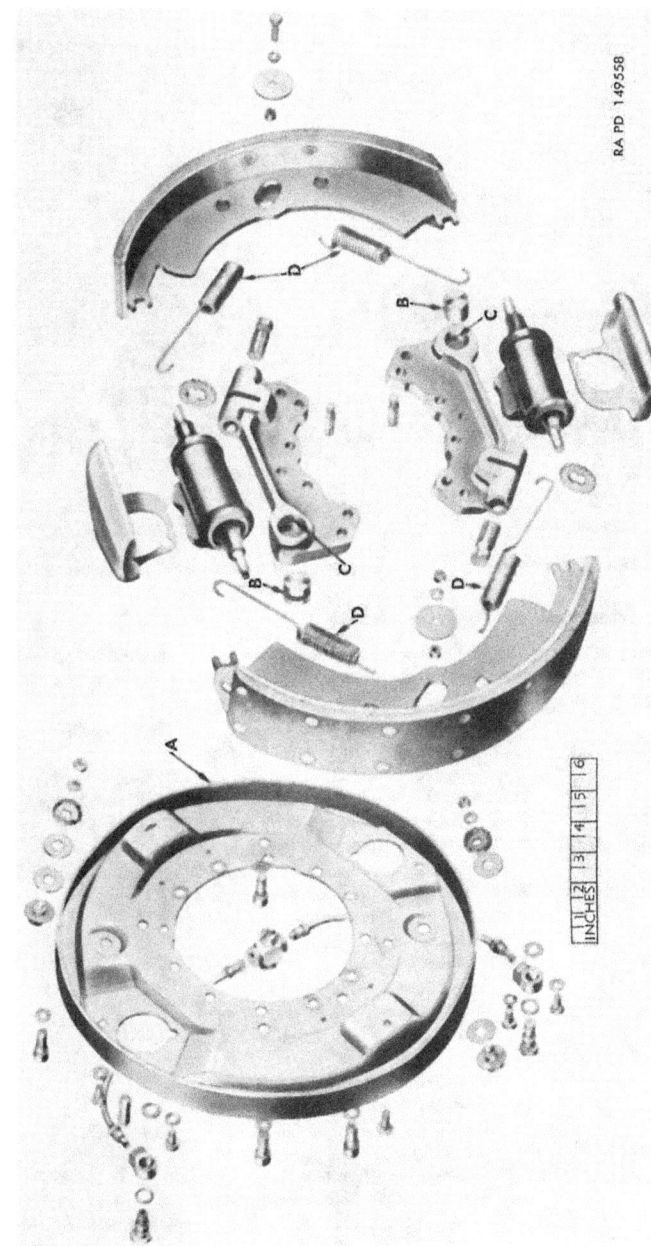

*Figure 249. Repair and rebuild standard points of measurement for service brakes.*

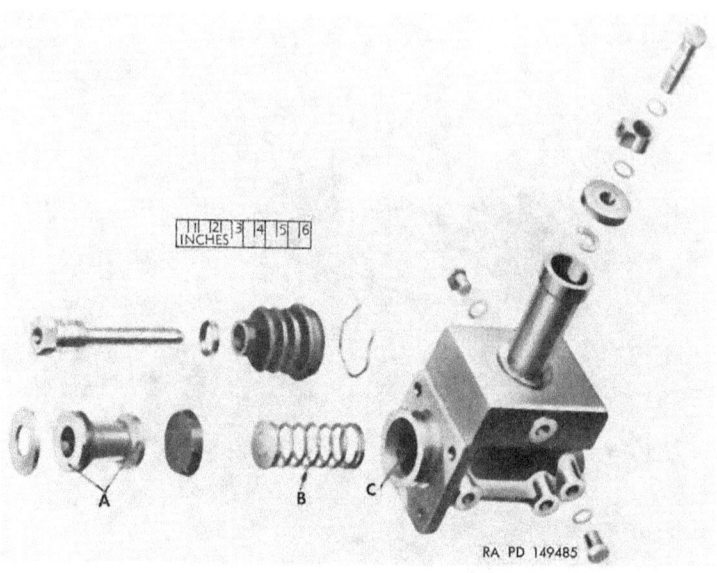

*Figure 250. Repair and rebuild standard points of measurement for master cylinder.*

## 351. Mechanical Parking Brake

| Fig. No. | Ref. letter | Point of measurement | Sizes and fits of new parts |
|---|---|---|---|
| 252 | A | Diameter of special pin | 0.996 to 0.998 |
| 252 | B | Inside diameter of lever bracket bearing | 0.999 to 1.001 |
| 252 | A–B | Fit of special pin in lever bracket bearing | 0.001L to 0.005L |
| 252 | C | Diameter of relay lever shaft | 0.997 to 1.000 |
| 252 | D | Inside diameter of relay lever bearing | 1.002 to 1.003 |
| 252 | C–D | Fit of relay lever shaft in relay lever bearing | 0.003L to 0.005L |
| 252 | E | Runout of drum braking surface not to exceed (total dial indicator reading) | 0.005 |
| 252 | F | Anchor spring: | |
| | | Approximate free length | $1\tfrac{1}{16}$ |
| | | Working load when compressed to ¾ | 11 to 15 lb |
| 252 | G | Release springs: | |
| | | Approximate free length | $2\tfrac{11}{16}$ |
| | | Working load when compressed to $1\tfrac{15}{32}$ | 25 to 35 lb |
| 252 | H | Tension spring: | |
| | | Approximate free length | 1.360 to 1.390 |
| | | Working load when compressed to ¼ | 600 to 800 lb |

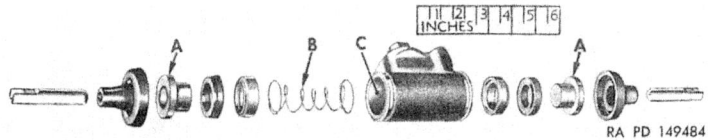

*Figure 251. Repair and rebuild standard points of measurement for wheel cylinders.*

## 352. Temporary (Electric) Parking Brake Valve

| Fig. No. | Ref. letter | Point of measurement | Sizes and fits of new parts |
|---|---|---|---|
| 178 | B | Check valve spring: | |
| | | Approximate free length | $1\frac{1}{16}$ |
| | | Working load when compressed to $1\frac{1}{32}$ | $2\frac{3}{4}$ to $3\frac{1}{4}$ lb |

## 353. Steering Gear Assembly

| Fig. No. | Ref. letter | Point of measurement | Sizes and fits of new parts |
|---|---|---|---|
| 253 | A | Inside diameter of Pitman shaft bearing (inside cover) | 1.1255 to 1.1260 |
| 253 | B | Outside diameter of Pitman arm shaft | 1.1235 to 1.1240 |
| 253 | B–A | Fit of Pitman arm shaft in bearing (in side cover) | 0.0015L to 0.0025L |
| 253 | C | Outside diameter of Pitman arm shaft | 1.3750 to 1.3755 |
| 253 | D | Inside diameter of Pitman shaft bearing (in housing) | 1.379 to 1.3795 |
| 253 | C–D | Fit of Pitman arm shaft in bearing (in housing) | 0.0035L to 0.0045L |
| 253 | E | Diameter of balls | $\frac{9}{32}$ |
| 253 | F–G | Clearance between end of adjusting screw and notch in Pitman arm shaft | 0.002 (max) |
| 253 | H | Shaft bearing spring: | |
| | | Free length | $\frac{3}{4}$ |
| | | Working load when compressed to $\frac{1}{2}$ to $\frac{5}{8}$ | 25 to 35 lb |
| | | Backlash between Pitman arm shaft teeth and worm ball nut teeth—pounds pull on steering wheel through center (par. 264d) | $2\frac{1}{2}$ to 3 lb |
| | | Worm bearing adjustment—pounds pull to keep steering wheel moving (par. 264c) | $1\frac{1}{2}$ to 2 lb. |

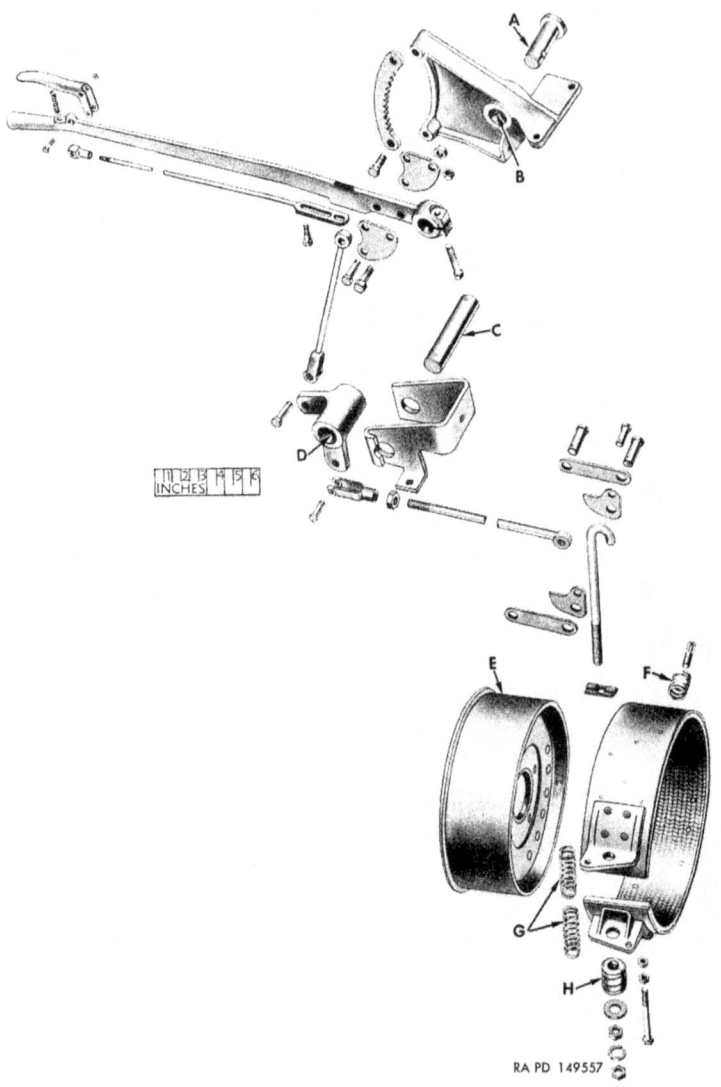

*Figure 252. Repair and rebuild standard points of measurement for mechanical parking brake.*

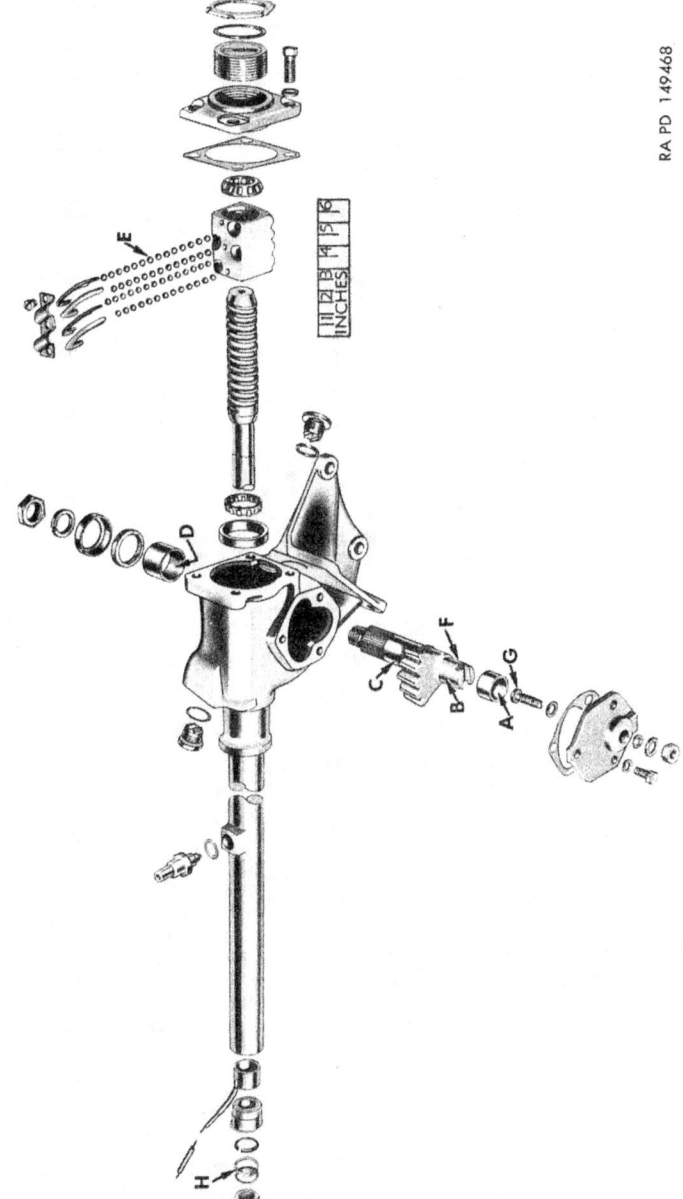

*Figure 253. Repair and rebuild standard points of measurement for steering gear assembly.*

### 354. Front Spring Suspension Components

| Fig. No. | Ref. letter | Point of measurement | Sizes and fits of new parts |
|---|---|---|---|
| 254 | A | Diameter of rear shackle bolt | 0.7465 to 0.7475 |
| 254 | B | Inside diameter of spring eye bearing | 0.749 to 0.754 |
| 254 | C | Outside diameter of shackle pin | 0.7465 to 0.7475 |
| 254 | D | Inside diameter of shackle bearing | 0.749 to 0.752 |
| 254 | A-D | Fit of shackle bolt in shackle bearing | 0.0015L to 0.0055L |
| 254 | C-B | Fit of shackle pin in spring eye bearing | 0.0015L to 0.0075L |
| | | Fit of shock absorber piston in cylinder | 0.0014L to 0.0025L |

### 355. Winch Assembly

| Fig. No. | Ref. letter | Point of measurement | Sizes and fits of new parts |
|---|---|---|---|
| 255 | A | Free length of clutch control lever ball spring | 11/16 |
| 255 | B | Diameter of drum shaft | 1.873 to 1.875 |
| 255 | C | Inside diameter of bushing type bearings | 1.877 to 1.880 |
| 255 | B-C | Fit of drum shaft in bearings | 0.002L to 0.007L |
| 255 | D | Automatic brake adjusting spring: Free lenth | 1½ |
| | | Working load when compressed to 1 | 50 lb |
| 255 | E | Thickness of drum thrust ring | 0.486 to 0.490 |
| 255 | F | Thickness of end frame bearing thrust ring (at outer edge) | 0.240 to 0.250 |
| 255 | G | Drag brake spring: Free length | 2 |
| | | Working load when compressed to 1 11/16 | 48 lb |

### 356. Winch Drive System

| Fig. No. | Ref. letter | Point of measurement | Sizes and fits of new parts |
|---|---|---|---|
| 256 | A | Diameter of shear pin hole in front fixed yoke | 0.312 to 0.314 |
| 256 | B | Inside diameter of bearings | 0.627 to 0.629 |
| 256 | C | Outside diameter of journal arms | 0.623 to 0.625 |
| 256 | B-C | Fit of bearings on journals | 0.002L to 0.006L |
| 256 | D | Width of grooves in slip yoke | 0.311 to 0.313 |
| 256 | E | Width of teeth on drive shafts | 0.308 to 0.310 |
| 256 | D-E | Blacklash between drive shaft and slip yoke splines | 0.001 to 0.005 |
| 256 | F | Drive shaft runout not to exceed (total dial indicator reading) | 0.010 |

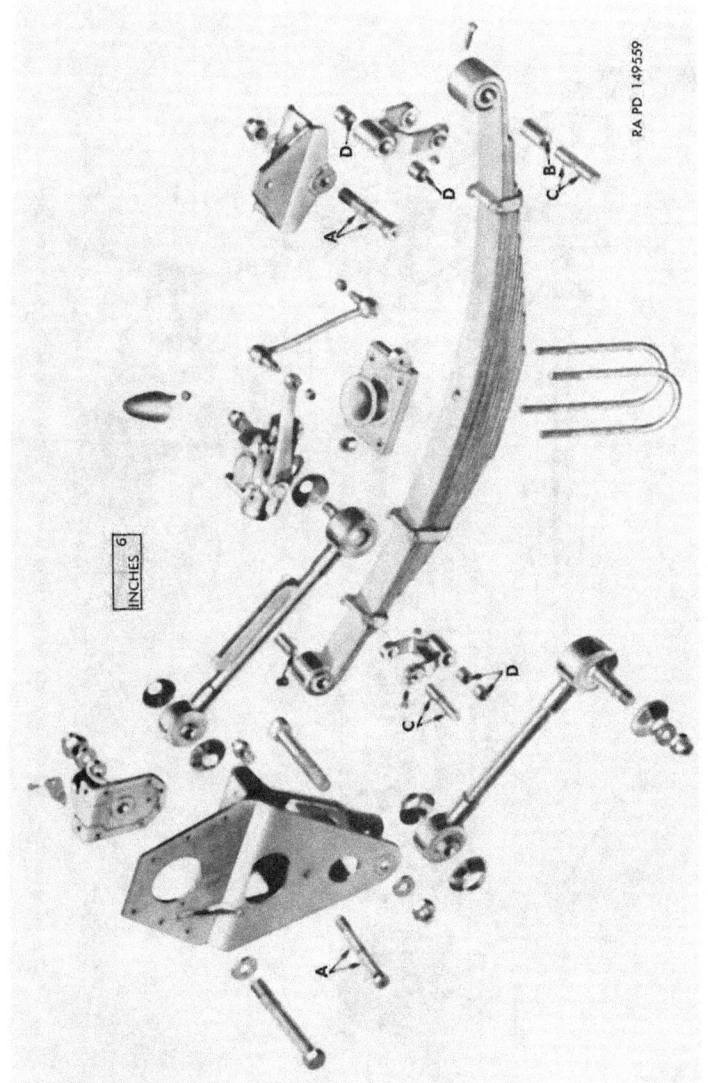

*Figure 254. Repair and rebuild standard points of measurement for front spring suspension components.*

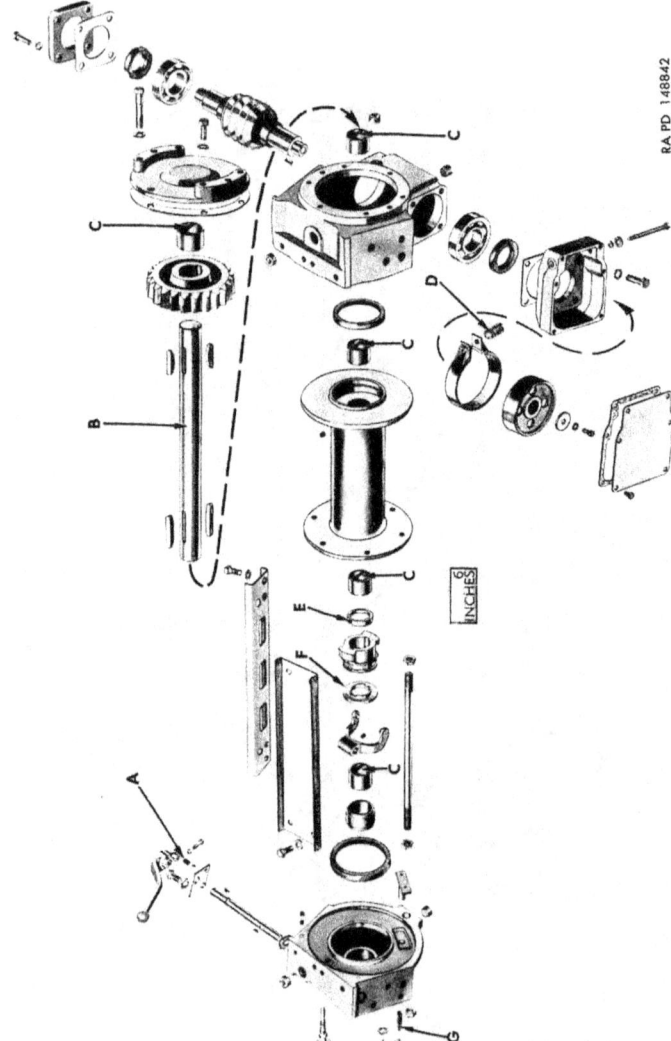

Figure 255. Repair and rebuild standard points of measurement for winch assembly.

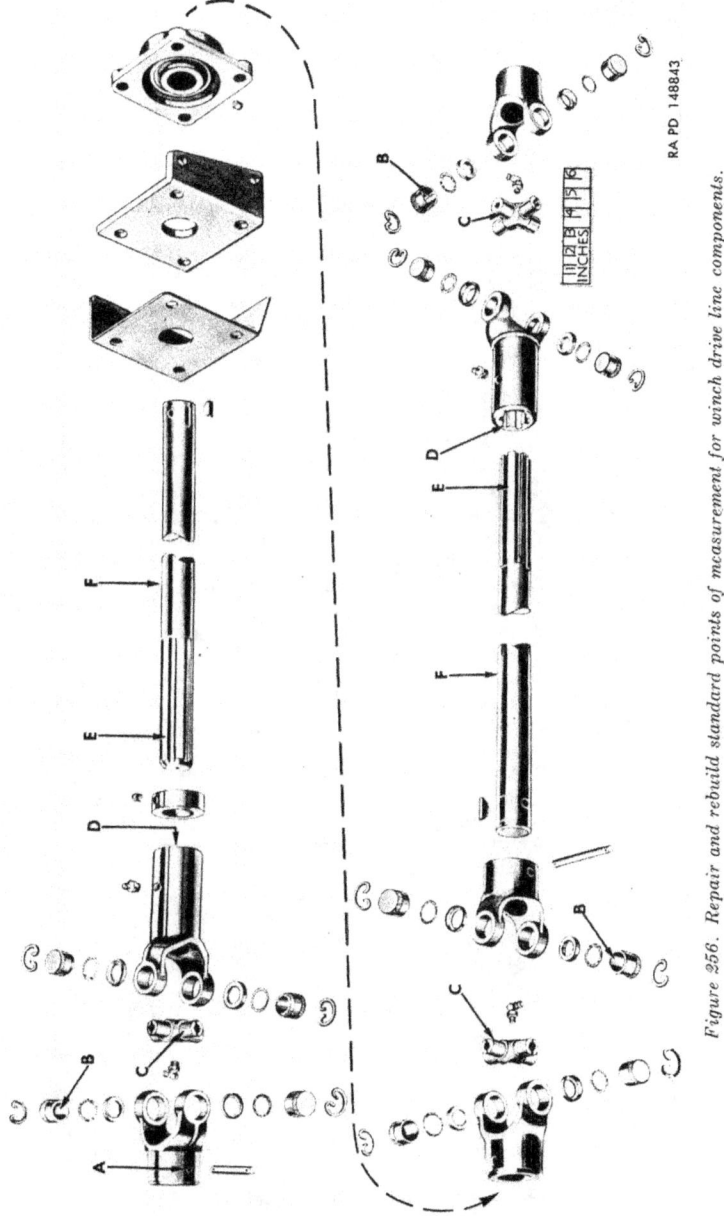

Figure 256. Repair and rebuild standard points of measurement for winch drive line components.

## 357. Torque Wrench Specifications

The following torque wrench specifications are itemized in the sequence appearing in text in chapters as indicated.

CHAPTER 4

| Fig. No. | Ref. letter | Location | Torque (lb-ft) |
|---|---|---|---|
| 33 | A | Torque rod bracket to cross member screw nuts | 33 to 43 |
| 35 | E | Secondary spring seat to frame bracket screw nuts | 33 to 43 |
| 35 | A | Secondary rear spring "U" bolt nuts | 375 to 400 |
| 39 | N | Rear spring seat adjusting nut | 60 to 75 |
| 39 | Q | Rear spring seat adjusting nut (lock) | 100 to 150 |
| 29 | D | Front spring "U" bolt nuts | 170 to 200 |
| 29<br>30 | B<br>F | Front axle upper torque bolt nut | 190 to 250 |
| 29 | H | Front axle upper torque rod pin nut | 350 to 400 |
|  |  | Shock absorber mounting screw nut | 63 to 84 |
| 29 | F | Shock absorber link pin nuts | 48 to 64 |
|  |  | Pillow block stud nuts | 48 to 64 |
|  |  | Rear axle torque rod pin nuts | 350 to 400 |
| 28 | A | Transfer mounting bolts | 60 to 85 |
|  |  | Transmission support and bracket to frame screw nuts | 48 to 64 |
|  |  | Transmission support bracket to frame screw nuts | 48 to 64 |
|  |  | Transfer reverse cross shaft right bracket to frame screw nuts | 20 to 27 |
|  |  | Transfer reverse cross shaft left bracket to frame screw nuts | 9½ to 13 |
| 36 | A & D | Transfer control cross shaft bracket screw nut | 20 to 27 |
| 36 | B & C | Parking brake relay lever shaft bracket to frame screw nuts | 20 to 27 |
|  |  | Wheel stud nuts | 300 to 350 |
|  |  | Winch drive shaft pilot bearing bracket screw nuts | 48 to 64 |
| 26 | C | Brake master cylinder to bracket screw nuts | 20 to 27 |
| 26 | A | Master cylinder brake line connector bolt | 20 to 30 |
|  |  | Air-hydraulic cylinder to end plate bolts | 9½ to 13 |
| 27 | F | Air-hydraulic cylinder rear bracket to frame screw nuts | 20 to 27 |
|  |  | Air-hydraulic cylinder to rear bracket stud nut | 9½ to 13 |
|  |  | Pitman arm nut | 115 to 155 |
| 25<br>124 | C<br>A | Drag link ball stud nuts | 75 to 100 |
| 25 | A & B | Steering gear to frame screw nuts | 48 to 64 |
|  |  | Air reservoir "U" bolt nuts | 9½ to 13 |
| 41 | P | Muffler support bracket to frame screw nuts | 9½ to 13 |
|  |  | Engine ground strap to frame screw nut | 5 to 7 |
|  |  | Rear exhaust pipe hanger strap screw nut | 20 to 27 |
|  |  | Steering column bracket cap screw nut | 20 to 27 |

| Fig. No. | Ref. letter | Location | Torque (lb-ft) |
|---|---|---|---|
| 198 | S | Steering wheel nut | 40 to 55 |
|  |  | Brake upper pedal to lower pedal screw nut | 20 to 27 |
| 17 | A, B, C, & F. | Winch support brackets to frame screw nuts | 95 to 127 |
|  |  | Spare wheel carrier bracket to frame screw nuts | 48 to 64 |
|  |  | Spare wheel to carrier swivel bracket stud nuts | 230 to 300 |
|  |  | Spare wheel carrier swivel bracket to lock bracket stud nut | 250 to 300 |
| 24 | B | Fuel tank support to frame screw nut | 20 to 27 |
| 37 | A | Rear bumper to frame screw nut | 48 to 64 |
|  |  | Pintle bracket to frame screw nuts | 165 to 220 |
| 37, 38 | B, A | Tail light bracket to frame screw nuts | 48 to 64 |
| 46 | D | Cargo body rear mounting screw nuts | 165 to 220 |

## CHAPTER 5

| Fig. No. | Ref. letter | Location | Torque (lb-ft) |
|---|---|---|---|
| 67 |  | Output gear rear bearing lock sleeve screw | 20 to 25 |
| 52 | AH | Front axle output shaft front bearing retainer cap screws | 20 to 25 |
| 52 | NN | Forward rear axle output shaft rear retainer cap screws | 20 to 25 |
| 52 | AS | Bearing lock nut (min) | 100 |
| 52 | KK | Forward rear axle output shaft (min) | 130 |
| 52 | AE | Front axle output shaft front bearing support cap screws | 20 to 25 |
| 52 | AL | Front axle output shaft flange nut (min) | 130 |
| 52 | BF & CC | Idler shaft bearing lock nuts (min) | 100 |
| 52 | AZ | Idler shaft front bearing retainer cap screws | 20 to 25 |
| 52 | BK | Speedometer driven gear assembly | 30 to 35 |
| 52 | BT | Input shaft front bearing retainer screws | 20 to 25 |
| 52 | BQ | Input shaft flange nut (min) | 130 |
| 54 | F | Sliding gear shifter fork bolt | 15 to 20 |
|  |  | Shifter shaft support bolts | 20 to 25 |
| 52 | FF | Output gear bearing retainer screws | 20 to 25 |
| 52 | P | Output gear flange nut (min) | 100 |
| 80 | B | Power take-off opening cover screws | 20 to 25 |
| 80 | C & B | Transfer case cover cap screws | 20 to 25 |

## CHAPTER 7

| Fig. No. | Ref. letter | Location | Torque (lb-ft) |
|---|---|---|---|
| 109 | P & N | Trunnion bearing nuts or spacers | 70 to 80 |
| 132 | AA | Differential and carrier assembly to housing stud nuts | 45 to 55 |
| 121, 148 | E, G | Axle housing cover to housing stud nuts | 45 to 55 |
| 130 |  | Tie rod end stud nuts | 175 to 200 |

## CHAPTER 8

| Fig. No. | Ref. letter | Location | Torque (lb-ft) |
|---|---|---|---|
| 132 | PP | Oil seal retainer cap screws | 160 to 180 |
| 132 | JJ | Drive pinion nuts (at axles) | 160 to 280 |
| 132 | W | Drive gear cap screws | 85 to 95 |
| 132 | D | Differential bearing cap screws | 130 to 160 |
| 132 | AA | Differential carrier stud nuts | 45 to 55 |
| 148 | G | Axle housing cover std nut | 45 of 55 |

## CHAPTER 9

| Fig. No. | Ref. letter | Location | Torque (lb-ft) |
|---|---|---|---|
| 154 | A | Pillow block shaft nuts | 175 to 225 |

## CHAPTER 10

| Fig. No. | Ref. letter | Location | Torque (lb-ft) |
|---|---|---|---|
| 182 | U | Brake drum to adapter stud nuts | 20 to 27 |
| 168 | G | Master cylinder filler pipe | 75 to 80 |
| 168 | N | Master cylinder filler cap bolt (max) | 30 |

## CHAPTER 11

| Fig. No. | Ref. letter | Location | Torque (lb-ft) |
|---|---|---|---|
| 178 | A | Parking brake valve outlet plug | 145 to 155 |

## CHAPTER 12

| Fig. No. | Ref. letter | Location | Torque (lb-ft) |
|---|---|---|---|
| 182 | V | Hub drive flange studs | 50 to 60 |
| 182 | D | Wheel bearing adjusting nuts | 60 to 75 |
| 182 | B | Wheel bearing adjusting nuts (lock) | 100 to 150 |
|   |   | Rear axle shaft or front hub drive flange stud nuts. | 55 to 65 |

## CHAPTER 13

| Fig. No. | Ref. letter | Location | Torque (lb-ft) |
|---|---|---|---|
| 198 | S | Steering wheel nut | 40 to 55 |
|   |   | Pitman arm nut | 115 to 155 |

## CHAPTER 15

| Fig. No. | Ref. letter | Location | Torque (lb-ft) |
|---|---|---|---|
| 214 | AA | Rear spring seat shaft safety nut | 650 to 700 |

# APPENDIX
# REFERENCES

## 1. Publication Indexes

The following publication indexes and lists of current issue should be consulted frequently for latest changes or revisions of references given in this appendix and for new publications relating to matériel covered in this manual:

Index of Administrative Publications_____ SR 310-20-5
Index of Army Motion Pictures and Film Strips and
  Kinescope Recordings_____ SR 110-1-1
Index of Blank Forms and Army Personnel Classification Tests_____ SR 310-20-6
Index of Technical Manuals, Technical Regulations,
  Technical Bulletins, Supply Bulletins, Lubrication
  Orders and Modification Work Orders_____ SR 310-20-4
Index to Tables of Organization and Equipment, Reduction
  Tables, Tables of Organization, Tables of Equipment, and
  Tables of Allowances_____ 310-20-7
Index of Training Publications_____ SR 310-20-3
Introduction and Index (supply catalogs)_____ ORD 1
Military Training Aids_____ FM 21-8
Ordnance Major Items and Major Combinations and Pertinent Publications_____ SB 9-1

## 2. Supply Catalogs

The following catalogs of the Department of the Army Supply Catalog pertain to this matériel:

*a. Destruction to Prevent Enemy Use.*
Land Mines and Fuzes, Demolition Material, and
  Ammunition for Simulated Artillery and Grenade Fire_____ ORD 11 SNL R-7

*b. Repair and Rebuild.*
Antifriction Bearings and Related Items_____ ORD 5 SNL H-12
Cleaners, Preservatives, Lubricants, Recoil Fluids,
  Special Oils, and Related Maintenance Materials_ ORD 3 SNL K-1

425

Electrical Fittings _____ ORD 5 SNL H-4
Items of Soldering, Metallizing, Brazing and Welding Materials: Gases and Related Items _____ ORD 3 SNL K-2
Lubricating Equipment, Accessories and Related Dispensers _____ ORD (*) SNL K-3
Lubricating Fittings, Oil Filters, and Oil Filter Elements _____ ORD 5 SNL H-16
Major Items and Major Combinations of Group G_ ORD 3 SNL G-1
Miscellaneous Hardware _____ ORD 5 SNL H-2
Oil Seals _____ ORD 5 SNL H-13
Pipe and Hose Fittings _____ ORD 5 SNL H-6
Shop Set, Auto Fuel and Electrical System, Field Maintenance _____ ORD 6 SNL J-8, Sec. 12
Shop Set, Contact and Emergency Repair, Field Maintenance _____ ORD 6 SNL J-8, Sec. 18
Shop Set, Engine and Power Train Rebuild Company (Armament), Depot Maintenance _____ ORD 6 SNL J-9, Sec. 8
Shop Set, Engine Rebuild Company (Automotive), Depot Maintenance _____ ORD 6 SNL J-9, Sec. 3
Shop Set, Headquarters and Service Company, Depot Maintenance, Automotive or Armament _____ ORD 6 SNL J-9, Sec. 2
Shop Set, Maintenance (Field) Automotive_ ORD 6 SNL J-8, Sec. 13
Shop Set, Power Train Rebuild Company (Automotive), Depot Maintenance _____ ORD 6 SNL J-9, Sec. 1
Shop Set, Tire Rebuild Company, Depot Maintenance _____ ORD 6 SNL J-9, Sec. 10
Standard Hardware _____ ORD 5 SNL H-1
Tool Set, Auto Fuel and Electrical System Repairman _____ ORD 6 SNL J-10, Sec. 8
Tool Set, Canvas and Leather Repairman __ ORD 6 SNL J-10, Sec. 15
Tool Set, General Mechanic's _____ ORD 6 SNL J-10, Sec. 4
Tool Set, Maintenance (Field), Motor Vehicle Assembly Company _____ ORD 6 SNL J-8, Sec. 7
Tool Set, Metal Body Repairman _____ ORD 6 SNL J-10, Sec. 7
Tool Set, Tire Rebuilder and Inspector _____ ORD 6 SNL J-10, Sec. 11
Tool Set, Vulcanizer's _____ ORD 6 SNL J-10, Sec. 12

*c. Vehicle.*

Truck, Cargo, 2½-Ton, 6 x 6, M135 _____ ORD (*) SNL G-749

---

*See ORD 1 for published catalogs of the ordnance section of the Department of the Army Supply Catalog.

## 3. Forms

The following forms pertain to this matériel:

| | |
|---|---|
| WD AGO Form 9-1, | Matériel Inspection Tag |
| DA Form 9-3, | Processing Record for Storage and Shipment (Tag) and Boxed Engines |
| WD AGO Form 9-4, | Vehicular Storage and Servicing Record (Card) |
| DA Form 9-68, | Spot Check Inspection Report for Wheeled and Half-Track Vehicles |
| WD AGO Form 9-71, | Locator and Inventory Control Card |
| WD AGO Form 9-72, | Ordnance Stock Record Card |
| DA Form 9-76, | Request for Work Order |
| DA Form 9-77, | Job Order Register |
| WD AGO Form 9-78, | Job Order |
| DA Form 9-79, | Parts Requisition |
| WD AGO Form 9-80, | Job Order File |
| WD AGO Form 9-81, | Exchange Part or Unit Identification Tag |
| DA Form 447, | Property Turn-In Slip |
| DA Form 460, | Preventive Maintenance Roster |
| DA Form 461, | Preventive Maintenance Service and Inspection for Wheeled and Half-Track Vehicles |
| DA Form 461-5, | Limited Technical Inspection |
| DA Form 446, | Issue Slip |
| DA Form 468, | Unsatisfactory Equipment Report |
| DA Form 478, | MWO and Major Unit Assembly Replacement Record and Organizational Equipment File |
| DA Form 811, | Work Request and Job Order |
| DA Form 811-1, | Work Request and Hand Receipt |
| WD AGO Form 865, | Work Order |
| WD AGO Form 866, | Consolidation of Parts |
| WD AGO Form 867, | Status of Modification Work Order |
| DD Form 6, | Report of Damaged or Improper Shipment |
| DD Form 317, | Preventive Maintenance Service Due (Sticker) |

## 4. Other Publications

The following explanatory publications contain information pertinent to this matériel and associated equipment:

*a. Camouflage.*
Camouflage, Basic Principles_____ FM 5-20

Camouflage of Vehicles _____ FM 5–20B
  *b. Decontamination.*
Decontamination _____ TM 3–220
Decontamination of Armored Force Vehicles _____ FM 17–59
Defense Against Chemical Attack _____ FM 21–40
  *c. Destruction to Present Enemy Use.*
Explosives and Demolitions _____ FM 5–25
Ordnance Service in the Feld _____ FM 9–5
  *d. General.*
Cooling Systems: Vehicles and Powered Ground Equipment _____ TM 9–2858
Disposal of Supplies and Equipment: Uneconomically
  Repairable Ordnance Vehicles _____ SR 755–105–5
Fuels and Carburetion _____ TM 10–550
Inspection of Ordnance Matériel in the Hands of Troops _ TM 9–1100
Lubrication Order _____
Military Vehicles _____ TM 9–2800
Motor Vehicles (Ordnance Corps Responsibility) _____ AR 700–105
Ordnance Field Maintenance _____ FM 9–10
Precautions in Handling Gasoline _____ AR 850–20
Prevention of Motor Vehicle Accidents _____ SR 385–155–1
Principles of Automotive Vehicles _____ TM 9–2700
Report of Accident Experience _____ SR 385–10–40
Storage Batteries, Lead-Acid Type _____ TM 9–2857
Supplies and Equipment: Unsatisfactory Equipment
  Report _____ SR 700–45–5
  *e. Repair and Rebuild.*
Abrasives, Cleaning, Preserving, Sealing, Adhesive, and
  Related Materials Issued for Ordnance Matériel _____ TM 9–850
Body and Components of 2½ Ton 6 x 6 Water tank
  Truck M50, Gasoline Truck M49, and Tractor Truck
  M48 _____ TM 9–1819D
Dump Body and Controls for 2½-Ton 6 x 6 Dump Truck
  M47 _____ TM 9–1819C
Electrical Equipment (Bendix-Scintilla) _____ TM 9–1825E
Ordnance Maintenance: Engine (GMC Model 302) __ TM 9–1819AA
Hand, Measuring, and Power Tools _____ TM 10–590
Instruction Guide: Care and Maintenance of Ball and
  Roller Bearings _____ TM 37–265
Lubrication _____ TM 9–2835
Lubrication Order _____ LO 9–819A
Maintenance and Care of Hand Tools _____ TM 9–867
Maintenance and Care of Pneumatic Tires and Rubber
  Treads _____ TM 31–200

Maintenance of Supplies and Equipment: Maintenance Responsibilities and Shop Operation_____ AR 750–5
Modification of Ordnance Matériel_____ SB 9–38
Ordnance Maintenance: Carburetors (Holley)_____ TM 9–1826D
Ordnance Maintenance: Electrical Equipment (Auto-Lite) _____ TM 9–1825B
Ordnance Maintenance: Electrical Equipment (Delco-Remy)_____ TM 9–1825A
Ordnance Maintenance: Fuel Pumps_____ TM 9–1828A
Ordnance Maintenance: Power Brake Systems_____ TM 9–1827A
Ordnance Maintenance: Speedometers, Tachometers, and Recorders_____ TM 9–1829A
Ordnance Maintenance: Vacuum Brake System_____ TM 9–1827B
Ordnance Maintenance: Vehicular Maintenance Equipment, Grinding, Boring, Valve Reseating Machines, and Lathes_____ TM 9–1834A
Ordnance Maintenance, Hydra-Matic Transmission (GMC Model 302M) _____ TM 9–1819AB
Painting Instructions For Field Use_____ TM 9–2851
Uneconomically Repairable Ordnance Vehicles_____ SR 755–105–5
Power Train, Body, and Frame for 20½-Ton 6 x 6 Cargo Truck M34, M35, M36, M44, M45, M46, M47, M48, M49, M50, M108, and M109_____ TM 9–1819B
Preparation of Ordnance Matériel for Deep-Water Fording_____ TM 9–2853
Preventive Maintenance of Electric Motors and Generators_____ TM 55–405
Wheeled and Half-Track Vehicles, Trailers, and Towed Artillery: Lubrication of Wheel Bearings_____ TB 9–2835–12

*f. Operation.*

2½-Ton 6 x 6 Cargo Truck M135_____ TM 9–819A

*g. Shipment and Stand-By or Long-Term Storage.*

Army Shipping Document_____ TM 38–705
Instruction Guide: Ordnance Packaging and Shipping (Posts, Camps, and Stations)_____ TM 9–2854
Marking and Packing of Supplies and Equipment: Marking of Oversea Supply_____ SR 746–30–5
Military Standard—Marking of Shipments_____ MIL–STD–129**
Ordnance Storage and Shipment Chart—Group G__ TB 9–OSSC–G
Preparation of Supplies and Equipment for Shipment: Processing of Unboxed and Uncrated Equipment for Oversea Shipment_____ AR 747–30

429

Processing of Motor Vehicles and Related Unboxed Matériel for Shipment and Storage_____ SB 9-4
Preservation, Packaging, and Packing of Military Supplies and Equipment_____ TM 38-230
Protection of Ordnance General Supplies in Open Storage_____ TB ORD 379
Shipment of Supplies and Equipment: Report of Damaged or Improper Shipment_____ SR 745-45-5
Standards for Oversea Shipment and Domestic Issue of Ordnance Matériel Other Than Ammunition and Army Aircraft_____ TB ORD 385

# INDEX

| | Paragraph | Page |
|---|---|---|
| Absorber, shock (*See* Shock abosrber) | | |
| Accessory drive, power take-off: | | |
|   Description | 142 | 153 |
|   Installation | 144 | 155 |
|   Rebuild: | | |
|     Dump truck accessory drive | 149–152 | 163 |
|     Tank truck accessory drive | 145–148 | 155 |
|   Removal | 143 | 154 |
|   Standards, repair and rebuild | 345 | 407 |
| Adjustments: | | |
|   Bearings, wheel | 256*f* | 291 |
|   Drive gear and pinion backlash (axle) | 202*g* | 225 |
|   Main spring seat bearing | 49 | 60 |
|   Steering gear | 264 | 311 |
|   Trunnion bearings, steering knuckle | 186*d* | 203 |
|   Turning angle, front wheel | 155 | 174 |
|   Ventilator, cowl | 328*d* | 383 |
| Alinement: | | |
|   Frame | 287 | 336 |
|   Front axle | 155 | 174 |
| Allocation, field and depot maintenance | 2 | 14 |
| Axle, front: | | |
|   Alinement | 155 | 174 |
|   Assembly of axle from subassemblies | 185–192 | 201 |
|   Data | 154 | 173 |
|   Data, alinement | 156 | 177 |
|   Description and operation | 153 | 171 |
|   Disassembly into subassemblies | 157–162 | 180 |
|   Installation | 50 | 62 |
|   Rebuild: | | |
|     Axle shaft and universal joint | 164–169 | 183 |
|     Housing | 174–179 | 194 |
|     Steering knuckle, support, trunnions, seals, and bearings | 170–173 | 189 |
|     Tie rod | 180–184 | 198 |
|   Removal | 36 | 49 |
|   Standards, repair and rebuild | 346 | 407 |
| Axles, rear: | | |
|   Assembly of axle from subassemblies | 205–209 | 230 |
|   Cleaning and inspection of axle housing and axle shafts | 203–204 | 227 |
|   Data | 194 | 209 |
|   Description and operation | 193 | 209 |
|   Disassembly of axle into subassemblies | 195–196 | 209 |
|   Installation | 53 | 64 |
|   Removal | 37 | 49 |

|  | Paragraph | Page |
|---|---|---|
| Axle shaft and universal joint: | | |
|     Assembly | 169 | 186 |
|     Cleaning | 166 | 185 |
|     Disassembly | 165 | 183 |
|     Inspection | 167 | 185 |
|     Installation | 190 | 207 |
|     Removal | 159 | 180 |
|     Repair | 168 | 186 |
| Bearings, hub (wheel) | 256 | 287 |
| Block, pillow (*See* Pillow block) | | |
| Body, cargo (*See* Cargo body) | | |
| Brake drums | 235 | 265 |
| Brakes: | | |
|     Parking: | | |
|         Data | 243 | 272 |
|         Description | 242 | 272 |
|         Rebuild of mechanical parking brake | 244–246 | 273 |
|         Rebuild of temporary (electric) parking brake valve | 247–252 | 278 |
|         Standards, repair and rebuild | 351 | 414 |
|     Service: | | |
|         Assembly of brake components | 233 | 260 |
|         Brake drums | 235 | 265 |
|         Cleaning and inspection of brake components | 231 | 258 |
|         Data | 227 | 251 |
|         Description | 226 | 248 |
|         Disassembly of brake components | 230 | 253 |
|         Installation of brake assembly | 234 | 263 |
|         Repair of brake components | 232 | 259 |
|         Rebuild: | | |
|             Brake shoes and drums | 228–235 | 251 |
|             Master cylinder | 236–238 | 265 |
|             Wheel cylinders | 239–241 | 270 |
|         Removal of brake assembly | 229 | 252 |
|         Standards, repair and rebuild | 348 | 412 |
| Bumper: | | |
|     Front: | | |
|         Installation | 74 | 80 |
|         Removal | 17 | 29 |
|     Rear: | | |
|         Installation | 77 | 81 |
|         Removal | 42 | 58 |
| Cab: | | |
|     Data | 321 | 371 |
|     Description | 320 | 370 |
|     Door | 330–332 | 384 |
|     Installation | 70 | 32 |
|     Rebuild | 322–329 | 371 |
|     Removal | 21 | 73 |
|     Repair | 322 | 371 |
|     Windshield | 333, 334 | 391 |
| Camber, front axle | 155 | 174 |
| Cargo body: | | |
|     Data | 336 | 400 |
|     Description | 335 | 395 |

|  | Paragraph | Page |
|---|---|---|
| Cargo body—Continued | | |
| Installation | 79 | 82 |
| Patching and welding | 339 | 400 |
| Removal | 16 | 28 |
| Repair of top paulin and end curtain | 341 | 402 |
| Repair of wood parts | 340 | 401 |
| Straightening | 338 | 400 |
| Caster, front axle | 155 | 174 |
| Common tools and equipment | 9 | 20 |
| Connectors, bayonet type | 293 | 342 |
| Curtain: | | |
| End, cargo body | 341 | 402 |
| Rear, cab | 323 | 372 |
| Data: | | |
| Alinement, front axle | 156 | 177 |
| Axle, front | 154 | 173 |
| Axle, rear | 194 | 209 |
| Brakes: | | |
| Parking | 243 | 272 |
| Service | 227 | 251 |
| Cargo body | 336 | 400 |
| Frame | 285 | 334 |
| Power take-off | 122 | 137 |
| Spring suspension: | | |
| Front | 268 | 316 |
| Rear | 279 | 331 |
| Steering gear | 258 | 295 |
| Transfer | 82 | 89 |
| Wheels and hubs | 254 | 286 |
| Winch and drive line | 299 | 348 |
| Differences between models | 5 | 18 |
| Differential carrier: | | |
| Assembly | 202 | 221 |
| Cleaning | 199 | 216 |
| Disassembly | 198 | 211 |
| Inspection | 200 | 218 |
| Installation: | | |
| Front | 187 | 206 |
| Rear | 206 | 230 |
| Removal: | | |
| Front | 162 | 182 |
| Rear | 196e | 211 |
| Repair | 201 | 219 |
| Standards, repair and rebuild | 347 | 411 |
| Doors, cab: | | |
| General | 330 | 384 |
| Replacement of doors | 331 | 384 |
| Replacement of door components | 332 | 387 |
| Drag link: | | |
| Description | 265 | 314 |
| Inspection and repair | 266 | 314 |
| Drive line, winch (See Winch drive line) | | |
| Drum, brake | 235 | 265 |

|  | Paragraph | Page |
|---|---|---|
| Electrical system: | | |
| Data | 292 | 341 |
| Description | 291 | 341 |
| Rebuild of horn | 295–297 | 344 |
| Repair of cable connections | 293, 294 | 242 |
| Forms, records, and reports | 3 | 15 |
| Frame: | | |
| Alinement | 287 | 336 |
| Data | 285 | 334 |
| Description | 284 | 334 |
| Rebuild of pintle | 289 | 339 |
| Repair | 288 | 337 |
| Front axle (*See* Axle, front) | | |
| Front spring suspension (*See* Spring suspension, front) | | |
| Fuel tank: | | |
| Installation | 76 | 80 |
| Removal | 23 | 40 |
| Gear, steering (*See* Steering gear) | | |
| Hood: | | |
| Installation | 72 | 78 |
| Removal | 19 | 30 |
| Horn: | | |
| Assembly | 297 | 344 |
| Description | 291b | 341 |
| Disassembly | 295 | 344 |
| Inspection | 296 | 344 |
| Housing: | | |
| Front axle: | | |
| Assembly | 179 | 197 |
| Cleaning | 176 | 194 |
| Disassembly | 175 | 194 |
| Inspection | 177 | 196 |
| Repair | 178 | 197 |
| Rear axle: | | |
| Cleaning | 203 | 227 |
| Inspection | 204 | 229 |
| Hubs (*See* Wheels and hubs) | | |
| Introduction: | | |
| Data | 6 | 19 |
| Description (of vehicles) | 4 | 16 |
| Differences between models | 5 | 18 |
| Field and depot maintenance allocation | 2 | 14 |
| Forms, records, and reports | 3 | 15 |
| Scope | 1 | 1 |
| Lining, brake shoe, replacement | 232 | 259 |
| Major components, removal, and installation: | | |
| Air-hydraulic cylinder: | | |
| Installation | 64 | 67 |
| Removal | 31 | 46 |
| Air reservoir: | | |
| Installation | 66 | 68 |
| Removal | 30 | 45 |

Major components, removal, and installation—Continued

| | Paragraph | Page |
|---|---|---|
| Axle, front: | | |
|   Installation | 50 | 62 |
|   Removal | 36 | 49 |
| Axles, rear: | | |
|   Installation | 53 | 64 |
|   Removal | 37 | 49 |
| Bumper, front: | | |
|   Installation | 74 | 80 |
|   Removal | 17 | 29 |
| Cab: | | |
|   Installation | 70 | 73 |
|   Removal | 21 | 32 |
| Cargo body: | | |
|   Installation | 79 | 82 |
|   Removal | 16 | 28 |
| Carrier, spare wheel: | | |
|   Installation | 75 | 80 |
|   Removal | 25 | 43 |
| Controls and brackets installation | 55–58 | 65 |
| Control shafts and brackets removal | 41 | 54 |
| Exhaust pipe, rear: | | |
|   Installation | 69 | 73 |
|   Removal | 24 | 42 |
| Fender and skirt: | | |
|   Installation | 71 | 78 |
|   Removal | 20 | 31 |
| Fuel tank: | | |
|   Installation | 76 | 80 |
|   Removal | 23 | 40 |
| General | 15, 44 | 28, 58 |
| Hood: | | |
|   Installation | 72 | 78 |
|   Removal | 19 | 30 |
| Inspection | 80 | 82 |
| Lines and wiring harnesses: | | |
|   Installation | 45 | 58 |
|   Removal | 43 | 58 |
| Main springs, rear: | | |
|   Installation | 49 | 60 |
|   Removal | 39 | 54 |
| Master cylinder: | | |
|   Installation | 62 | 67 |
|   Removal | 28 | 45 |
| Muffler: | | |
|   Installation | 67 | 69 |
|   Removal | 24 | 42 |
| Power plant: | | |
|   Installation | 68 | 69 |
|   Removal | 22 | 37 |
| Propeller shaft: | | |
|   Installation | 59 | 66 |
|   Removal | 32 | 46 |
| Rear bumpers, tow hooks, and pintle removal | 42 | 58 |

| Major components, removal, and installation—Continued | Paragraph | Page |
|---|---|---|
| Rear tow hooks, bumpers, and pintle installation | 77 | 81 |
| Secondary spring: | | |
|     Installation | 48 | 60 |
|     Removal | 40 | 54 |
| Shock absorber: | | |
|     Installation | 52 | 63 |
|     Removal | 35 | 48 |
| Spare wheel and carrier: | | |
|     Installation | 75 | 80 |
|     Removal | 25 | 43 |
| Springs, front: | | |
|     Installation | 47 | 59 |
|     Removal | 38 | 53 |
| Steering gear: | | |
|     Installation | 65 | 68 |
|     Removal | 26 | 43 |
| Transfer: | | |
|     Installation | 54 | 65 |
|     Removal | 33 | 46 |
| Winch: | | |
|     Installation | 73 | 79 |
|     Removal | 18 | 29 |
| Winch drive line: | | |
|     Installation | 61 | 66 |
|     Removal | 29 | 45 |
| Wheel: | | |
|     Installation | 60 | 66 |
|     Removal | 34 | 48 |
| Master cylinder: | | |
|     Assembly | 238 | 268 |
|     Cleaning, inspection, and repair | 237 | 267 |
|     Description | 226b | 248 |
|     Disassembly | 236 | 265 |
|     Installation | 62 | 67 |
|     Removal | 28 | 45 |
|     Standards, repair and rebuild | 349 | 412 |
| Muffler: | | |
|     Installation | 67 | 69 |
|     Removal | 24 | 42 |
| Parts | 8 | 20 |
| Paulin top: | | |
|     Cab | 323 | 372 |
|     Cargo body | 341 | 402 |
| Pillow block: | | |
|     Assembly | 225 | 246 |
|     Cleaning, inspection, and repair | 224 | 241 |
|     Description | 222 | 240 |
|     Disassembly | 223 | 241 |
| Pintle: | | |
|     Installation | 77 | 81 |
|     Rebuild | 289 | 339 |
|     Removal | 42 | 58 |

|  | Paragraph | Page |
|---|---|---|
| Power plant: | | |
|   Installation | 68 | 69 |
|   Removal | 22 | 37 |
| Power take-off: | | |
|   Accessory drive | 142–152 | 153 |
|   Assembly | 136–141 | 145 |
|   Data | 122 | 137 |
|   Description | 121 | 133 |
|   Disassembly | 123–128 | 137 |
|   Rebuild of components | 129–135 | 143 |
|   Standards, repair and rebuild | 344 | 407 |
| Propeller shafts and universal joints: | | |
|   Arrangment of propeller shafts | 210 | 232 |
|   Assembly | 217 | 236 |
|   Cleaning | 216a | 234 |
|   Description | 210–213 | 232 |
|   Disassembly | 215 | 234 |
|   Inspection | 216b | 234 |
|   Installation | 59 | 66 |
|   Removal | 32 | 46 |
| Propeller shaft, transmission-to-transfer: | | |
|   Assembly | 221 | 240 |
|   Cleaning, inspection, and repair | 220 | 237 |
|   Description | 218 | 237 |
|   Disassembly | 219 | 237 |
| Rear axle (*See* Axles, rear) | | |
| Rear spring suspension (*See* Spring suspension, rear) | | |
| Riveting (frame) | 288d | 338 |
| Scope (of manual) | 1 | 1 |
| Seats, cab | 324 | 375 |
| Shaft, propeller (*See* Propeller shafts *and* universal joints) | | |
| Sheet metal | 329 | 384 |
| Special tools and equipment | 10 | 20 |
| Specifications, torque wrench | 357 | 422 |
| Spring suspension: | | |
|   Front: | | |
|     Description | 267 | 315 |
|     Inspection | 269 | 316 |
|     Repair | 270–272 | 317 |
|     Shock absorbers | 273–277 | 321 |
|     Spring installation | 47 | 59 |
|     Spring removal | 38 | 53 |
|     Standards, repair and rebuild | 354 | 418 |
|   Rear: | | |
|     Data | 279 | 331 |
|     Description | 278 | 329 |
|     Inspection | 281 | 332 |
|     Main spring: | | |
|       Installation | 49 | 60 |
|       Removal | 39 | 54 |
|     Repair | 282 | 332 |
|     Secondary spring: | | |
|       Installation | 48 | 60 |
|       Removal | 40 | 54 |

| | Paragraph | Page |
|---|---|---|
| Spring suspension—Continued | | |
|   Rear—Continued | | |
|     Spring seat shaft replacement | 283 | 332 |
| Shock absorbers: | | |
|   Assembly | 277 | 326 |
|   Cleaning and inspection | 276 | 324 |
|   Construction | 273 | 321 |
|   Disassembly | 275 | 323 |
|   Installation | 52 | 63 |
|   Operation | 274 | 322 |
|   Removal | 35 | 48 |
| Standard, repair and rebuild: | | |
|   Accessory drive, power take-off | 345 | 407 |
|   Axle, front | 346 | 407 |
|   Brake, parking: | | |
|     Electric | 352 | 415 |
|     Mechanical | 351 | 414 |
|   Brakes, service | 348 | 412 |
|   Differential, axle | 347 | 411 |
|   Master cylinder | 349 | 412 |
|   Power take-off | 344 | 407 |
|   Spring suspension, front | 354 | 418 |
|   Steering gear | 353 | 415 |
|   Torque wrench specifications | 357 | 422 |
|   Transfer | 343 | 403 |
|   Wheel cylinders | 350 | 412 |
|   Winch | 355 | 418 |
|   Winch drive line | 356 | 418 |
| Steering gear: | | |
|   Adjustments | 264 | 311 |
|   Assembly | 263 | 306 |
|   Cleaning | 261 | 301 |
|   Data | 258 | 295 |
|   Description | 257 | 293 |
|   Disassembly | 268 | 316 |
|   Inspection and repair | 262 | 301 |
|   Installation | 65 | 68 |
|   Removal | 26 | 43 |
|   Standards, repair and rebuild | 353 | 415 |
| Steering knuckle support: | | |
|   Bearing adjustment | 186d | 203 |
|   Inspection | 172 | 189 |
|   Installation | 186c | 203 |
|   Removal | 161 | 182 |
|   Repair | 173 | 191 |
| Steering wheel: | | |
|   Installation | 263k | 311 |
|   Removal | 259c | 296 |
| Tank, fuel: | | |
|   Installation | 76 | 80 |
|   Removal | 23 | 40 |
| Tie rod, front axle: | | |
|   Assembly | 184 | 200 |
|   Cleaning | 182 | 200 |

|                                                              | Paragraph | Page |
|---|---|---|
| Tie rod, front axle—Continued | | |
|     Disassembly | 181 | 200 |
|     Installation | 189 | 206 |
|     Inspection | 183 | 200 |
|     Removal | 160 | 182 |
| Tools: | | |
|     Common | 9 | 20 |
|     Special | 10 | 20 |
| Torque rods: | | |
|   Front: | | |
|     Description | 267c | 315 |
|     Inspection | 269g | 317 |
|     Repair | 272 | 320 |
|   Rear: | | |
|     Description | 278d | 321 |
|     Inspection | 281c | 332 |
| Torque wrench specification | 357 | 422 |
| Transfer: | | |
|   Assembly of transfer from subassemblies | 114–120 | 126 |
|   Data | 82 | 89 |
|   Description | 81 | 84 |
|   Disassembly into subassemblies | 83–91 | 89 |
|   Installation | 54 | 65 |
|   Rebuild: | | |
|     Case and covers | 112, 113 | 125 |
|     Forward rear axle output shaft components | 104–106 | 115 |
|     Front axle output shaft components | 101–103 | 110 |
|     Idler shaft components | 107–109 | 118 |
|     Input shaft and components | 92–94 | 99 |
|     Output gear bearing retainer and components | 98–100 | 107 |
|     Output gear and components | 95–97 | 103 |
|     Shifting mechanism | 110, 111 | 123 |
|   Removal | 33 | 46 |
|   Standards, repair and rebuild | 343 | 403 |
| Trouble shooting: | | |
|   Bent brake backing plate | 14 | 27 |
|   Excessive front axle caster | 12 | 27 |
|   Incorrect front wheel alinement | 13 | 27 |
|   Purpose | 11 | 27 |
| Turning angle, front wheel | 155 | 174 |
| Universal joint (*See* Propeller shafts and universal joints) | | |
| Valve, electric parking brake: | | |
|   Assembly | 252 | 281 |
|   Cleaning and inspection | 251 | 280 |
|   Disassembly | 250 | 280 |
|   Operation | 248 | 278 |
|   Standards, repair and rebuild | 352 | 415 |
|   Testing | 249 | 280 |
| Welding: | | |
|   Cargo body | 339 | 400 |
|   Frame | 288e | 338 |

|  | Paragraph | Page |
|---|---|---|
| Wheel cylinders: | | |
| Assembly | 241 | 270 |
| Cleaning, inspection, and repair | 240 | 270 |
| Description | 226d | 249 |
| Disassembly | 239 | 270 |
| Standards, repair and rebuild | 350 | 412 |
| Wheels: | | |
| Installation | 60 | 66 |
| Removal | 34 | 48 |
| Wheels and hubs: | | |
| Bearing adjustment | 256f | 291 |
| Data | 254 | 286 |
| Description | 253 | 283 |
| Hubs and bearings, rebuild | 256 | 287 |
| Wheels, rebuild | 255 | 286 |
| Winch: | | |
| Assembly of winch from subassemblies | 313–315 | 362 |
| Data | 299 | 348 |
| Description | 298 | 346 |
| Disassembly into subassemblies | 300–302 | 348 |
| Installation | 73 | 79 |
| Rebuild: | | |
| Drive system | 316–319 | 365 |
| Drum | 311, 312 | 362 |
| End frame | 303–305 | 352 |
| Gear case and automatic brake | 306–310 | 355 |
| Removal | 18 | 29 |
| Standards, repair and rebuild | 355 | 418 |
| Winch drive system: | | |
| Installation | 61 | 66 |
| Rebuild | 316–319 | 365 |
| Removal | 29 | 45 |
| Standards, repair and rebuild | 356 | 418 |
| Windshield: | | |
| General | 333 | 391 |
| Replacement of windshield components | 334 | 391 |
| Wrench, torque, specifications | 357 | 422 |

©2013 Periscope Film LLC
All Rights Reserved
ISBN#978-1-940453-18-7
www.PeriscopeFilm.com

www.ingramcontent.com/pod-product-compliance
Lightning Source LLC
Chambersburg PA
CBHW051822230426
43671CB00008B/808